Introduction to the Physics of Gyrotrons

Johns Hopkins Studies in Applied Physics

Introduction to the Physics of Gyrotrons

Gregory S. Nusinovich

Foreword by Victor Granatstein and Richard Temkin

The Johns Hopkins University Press
Baltimore and London

Printed in the United States of America on acid-free paper
2 4 6 8 9 7 5 3 1

The Johns Hopkins University Press
2715 North Charles Street
Baltimore, Maryland 21218-4363
www.press.jhu.edu

Library of Congress Cataloging-in-Publication Data

Nusinovich, G. S.
Introduction to the physics of gyrotrons / by Gregory S. Nusinovich ; foreword by Victor Granatstein and Richard Temkin.
p. cm.
Includes bibliographical references and index.
ISBN 0-8018-7921-3 (hardcover : acid-free paper)
1. Gyrotrons. 2. Plasma radiation. 3. Dynamics of a particle. I. Title.
QC718.5.G95N87 2004
530.4'4—dc22 2003024182

A catalog record for this book is available from the British Library.

To my teachers:

A. V. Gaponov-Grekhov, M. I. Petelin, and V. K. Yulpatov

Contents

Foreword

It is a great pleasure to write a foreword for this book, *Introduction to the Physics of Gyrotrons,* by Gregory Nusinovich. Dr. Nusinovich has been a pioneer in the field of gyrotron research for over three decades. His journey in this world of science and engineering took him from the Institute of Applied Physics in Nizhny Novgorod to his present home at the Institute for Research in Electronics and Applied Physics at the University of Maryland, College Park. Along the way, he has been a leading scientist in the gyrotron research program, first in Russia and today in the United States. This journey has given him a unique perspective on the exciting, developing field of gyrotron research. We believe that it is accurate to say that Dr. Nusinovich is today the single leading authority in the field of gyrotron research.

The gyrotron was developed to help bridge the gap between the realm of conventional microwave tubes at low frequency and the realm of lasers at high frequency. In the past few years, gyrotrons have been demonstrated at average power levels near one megawatt and at wavelengths of a few millimeters, resulting in unprecedented levels of power density over an extensive portion of the electromagnetic spectrum. This recent progress more than justifies the writing of a new, specialized book explaining the physics and engineering of gyrotron devices. This new progress has built on a long history of intensive research in the field. One may count over two thousand papers published in the field of gyrotrons in the past thirty years. It is remarkable that Dr. Nusinovich is an author or co-author of about two hundred of them, surely indicating his impact on the field. Dr. Nusinovich and colleagues were the first to calculate the generalized, nonlinear efficiency of the gyrotron and the first to explain and estimate the effects of mode competition, to name just two early contributions. More recently, he has pioneered research on gyrotron amplifiers at both the fundamental cyclotron resonance and at cyclotron harmonics.

The present book is a remarkable achievement. It begins with material introducing the gyrotron, a device that can be built in many forms: oscillator, amplifier, multicavity, traveling wave, etc. The great strength of this book lies in the subsequent analysis that provides rigorous derivations of the key equations of the gyrotron interaction. This rigorous derivation is very useful for scientists working in the field of gyrotron research. It also forms a model for developing the theory of any vacuum electron device. The book provides detailed results for all of the various forms of the gyrotron that are being studied today. Numerous excellent examples make the material much easier to understand.

Scientists and engineers working in the field of gyrotrons and related devices owe Dr. Nusinovich a debt of gratitude for producing this excellent volume. It will help young people entering the field and will serve as a valuable reference work for more experienced scientists. We believe that the field of gyrotron research will progress more rapidly and vigorously as a result of the publication of this excellent volume.

Victor Granatstein, *College Park, Maryland*
Richard Temkin, *Cambridge, Massachusetts*

Preface

Gyrotrons are well recognized as high-power sources of coherent electromagnetic radiation. In the millimeter- and submillimeter-wavelength regions, the power that gyrotrons can radiate in continuous-wave and long-pulse regimes exceeds the power of classical microwave tubes (klystrons, magnetrons, traveling-wave tubes, backward-wave oscillators, etc.) by many orders of magnitude.

This gyrotron superiority stems from the physics of gyrotron operation. Classical microwave tubes are based on such kinds of electron coherent radiation (Cherenkov radiation or transition radiation) that require microwave structures with elements smaller than a wavelength. For instance, traveling-wave tubes and backward-wave oscillators are based on the Cherenkov synchronism between electrons and slow waves whose phase velocity should be close to the electron velocity. Such waves can be excited in periodic slow-wave structures whose period should be smaller than a wavelength. The distance between electrons and walls of these structures should also be much smaller than a wavelength, because slow waves are localized near the structure walls. All these factors cause rapid miniaturization of the interaction space with the wavelength shortening even at millimeter wavelengths. Correspondingly, the power that can be handled by such structures decreases drastically.

Gyrotrons, however, are based on the mechanism of coherent cyclotron radiation from electrons gyrating in a constant magnetic field. In these devices, the electrons can resonantly interact with fast waves, which, in principle, can propagate even in free space. Therefore, the interaction space in gyrotrons can be much larger than in classical microwave tubes operating at the same wavelength.

The arrangement of a simplest gyrotron is shown schematically in Fig. P.1. Here, a magnetron-type electron gun is shown on the left. The voltage applied to the anode creates the electric field at the cathode. This field has both the perpendicular and parallel components with respect to the lines of the magnetic field produced by a solenoid. Thus, electrons emitted from the cathode acquire there both the orbital and axial velocity components. Then,

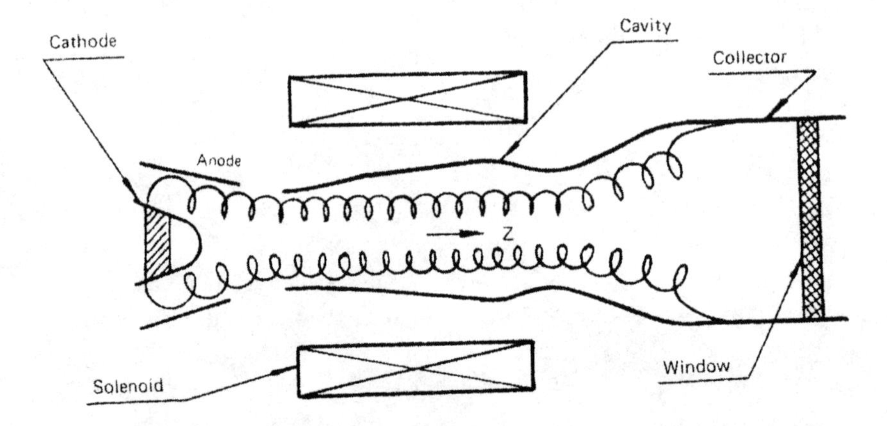

Fig. P.1. Gyrotron arrangement

the electrons move toward the cavity in the growing magnetic field, in which the electron flow undergoes the adiabatic compression and the electron orbital momentum increases. In the region of the uniform magnetic field, the electrons interact with the eigen-mode of the cavity and transform a part of their kinetic energy into the microwave energy. Then, the spent beam exits from the axially open cavity, undergoes decompression in the decreasing magnetic field and settles on the collector. The latter is also functioning as an oversized output waveguide, which directs the outgoing radiation toward the output window shown on the right.

The coherent cyclotron radiation in gyrotrons is caused by the cyclotron maser instability. This instability was discovered almost simultaneously in the late 1950s by several scientists working independently. Then, in the mid-1960s, the most useful and promising gyrotron configurations were invented at the Radiophysical Research Institute (Gorky, USSR). Since then, gyrotrons have dominated the millimeter-wave region at a megawatt power level and successfully entered the submillimeter-wavelength region. These devices became ubiquitous tools in various plasma experiments at research laboratories around the globe. In particular, they are of great importance for controlled fusion experiments based on the so-called electron-cyclotron plasma heating. Gyrotrons are also used in many other applications.

In spite of this active development for more than 40 years, so far there is a lack of a single comprehensive book on gyrotrons addressed to a general audience. The absence of such a book was one of my motivations to write the present one. This book was also motivated by the fact that many of the pioneering studies done by Russian inventors of gyrotrons are still unavailable for Western readers. Moreover, there are many important studies, results of which were not published even in Russia. Note that the gyrotron theory was developed in Russia from the very beginning in a general form. To deliver these fundamentals of gyrotron theory to all physicists who are interested in gyrotrons is important and challenging. For a number of reasons, I decided to accept this challenge.

The book consists of three parts. The first part is an introductory one. It contains two chapters: the first chapter is on the basic physics of gyrodevices and the second chapter describes their key components. The second part of the book is the core one. It contains seven chapters (Chapters 3–9) describing the fundamentals of the theory of gyrotron oscillators and amplifiers. This part is supplemented with the three appendixes found at the back of the book, in which the derivation of the most important equations is reproduced. The third part contains three chapters (Chapters 10–12) that describe the state-of-the-art in the development of gyrodevices. Each of these chapters starts with a historical introduction that helps one understand the genesis of the gyrotron development. Part III ends with Chapter 13, in which a number of gyrodevices not considered in the main part of the book are described. Many of the chapters contain some problems, the solving of which may help one better understand the topics under consideration. The book ends with the Summary and References. Although it is impossible, of course, to make references to all contributions to the field, I have tried to mention the pioneering papers as well as the papers that can be useful for more detailed studies of the topics discussed in the book at the introductory level.

I have dedicated this book to my three teachers: Andrei Victorovich Gaponov-Grekhov, Mikhail Ivanovich Petelin, and Valeriy Konstantinovich Yulpatov. I had the honor to work with them for more than 20 years, first at the Radiophysical Research Institute, and then at the Institute of Applied Physics of the Academy of Sciences of the USSR. During this time, I also greatly benefited from working with Mark Moiseev, Vladimir Bratman, Naum Ginzburg, and Nikolay Kovalev.

Since January 1991, when I immigrated to the United States, I have been working at the Institute for Research in Electronics and Applied Physics (IREAP), formerly the Institute for Plasma Research, at the University of Maryland. At this institute, I have actively and fruitfully collaborated with Thomas Antonsen, Jr., Victor Granatstein, Baruch Levush, Peter Latham, and Alexander Vlasov.

The gyrotron-oriented work at IREAP has been supported by T. V. George of the Office of Fusion Energy Sciences, U.S. Department of Energy; Jack

Agee and Robert Barker of the Air Force Office of Scientific Research; David Sutter of the Division of High Energy Physics, U.S. Department of Energy; and Robert Parker of the Vacuum Electronics Branch, Naval Research Laboratory. I gratefully acknowledge this support.

I am indebted to Olexander Sinitsyn, Janet Wofsheimer, and Dorothea Brosius for their help in the preparation of the manuscript and to Jagadishvar Sirigiri for his comments on the original manuscript.

PART I

Introduction to Gyrodevices

▪ CHAPTER 1 ▪

Introduction

The history of the development of CRM and gyrotron oscillators and amplifiers will be described later in Chapters 10–12, which are devoted to specific types of devices. In this chapter, we will focus on the physical issues that are the most important for gyrotron operation.

1.1 Relativistic Dependence of Electron Cyclotron Frequency on Electron Energy

It is well known that the frequency of electron gyration Ω in a constant magnetic field $\vec{H}_0$ depends on the electron energy $\mathcal{E}$ through the equation $\Omega = eH_0c/\mathcal{E}$ (here e is the electron charge and c is the speed of light). However, it is less obvious that this relativistic effect can be important even in the case of weakly relativistic electrons. (These are electrons whose kinetic energy is much smaller than their rest energy, or, in other words, electrons accelerated by voltage, which is much smaller than 511 kV.)

To illustrate this statement, let us consider a ring of electrons gyrating in a constant magnetic field around the same guiding center and interacting with an electromagnetic (EM) wave whose frequency ω is close to Ω:

$$\omega \approx \Omega. \tag{1.1}$$

Assume, first, that the wave amplitude is constant and small in comparison with the external magnetic field and that the electric field of the wave rotates in the plane of electron gyration in the same direction as the electrons. Then, in accordance with the general equation for electron energy,

$$\frac{d\mathcal{E}}{dt} = -e(\vec{v} \cdot \vec{E}), \tag{1.2}$$

the changes in electron energy, $\Delta\mathcal{E}$, can be evaluated as

$$|\Delta\mathcal{E}| \sim ev_{\perp}E\tau, \tag{1.3}$$

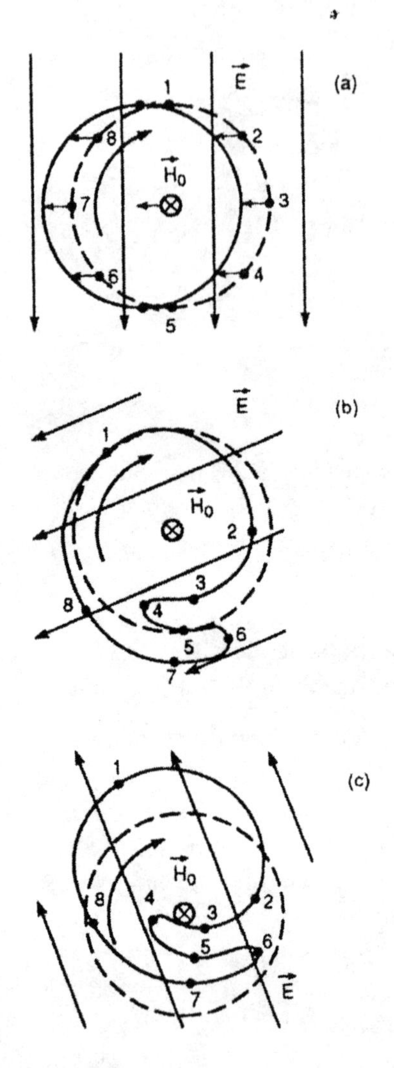

Fig. 1.1. Motion of gyrating relativistic electrons near the fundamental cyclotron resonance with a circularly polarized wave. A dotted line describes the initial distribution of electrons: (a) initial modulation, (b) electron bunching, (c) deceleration of the bunch.

where $v_\perp$ is the electron orbital velocity and τ is the interaction time. It follows from (1.2) that the sign of $\Delta\mathcal{E}$ depends on the initial electron gyrophase with respect to the wave: $(\vec{v} \cdot \vec{E}) < 0$ corresponds to the initially accelerating phase $(\Delta\mathcal{E} > 0)$ while $(\vec{v} \cdot \vec{E}) > 0$ corresponds to the decelerating phase $(\Delta\mathcal{E} < 0)$. This initial modulation of electron energies and the corresponding electron displacement from a stationary orbit are shown in Fig. 1.1a.

Since the electron gyrofrequency depends on the electron energy, the energy modulation causes changes in the frequency of gyration:

$$\Delta\Omega = -\Omega_0 \left(\frac{\Delta\mathcal{E}}{\mathcal{E}_0}\right). \tag{1.4}$$

(Here the subscript "0" designates initial values of variables.) Correspondingly, the accelerated particles start to gyrate more slowly, while the decelerated particles start to gyrate faster. (Note that this effect is also responsible for the negative mass instability; see Nielsen and Sessler 1959). The slippage of the electron phase with respect to the phase of initial gyration can be evaluated as

$$|\Delta\theta| \sim |\Delta\Omega|\,\tau \sim \Omega_0\tau \left|\frac{\Delta\mathcal{E}}{\mathcal{E}_0}\right|. \tag{1.5}$$

It is important to note that, after an initial energy modulation, the phase slippage may also proceed in the absence of the EM wave, e.g., in some drift regions, which are free of microwaves. Thus, such a phase slippage, which may result in the *orbital bunching* of electrons, has certain inertia, which is also typical for electron *axial bunching* in conventional klystrons. The formation of an electron bunch on the orbit is shown in Fig. 1.1b. Let us emphasize that, in accordance with (1.3) and (1.5), the phase bunching proceeds as τ^2, which is why it is sometimes called "quadratic bunching" (Gaponov, Petelin, and Yulpatov 1967). Note, however, that in multistage devices with separate sections for electron energy modulation and electron phase bunching, the variable τ in (1.3) and (1.5) has different meanings, thus, in such cases it can be designated as τ_{mod} and τ_{bunch} in (1.3) and (1.5), respectively.

When the initial gyrofrequency of the electrons exceeds the wave frequency ω, decelerated electrons, which increase their gyrofrequency, soon shift from the decelerating phase into the accelerating phase. At the same time, initially accelerated particles slightly decrease their gyrofrequency, and thus remain in this phase. As a result, when $\Omega_0 > \omega$, the bunch is formed in the accelerating phase. On the other hand, when $\Omega_0 < \omega$, initially decelerated particles stay in the same phase, while initially accelerated particles decrease their gyrofrequency, which leads to their slippage into the decelerating phase. Thus, when $\Omega_0 < \omega$, the bunch is formed in the decelerating phase. Then, further deceleration leads to extraction of the orbital momentum from electrons. This process is shown in Fig. 1.1c.

So far, we were considering only the case of the fundamental cyclotron resonance given by (1.1). In principle, of course, the frequency of the EM field can be close to one of the cyclotron harmonics:

$$\omega \approx s\Omega_0, \tag{1.6}$$

where s is the cyclotron harmonic number. In order to describe the interaction at harmonics, it is expedient to use the polar coordinate system (r, θ) with the origin on the gyration axis and to expand the electric field, $\vec{E} = Re\{\vec{E}(r,\theta)\exp(i\omega t)\}$ into the Fourier series in the angular variable θ:

$$\vec{E}(r,\theta) = \sum \vec{E}_\ell(r)\exp(-il\theta).$$

It follows from the cyclotron resonance condition given by (1.6) that the largest cumulative effect in electron motion is caused by the synchronous sth harmonic. This harmonic represents the field azimuthally rotating with an angular frequency ω/s. As shown by Flyagin et al. (1977), when the gyroradius is much smaller than the wavelength, this field has a quasistatic structure on the electron orbit and represents the field of a rotating sth order multipole. (This issue is analyzed in one of the problems at the end of the chapter.) The resonances with such fields at the first three cyclotron harmonics are shown in Fig. 1.2.

Our simple consideration of interaction between gyrating electrons and the synchronously rotating component of the electric field can yield even more results once we postulate that the cyclotron resonance is maintained as long as the phase shift of gyrating electrons with respect to the wave is not too large. Assume, for simplicity, that this restriction is

$$\Theta = (\omega - s\Omega)T < 2\pi, \tag{1.7}$$

where $T = L/v_z$ is the electron transit time, L is the interaction length, and v_z is the electron axial velocity. Following Bratman et al. (1981), let us represent

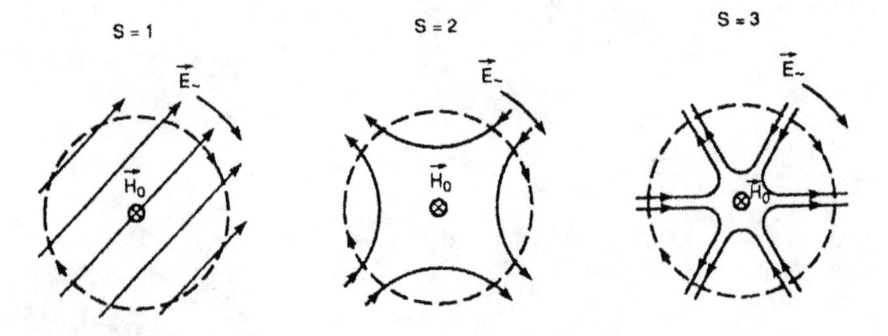

Fig. 1.2. Electron motion in the case of resonance at cyclotron harmonics. A dashed line shows the initial trajectory of electrons.

Ω as $\Omega_0 + \Delta\Omega$. Then, the transit angle Θ can be represented as a sum of the kinematic shift $\Theta_{\text{kin}} = (\omega - s\Omega_0)T$, which is associated with the initial cyclotron resonance mismatch and the dynamic shift $\Theta_d = -s\Delta\Omega T$, which is essentially the same phase slippage that was evaluated above by (1.5). In the simplest case, in order to fulfill the limitation given by (1.7), absolute values of both Θ_{kin} and Θ_d should not exceed 2π. The first condition, $|\Theta_{\text{kin}}| < 2\pi$, has an obvious meaning: the longer the interaction is, the closer the wave frequency should be to the resonant cyclotron harmonic. The second limitation, $|\Theta_d| < 2\pi$, restricts the energy that can be withdrawn by the EM field from electrons without violating the cyclotron resonance. Taking into account (1.5) and the relation $\Omega_0 T = 2\pi N$ between the interaction time T and the number of electron orbits in the interaction space N, this limitation can be rewritten as

$$\frac{|\Delta\mathcal{E}|}{\mathcal{E}_0 - mc^2} \le \frac{1}{sN\left(1 - \gamma_0^{-1}\right)}. \tag{1.8}$$

Here the left-hand side describes the ratio of the withdrawn energy to the initial kinetic energy (mc^2 is the rest energy). This ratio is called the single-particle efficiency, η_{sp}. On the right-hand side, $\gamma_0 = \mathcal{E}_0/mc^2 = 1 + eV_b/mc^2$ is the initial Lorentz factor, where V_b is the electron beam voltage.

Since a real resonance corresponds to a large number of electron orbits, (1.8) predicts a paradoxical result. It states that one should expect a high efficiency of the energy extraction mechanism based on a *purely relativistic* effect only in the case of *weakly relativistic* electrons! Indeed, as follows from (1.8), to realize $\eta_{\text{sp}} \sim 1$ one should have $V_b < (mc^2/e)/sN$, where $mc^2/e = 511$ kV. This result follows from the fact that, in the case of relativistic electrons, in accordance with (1.4), even rather small changes in electron energies disturb the cyclotron resonance condition given by (1.7).

Next, let us note that it is possible for the transit angle of (1.7) to remain small even if Θ_{kin} and Θ_d are both large, provided that these two terms compensate each other. In principle, for instance, one can taper the external magnetic field in such a way that the changes in the kinematic shift will compensate for the dynamic changes of the decelerating bunch. As a result, the transit angle determined by (1.7) will be small enough even when each of the two shifts is large, so that one can overcome the limitation given by (1.8). Such gyrodevices with tapered parameters have been studied elsewhere (Ginzburg 1987, Nusinovich 1992). Note that this principle of tapering in gyrodevices is quite similar to the known tapering of parameters in TWTs (Meeker and Rowe 1962) and FELs (Kroll et al. 1980, Sprangle et al. 1980).

1.2 Quantum Interpretation of Induced Cyclotron Radiation

Above, we outlined the process of induced electron cyclotron radiation in terms of the classical theory. Starting from the very first studies of CRMs (Twiss 1958, Schneider 1959, Gaponov 1960), this process was also interpreted in terms of the quantum theory. Indeed, as is known (see, e.g., Landau and Lifshitz 1958), an electron placed in a constant magnetic field has a discrete energy spectrum, and these energy levels,

$$W_n = mc^2[1 + (2n + 1)\hbar\Omega_r/mc^2]^{1/2} - mc^2, \tag{1.9}$$

are nonequidistant. (Here Ω_r is the rest electron gyrofrequency; also in (1.9) we ignored the spin and the motion of electrons along the magnetic field.) Therefore, a beam of such electrons having nonzero initial orbital velocities can be considered as an active medium with an inverse population of the energy levels, which are often called *Landau levels*. These levels are shown in Fig. 1.3. Since the gyrofrequency decreases with increasing energy, the transition from the nth level to the $(n - 1)$st level emits a higher energy photon than would be absorbed in the transition from the nth to the $(n + 1)$st level. Therefore, in each elementary act of radiation an electron loses more energy than in each elementary act of absorption. In general, the net transfer of power from the nth level, according to Schneider (1959), is equal to

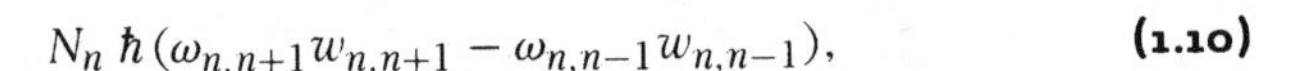

$$N_n\,\hbar\,(\omega_{n,n+1}w_{n,n+1} - \omega_{n,n-1}w_{n,n-1}), \tag{1.10}$$

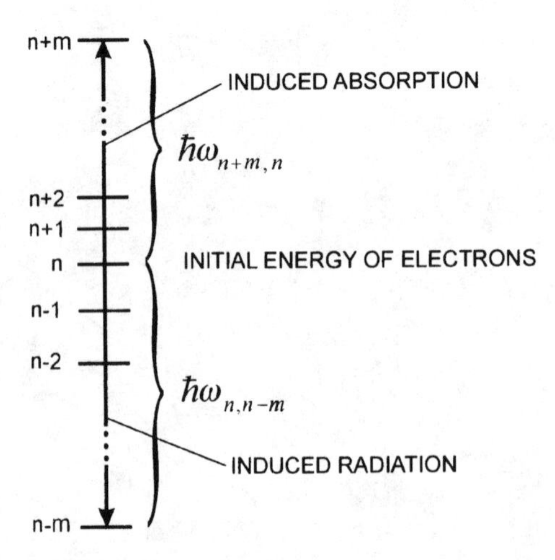

Fig. 1.3. Energy spectrum (Landau levels) and induced radiation and absorption processes in a system of gyrating relativistic electrons.

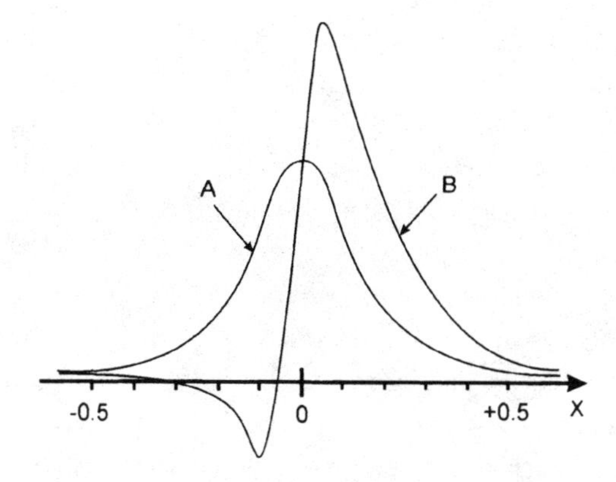

Fig. 1.4. Cyclotron resonance absorption lines of a nonrelativistic (A) and relativistic (B) electron. (Reproduced from Schneider 1959)

where N_n is the electron population at the nth level, $\omega_{n,n+1}$ and $\omega_{n,n-1}$ are the frequencies of transitions from the populated nth level to the $(n+1)$st and $(n-1)$st levels, respectively, and $w_{n,n+1}$ and $w_{n,n-1}$ are the transition probabilities for corresponding transitions. (So, a positive value of the net power given by (1.10) implies the absorption of the wave energy by electrons.) As is known, these probabilities are proportional to $n+1$ and n, respectively. Since $\omega_{n,n-1}$ is larger than $\omega_{n,n+1}$, this power can be negative, and thus power can be transferred from the electrons to the wave. This statement is illustrated with Fig.1.4 reproduced from Schneider (1959). Here the absorption lines for nonrelativistic (A) and relativistic (B) electrons are shown. The latter case means that the relativistic dependence of Ω on energy is taken into account. The argument $x = (\omega_{n,n+1} - \omega)\tau$ is the cyclotron resonance mismatch between the wave frequency ω and electron transition frequency, and τ is the interaction time. As was found by Schneider (1959), a net stimulated emission becomes possible when $\Omega\tau(W/mc^2) > 1$ (here W is the kinetic energy associated with electron gyration).

It is necessary to emphasize that the radiation linewidth, which is inversely proportional to the electron transit time through the interaction region, is much larger than the difference between the frequencies $\omega_{n,n-1}$ and $\omega_{n,n+1}$. As follows from (1.9),

$$\omega_{n,n-1} - \omega_{n+1,n} = [(W_n - W_{n-1}) - (W_{n+1} - W_n)]/\hbar \approx \Omega_0(\hbar\Omega_0/\mathcal{E}_0),$$

where $\Omega_0 = \Omega_r/\gamma_0$ is the initial cyclotron frequency. Therefore, the processes of the induced absorption and radiation are both present at the same time.

In principle, their separation can be realized when the linewidth of radiation, $\Delta\omega$, is extremely narrow; for instance, in the case of radiation at the wavelength of about 1 cm, it would require $\Delta\omega/\omega << 10^{-10}$. Note, however, that a hypothetical device with such extreme selective properties would limit one to using only one transition from, let us say, the nth level to $(n-1)$st level.

At this point we have arrived at the most important distinction between lasers and CRMs, which determines the preference of CRMs in the microwave and millimeter-wave regions. The active medium in lasers (at least, in their simple versions) can usually be treated as a three-level quantum object with a *strongly nonequidistant energy spectrum*. Therefore, it is only possible to emit a single photon from each excited element of the active medium. Since with the transition from optics to microwaves (or even to the infrared region) the energy of each photon decreases, this makes lasers extremely inefficient at long wavelengths. On the contrary, CRMs have an active medium with a *weakly nonequidistant energy spectrum* that allows one to radiate a huge number of photons from each electron. Indeed, even at very high magnetic fields required for operation at the fundamental cyclotron resonance at, let's say, a wavelength of 1 mm, the energy $\hbar\Omega_r$ is on the order of 10^{-3} eV, while typical high-power microwave sources operate at voltages higher than 10 kV. Therefore, such devices have initially populated levels $n > 10^7$, and, correspondingly, for their efficient operation it is necessary to use about 10^7 elementary acts of radiation emission from each electron. Note that since the nonequidistance of the levels is very weak, the lines of induced absorption and radiation in such an active medium are located at closely spaced frequencies, in contrast to lasers, where these lines are well separated, as is shown in Fig.1.5.

Let us emphasize that this nonequidistance, although it is very weak, plays an extremely important role because an active system with an *equidistant* energy spectrum, whose excited particles are immersed in a homogeneous EM field, cannot exhibit induced coherent radiation (see, e.g., W. Lamb, Jr., 1965).

1.3 Autoresonance

So far we have considered the interaction of an EM wave with a ring of gyrating electrons while ignoring the electron axial motion along the external magnetic field $\vec{H}_0 = H_0\vec{z}$. In reality, electrons immersed in the interaction space acquire not only an orbital velocity, $v_\perp$, but also an axial velocity, v_z. Therefore, the cyclotron resonance condition given earlier by (1.1) should be rewritten to

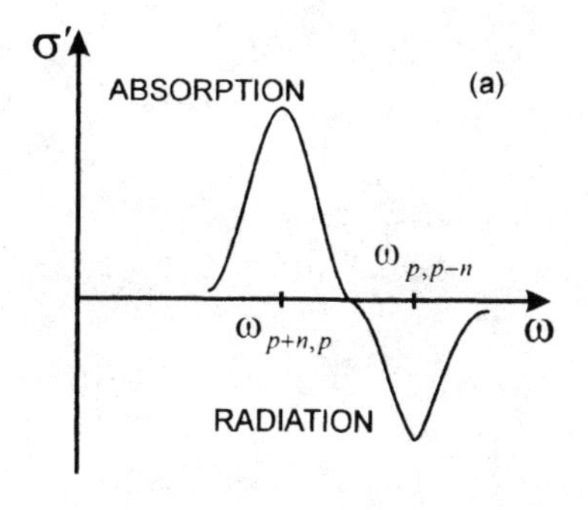

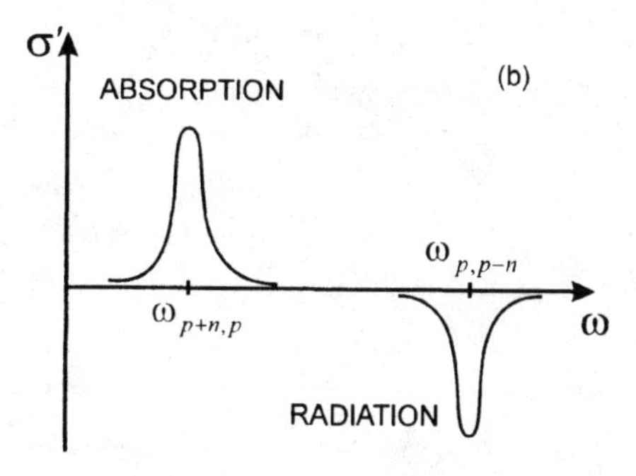

Fig. 1.5. Absorption and radiation lines for a system of gyrating electrons (a) and an active medium of a solid-state laser (b) (σ' is the real part of the conductivity of an active medium.)

account for the Doppler shift as

$$\omega - k_z v_z \approx s\Omega, \tag{1.11}$$

where k_z is the axial wavenumber of an EM wave, whose electric field $\vec{E}$ can be represented as $\vec{E} = \mathrm{Re}\{\vec{E}(\vec{r}) \exp[i(\omega t - k_z z)]\}$. It is obvious that the changing of the electron energy causes changes not only in the cyclotron frequency but also in the axial velocity, and that these changes have opposite signs. For instance, electron deceleration increases the cyclotron frequency but decreases the axial velocity, and, correspondingly, the Doppler term in (1.11). As a result, these two changes can, in principle, compensate for each other.

This compensation was analyzed a long time ago by V. Ya. Davydovsky (1962) and A. A. Kolomensky with A. N. Lebedev (1963) (see also Roberts and Buchsbaum 1964). These authors showed that, in the case of an EM wave propagating along an external magnetic field with a phase velocity, $v_{\text{ph}} = \omega/k_z$, equal to the speed of light c (i.e., $k_z = \omega/c$), the two changes exactly compensate each other. So, if the cyclotron resonance condition is initially fulfilled in such a system, then this resonance will automatically be fulfilled later, no matter how large the changes in the electron energies are. This remarkable effect is called *autoresonance*. Initially, the above-mentioned authors proposed to use this effect for electron acceleration. (Note that this issue is still under consideration; see, e.g., Hirshfield et al. 1996.) Only later, as will be discussed, were the suggestions to use this effect for improving the CRM efficiency formulated. Let us emphasize that this unique effect is a consequence of a quite specific dependence of electron cyclotron frequency on the energy. In other devices, such as, for instance, FELs, the frequency of electron oscillations in periodic external fields does not have such dependence; thus there is no autoresonance effect.

In a more general case of $k_z \neq \omega/c$, the compensation of these two changes is not complete. Following Bratman et al. (1981), the phase slippage in this general case can be analyzed in a fashion quite similar to the treatment done in Section 1.1. In accordance with the cyclotron resonance condition given by (1.11), let us introduce, instead of (1.7), the transit angle of electron gyration with respect to the Doppler shifted wave frequency,

$$\Theta = (\omega - k_z v_z - s\Omega)T. \tag{1.12}$$

Again, this angle can be represented as the sum of the kinematic phase shift $\Theta_{\text{kin}} = (\omega - k_z v_{z0} - s\Omega_0)T$, which is proportional to the initial cyclotron resonance mismatch, and the dynamic shift $\Theta_d = (-k_z \Delta v_z - s\Delta\Omega)T$. Here $\Delta\Omega$ is given by (1.4). In order to evaluate the changes in the axial velocity Δv_z, one can take into account that the radiation of one photon not only reduces the electron energy by $\hbar\omega$, but also, due to the recoil effect, changes the axial momentum, p_z, by $\Delta p_z = \hbar k_z$ (Gaponov 1960). Correspondingly, the changes $\Delta\mathcal{E}$ and Δp_z are related by

$$\Delta\mathcal{E} = v_{\text{ph}} \Delta p_z. \tag{1.13}$$

Equation (1.13) clearly shows that in the case of electron deceleration by a forward wave the electron axial momentum decreases. Correspondingly, in the case of the electron deceleration by a backward ($k_z < 0$, so $v_{\text{ph}} < 0$) wave this momentum increases, and in the case of perpendicular propagation of the wave, when $k_z = 0$, and hence $v_{\text{ph}} \to \infty$, this momentum remains constant. As will be discussed later in more detail, the simultaneous decrease in the energy of the radiating electron and the increase in its axial momentum indicates that the radiating energy is taken from electron orbital motion. (This issue will also be considered based on classical equations.)

Since the axial momentum is equal to $m\gamma v_z$, the change in it can be represented as $\Delta p_z = m(\gamma_0 \Delta v_z + v_{z0}\Delta\gamma)$. Correspondingly, the change in the axial velocity, in accordance with (1.13), is

$$\Delta v_z = \left(c^2/v_{ph} - v_{z0}\right)(\Delta\gamma/\gamma_0). \tag{1.14}$$

Substituting (1.14) into the definition of Θ_d and using the cyclotron resonance condition given by (1.11) yield the following expression for Θ_d:

$$\Theta_d = 2\pi s N(\Delta\gamma/\gamma_0)[(1-n^2)/(1-n\beta_{z0})]. \tag{1.15}$$

Here we introduced the refractive index $n = c/v_{ph} = ck_z/\omega$. In the case of the exact autoresonance, $k_z = \omega/c$ ($n = 1$). Therefore, there is no dynamic phase shift for arbitrary changes in the energy: $\Theta_d = 0$. At the same time, as was stated by M. I. Petelin (1974), it is possible to operate near exact autoresonance when a small difference $|1-n| << 1$ allows one to simultaneously realize a substantial extraction of electron energy and a shift of the electron bunch into a proper decelerating phase. Note that M. I. Petelin considered a limiting case of initially ultrarelativistic electrons and showed that to extract a substantial part of their kinetic energy one should operate with waves having n close to 1. Later, Ginzburg, Zarnitsyna, and Nusinovich (1981) analyzed a more general case of arbitrary initial energies. They showed that, as follows from (1.15), the refractive index can always be optimized with respect to initial energy in order to make it possible to completely decelerate one electron. This kind of operation will be considered in more detail later.

1.4 Normal and Anomalous Doppler Effects

When electrons interact with fast waves whose phase velocities always exceed the speed of light, this interaction obeys the condition of the normal Doppler effect, $v_z < v_{ph}$. However, in the case of electron interaction with slow waves having $v_{ph} = \omega/k_z < c$, both the normal and anomalous Doppler effects may take place (Ginzburg 1959, Nezlin 1976). These two effects are separated by the condition of the Cherenkov synchronism between the wave and electrons, $v_z = v_{ph}$. In terms of the cyclotron resonance condition given by (1.11), the interaction with fast waves occurs at positive cyclotron harmonics ($s > 0$), while the anomalous Doppler effect, which corresponds to $v_z > v_{ph}$, can be realized only at negative harmonics ($s < 0$). These two situations are illustrated with

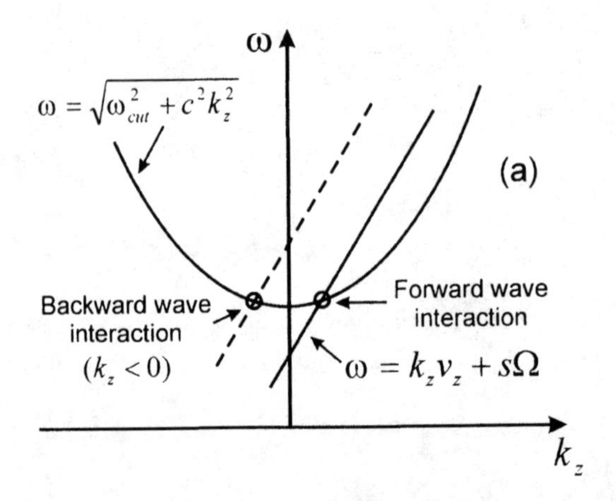

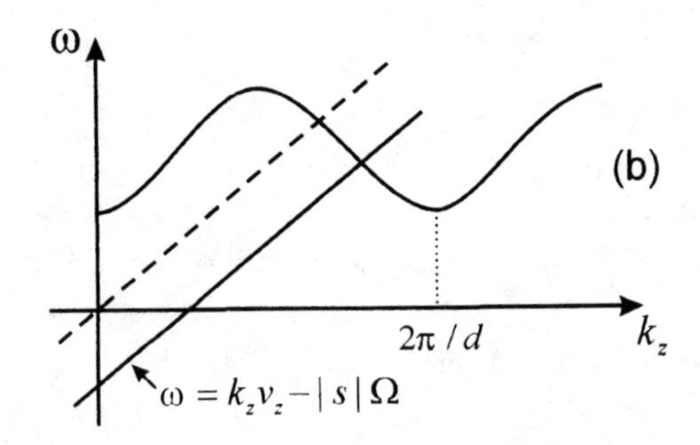

Fig. 1.6. Dispersion diagrams for the cases of fast-wave (a) and slow-wave (b) interactions.

Fig. 1.6. Fig. 1.6a shows the dispersion diagram with some possible resonances between gyrating electrons and fast waves. The dispersion equation for the fast waves in a waveguide has a standard form

$$\omega^2 = \omega_{\text{cut}}^2 + c^2 k_z^2, \tag{1.16}$$

where ω_{cut} is the waveguide cutoff frequency. Fig.1.6b shows a resonance with a slow wave at negative cyclotron harmonic: the dashed line shows the Cherenkov resonance ($s = 0$, i.e., $\omega = k_z v_z$), and the schematic dispersion curve is shown for a hypothetical slow-wave structure with the period d. (Of course, the resonance with slow waves can also take place at positive harmonics.) Recall that, in accordance with Floquet theorem, the periodicity of a slow-wave structure causes the presence of spatial harmonics in k_z (i.e., $k_{z,n} = k_{z,0} + n2\pi/d$), and hence, in the wave field (see, e.g., Sec. 8.8 in Collin 1966).

When an electron radiates under the condition of the normal Doppler effect, its energy and momentum components experience the changes that we already discussed in Sec. 1.3. The radiation under the condition of the anomalous Doppler effect is quite unusual. Therefore, we will briefly describe it here, in spite of the fact that this radiation has no direct relation to gyrotron operation.

In the case of the anomalous Doppler effect discussed by V. L. Ginzburg (1959) and M. V. Nezlin (1976), electrons move with a "superluminal" speed, which means that $v_z > v_{ph}$. In terms of the quantum theory used by these authors, such an electron can simultaneously radiate photons and experience transitions to higher (!) energy levels. Certainly, this energy is withdrawn from electron axial motion. (In a certain sense, this situation is opposite to the case of electron interaction with the backward wave under the condition of the normal Doppler effect, which was discussed in Sec. 1.3. If in that case radiating electrons lose their orbital momentum but gain axial momentum, they now lose axial momentum but increase their orbital momentum.) It is remarkable that in such a case one can observe coherent cyclotron radiation even from electrons with zero initial orbital velocity (Petelin 1974, Bratman and Tokarev 1974). In principle, this looks like an interesting opportunity, since often it is easier to form high-quality linear electron beams than beams of gyrating electrons. However, the necessity to significantly slow down the phase velocity of the wave reintroduces all of the limitations of slow-wave devices, discussed above in the Preface.

To complete this brief consideration of very interesting physical effects, let me mention that, since decelerating electrons lose their axial momentum, they move in the process of deceleration to the border $\beta_z = \beta_{ph}$ between the anomalous and normal Doppler effects. Then, at $\beta_z < \beta_{ph}$, their further deceleration is quite similar to the previously considered deceleration under the normal Doppler effect, i.e., they lose both components of their momentum, and thus can be completely decelerated (for more details see Ginzburg 1979, Vomvoridis 1997).

1.5 Electron Deceleration

Each of the preceding sections in this chapter ended with some comments regarding the range of possibilities for electron deceleration. This is not surprising because, for practically all sources of coherent radiation, the efficiency is one of the major concerns. Certainly, the real efficiency of any source should be evaluated by calculating the efficiency of one electron and averaging it over all initial distributions of electrons in the coordinate and momentum space. This will be discussed later. In the present section we will briefly summarize the results of the analysis of single-particle efficiency.

Our consideration will be based on the use of (1.13), which establishes the relation between changes in the electron energy and axial momentum. This equation can be rewritten as

$$p_{z0} - p_z = \frac{\mathcal{E}_0 - \mathcal{E}}{v_{ph}}. \qquad \textbf{(1.17)}$$

As follows from (1.17), in order to stop the electron axial motion and simultaneously extract all kinetic energy that corresponds to $p_{z,\mathrm{final}} = 0$ and $\mathcal{E}_{\mathrm{final}} = mc^2$, the wave should propagate along $\vec{H}_0$ with the optimum phase velocity (Ginzburg, Zarnitsyna, and Nusinovich 1981)

$$\beta_{\mathrm{ph,opt}} = (\gamma_0 - 1)/\gamma_0\beta_{z0}. \qquad \textbf{(1.18)}$$

One can easily verify that the electron orbital momentum, which, in accordance with the general relation

$$\mathcal{E}^2 = m^2c^4 + c^2\left(p_\perp^2 + p_z^2\right) \qquad \textbf{(1.19)}$$

and (1.17), can be represented as

$$p_\perp^2 = p_{\perp 0}^2 - 2(1 - n\beta_{z0})\gamma_0(\gamma_0 - \gamma) + (1 - n^2)(\gamma_0 - \gamma)^2, \qquad \textbf{(1.20)}$$

also becomes equal to zero when $n = 1/\beta_{\mathrm{ph,opt}}$. (In (1.20) the orbital momentum is normalized to mc.) So, the condition given by (1.18) is the condition of the complete deceleration of one electron, which is the case when the single-particle efficiency equals 100%.

In practice, the electron beam parameters are very often given in terms of the beam voltage, which relates to γ_0 as $V_b = (\gamma_0 - 1)(mc^2/e)$, and the orbital-to-axial velocity ratio, $\alpha = v_{\perp 0}/v_{z0}$. As follows from (1.19), in these terms the axial velocity can be determined as

$$\beta_{z0}^2 = \frac{\gamma_0^2 - 1}{\gamma_0^2(1 + \alpha^2)}. \qquad \textbf{(1.21)}$$

Thus, (1.18) can be rewritten (Nusinovich, Latham, and Dumbrajs 1995) as

$$\beta_{\mathrm{ph,opt}} = [(1 + \alpha^2)(\gamma_0 - 1)/(\gamma_0 + 1)]^{1/2}. \qquad \textbf{(1.22)}$$

In the Preface I pointed out that new trends in the development of high-power sources of millimeter wave radiation were focused on the use of electron interaction with fast waves, which propagate with $\beta_{ph} > 1$. Eq. (1.22) allows us to make some corrections to this general statement in our specific case. As follows from (1.22), the interaction with fast waves is optimal (in the sense of the single-particle efficiency) only at large enough velocity ratios: $\alpha^2 > 2/(\gamma_0 - 1)$. When the beam velocity ratios are in the range $1/\gamma_0 < \alpha^2 < 2/(\gamma_0 - 1)$, the optimal phase velocity corresponds to the interaction of electrons with slow waves ($\beta_{ph} < 1$) under the condition of the normal Doppler effect, which is $\beta_{ph} > \beta_{z0}$. Finally, in the case of small initial orbital velocities ($\alpha^2 < 1/\gamma_0$), it is necessary to slow down the wave even more for realizing $\beta_{ph} < \beta_{z0}$. This is the case of the "superluminal" velocities of electrons, when the anomalous Doppler effect, which was discussed in Sec. 1.4, takes place. This tendency to slow down the wave phase velocity as the electron orbital-to-axial velocity decreases follows from the fact that slow waves, in accordance with (1.17), allow one to extract more energy from electron axial motion.

Of course, for high-power sources of coherent radiation not only the efficiency, but also the power, which can be handled by a microwave structure, is important. Thus, all power limitations of slow-wave devices discussed in the Preface play a big role.

1.6 Optimum Choice of Parameters

The simple equations that were given above allow us to make "zero-order" estimates for the optimum values of some important parameters or, at least, to discuss the methods by which they can be estimated.

The external magnetic field should be chosen to yield a kinematic phase shift Θ_{kin} that will provide for the displacement of an electron bunch into the decelerating phase, i.e., $\Theta_{kin} \sim \pi$. In the case of a large number of electron orbits, $N >> 1$, this yields the following estimate for the initial cyclotron frequency of electrons:

$$\Omega_0 \sim (\omega/s)(1 - n\beta_{z0})[1 - (sN)^{-1}]. \qquad \textbf{(1.23)}$$

Here the factor $1/sN$ characterizes the mismatch of the cyclotron resonance with respect to the Doppler-shifted wave frequency $\omega(1 - n\beta_{z0})$. This mismatch is inversely proportional to the cyclotron harmonic number, s, and the number of electron orbits, N.

The number of electron orbits is, certainly, proportional to the interaction length. This number can be chosen in accordance with (1.15). As follows from this equation, the dynamic phase shift is on the order of 2π when the changes in electron energy are on the order of

$$\Delta\gamma/\gamma \sim (sN)^{-1}[(1 - n\beta_{z0})/(1 - n^2)]. \qquad \textbf{(1.24)}$$

Here the number N can be large when the operation is close enough to autoresonance. For changes in energy on the order of the initial kinetic energy, this means

$$|1 - n^2| << (1 - n\beta_{z0})/\left(1 - \gamma_0^{-1}\right). \tag{1.25}$$

Note that, from applying (1.24) to the last two terms in (1.20), it readily follows that, when $N >> 1$, the last term on the right-hand side of (1.20) is small, and therefore, the orbital momentum is approximately equal to

$$p_\perp \approx p_{\perp 0}\left\{1 - \frac{2}{\beta_{\perp 0}^2}(1 - n\beta_{z0})\left(1 - \frac{\gamma}{\gamma_0}\right)\right\}^{1/2}. \tag{1.26}$$

When the refractive index n is smaller than n_{opt} given by (1.18), a decelerating electron will first lose its orbital momentum prior to stopping its axial motion. In this case, for the final value of $p_\perp = 0$, the single-particle efficiency, as follows from (1.26), is equal to

$$\eta_{\text{sp}} = \frac{\beta_{\perp 0}^2}{2(1 - n\beta_{z0})\left(1 - \gamma_0^{-1}\right)}. \tag{1.27}$$

Nusinovich, Latham, and Dumbrajs (1995) discussed modification of this equation for the case of a small number of electron orbits. It should be noted, however, that in the case of a small N, the electric field of the wave, which decelerates the electrons, should be strong. At the same time, this amplitude can often be restricted either by the breakdown limit (in the case of pulsed operation) or by the possibilities to dissipate the power of ohmic losses in the microwave structure walls (in the continuous wave or high average power regimes).

The wave electric field can be estimated in accordance with (1.3). In the case of resonance at an arbitrary cyclotron harmonic, s, the synchronous electric field E_s, which has a structure of a rotating multipole, can be represented as $E_s \sim (a/\lambda)^{s-1}E$, where $a = v_\perp/\Omega_0$ is the Larmor radius. When the orbital-to-axial velocity ratio is large, the total length of the electron trajectory in the interaction space can be estimated as $2\pi aN$. Therefore,

$$|\Delta\mathcal{E}| \sim e(a/\lambda)^{s-1}E2\pi aN. \tag{1.28}$$

Combining (1.28) with (1.24) and taking into account the cyclotron resonance condition given by (1.11) yields (Bratman et al. 1981)

$$\frac{eE}{mc\omega\gamma_0}\left(\frac{a}{\lambda}\right)^{s-1}\beta_{\perp 0} \sim \frac{1}{2\pi(sN)^2}\frac{(1 - n\beta_{z0})^2}{1 - n^2}. \tag{1.29}$$

This equation shows that the wave amplitude that is required for electron deceleration is inversely proportional to N^2. This is the consequence of the "quadratic" bunching discussed earlier after (1.5). Also note that in the left-hand side of (1.29) $a/\lambda \approx s\beta_{\perp 0}/2\pi(1 - n\beta_{z0})$, so the wave amplitude is also inversely proportional to $\beta^{s}_{\perp 0}$.

So far, in this section, we have analyzed electron deceleration by a wave of given amplitude. In any real device, the EM field is excited (or amplified, in the case of amplifiers) by an electron beam. Correspondingly, the field amplitude depends on the beam parameters. In the simplest case of oscillations in a single cavity, this dependence can be determined by a balance equation.

The power balance equation is valid for stationary regimes. In these regimes, the microwave power withdrawn from the beam is equal to the power of the microwave losses in a cavity having a finite quality factor Q:

$$\eta P_b = (\omega/Q)W. \tag{1.30}$$

Here $P_b = V_b I_b$ is the beam power, η is the device interaction efficiency, and W is the microwave energy stored in the cavity. This energy can be determined as$V < |E|^2 > /8\pi$, where V is the cavity volume, which depends on the operating wavelength and on the operating mode, and the angular brackets designate the averaging of the intensity of this field over the cavity volume.

1.7 Problems and Solutions

Problems

1. Assume that a gyrating electron is initially accelerated by a 100 keV potential. Find the changes in its cyclotron frequency (in percent) when in the interaction process it (a) loses 50% of its initial kinetic energy, (b) gains 50% of its initial kinetic energy.
2. Consider an electron with the initial normalized energy $\gamma_0 = 2$ and the velocity ratio $\alpha = 1$. Assume that this electron interacts with a plane wave having the normalized phase velocity $\beta_{\text{ph}} = 2$ and find the single particle efficiency for the instant of time when this electron loses all its orbital momentum. Also find the ratio of the electron axial momentum at this instant of time to the initial axial momentum.
3. Show that the component of the RF field, which is synchronous with gyrating electrons at the sth cyclotron harmonic, has a structure of the s-order rotating multipole.

Solutions

1. A particle that loses energy will increase its cyclotron frequency by about 8.9%, while a particle that gains energy will decrease the cyclotron frequency by about 7.56%.

2. As follows from (1.20), the change in the electron energy, which corresponds to $p_\perp = 0$, for given parameters of the problem is equal to $\Delta\gamma/\gamma_0 = 0.3285$. Therefore, the single-particle efficiency, $\eta_{sp} = \Delta\gamma/(\gamma_0 - 1)$, is close to 65.7%. Correspondingly, the ratio of final axial momentum to its initial value, as follows from (1.17), is close to 0.73.
3. In any system of a cylindrical symmetry (i.e., a system, for which $\partial/\partial z = 0$), the transverse structure of the field of any mode/wave can be determined by the membrane function Ψ, which obeys the Helmholtz equation $\Delta_\perp \Psi + k_\perp^2 \Psi = 0$. Let us take an arbitrary point (x_0, y_0) as a center of local polar coordinates r, θ, and expand the membrane function in the vicinity of this point into Fourier series $\Psi = \sum_{l=-\infty}^{l=\infty} \Psi_l(r) e^{-il\theta}$. Since we consider the RF fields proportional to $e^{i\omega t}$, each summand in this series represents a rotating field. Now we should determine the structure of $\Psi_l(r)$.

   In general, in polar coordinates, the radial structure of the membrane function, as follows from the Helmholtz equation, can be described by the Bessel function $\Psi_l(r) = L_l J_{|l|}(k_\perp r)$. (Coefficients L_l will be determined below.) At small r's, i.e., in the vicinity of the local center of coordinates, $J_{|l|}(k_\perp r) \simeq (1/|l|!)(k_\perp r/2)^{|l|}$. This corresponds to reducing the Helmholtz equation to the Laplace equation $\Delta_\perp \Psi = 0$. In this limit, the Fourier series of the membrane function, which was given above, turns into the Taylor series,

$$\Psi = \Psi(x_0, y_0) + \sum_{l=-\infty}^{l=-1} \frac{L_l}{|l|!} \left(\frac{\rho}{2}\right)^{|l|} + \sum_{l=1}^{\infty} \frac{L_l}{|l|!} \left(\frac{\rho^*}{2}\right)^{l}, \qquad (1.31)$$

   where $\rho = k_\perp[(x - x_0) + i(y - y_0)]$.

   Thus, the synchronous field, for which the RF field frequency ω is close to $s\Omega$ (i.e., $\omega t - s\theta$ is a slowly variable phase difference), is proportional to $|\rho|^s$.

   To determine the coefficients L_l, one can use the lemma that was originally derived by Schwinger (Schwinger and Saxon 1968) with the use of integral representations for the membrane and Bessel functions. In accordance with this lemma, the action of the angle operator $\hat{\nabla} = \frac{\partial}{\partial(k_\perp x)} + i\frac{\partial}{\partial(k_\perp y)}$ upon the membrane function (1.31), with the account for obvious relations $\hat{\nabla}\rho = 0$ and $\hat{\nabla}^*\rho = 2$, yields $L_l = \hat{\nabla}^{*|l|}\Psi(x, y)$ for $l < 0$ and $L_l = \hat{\nabla}^{|l|}\Psi(x, y)$ for $l > 0$. This procedure is described in Appendix 1 in more detail.

▪ CHAPTER 2 ▪

Gyrotron Arrangement

2.1 Velocity Spread and Inhomogeneous Doppler Broadening Operation Near Cutoff

As is known, there are two key components in any high-power microwave source: an electron beam and a microwave structure in which this beam excites an EM field. In principle, cyclotron resonance masers, being treated as sources of induced coherent radiation from classical excited oscillators (Gaponov, Petelin, and Yulpatov 1967), form so large a class of devices that it was difficult (if not impossible) to determine from the very beginning which of these devices has the best potentials. The only obvious argument was that this device should operate at fast waves, so the microwave structure should not contain any small-scale elements. Thus, it should be either a smooth-wall waveguide or a large-size resonator. When the first CRM experiments were started, lasers had just been invented. So it was clear that in CRMs one could use open resonators with sizes greatly exceeding the operating wavelength. The choice of an electron gun was much less obvious.

Initially two kinds of electron beams were considered (Gaponov 1961): electrons moving along the helical trajectories in a constant external magnetic field (see Fig. 2.1a) and electrons moving along trochoidal trajectories in crossed electric and magnetic fields (Fig. 2.1b). In the first case, it was also understood from the very beginning (Bott 1964, Gaponov et al. 1965) that it makes sense to produce electrons in a weak magnetic field and then let them move into a strong magnetic field region. Due to the adiabatic invariance of the magnetic momentum

$$p_{\perp}^{2}/B = \text{const}, \qquad (2.1)$$

the electron motion in the increasing magnetic field leads to adiabatic compression of the beam and allows one to significantly increase the electron

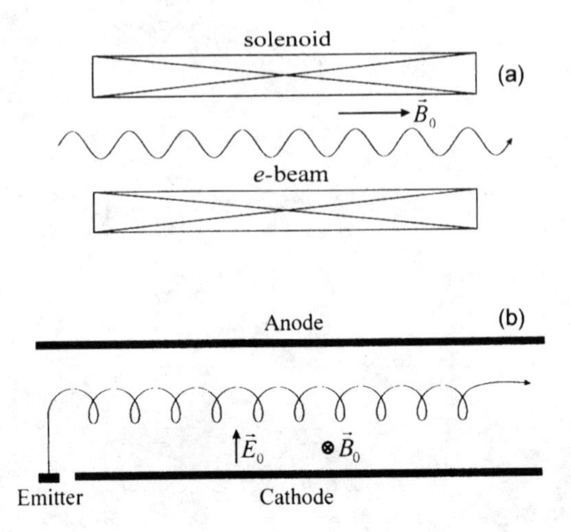

Fig. 2.1. Schematic representation of (a) helical trajectories of electrons moving in a constant magnetic field, and (b) trochoidal trajectories of electrons moving in crossed electric and magnetic fields.

orbital momentum. To produce such electron beams, the modified magnetron injection guns were used (Gaponov et al. 1965, Schriever and Johnson 1966). Less realized from the beginning was the fact that beams generated by such electron guns have a certain spread in electron axial velocities, and that this spread may play a crucial role in the electron interaction with the waves. Indeed, as it follows from the cyclotron resonance condition given by (1.11), the spread in electron axial velocities causes inhomogeneous Doppler broadening of the cyclotron resonance line, which is similar to the Doppler broadening known in gas lasers. In lasers (see, e.g., Lamb 1965), this effect is not so crucial for obtaining coherent EM radiation, because the lines of induced radiation and absorption are well separated, as we already discussed in Sec. 1.2. However, in CRMs these lines are not separated at all and, at the same time, the absolute value of the absorption line, as will be shown later, is larger than that of the radiation line. Thus, the broadening of the resonance may cause a substantial reduction of the radiation line and even its complete disappearance. The deteriorating effect of the axial velocity spread on the operation of CRMs with traveling waves was seen in the first experiments carried out by the group led by A. V. Gaponov in 1959 (Gaponov 1959c).

On the other hand, in trochoidal electron beams the axial (drift) velocity of all electrons moving in the crossed fields

$$\vec{v}_{\mathrm{dr}} = c[\vec{E} \times \vec{H}]/H^2, \tag{2.2}$$

is the same when the beam space charge field is negligibly small. Therefore, the first experiments with CRMs driven by such beams at relatively long wavelengths were quite successful (Antakov et al. 1960, 1966). Nevertheless, these devices did not have the potential for substantial increase in the operating frequency, as was understood from the very beginning (see, e.g., Pantell 1959). Indeed, for avoiding the increase of the space charge effects caused by the electron density $\rho = j/v_z$, it is necessary to keep the drift velocity large enough. However, as the operating frequency increases, the magnetic field increases proportionally to it, so the electric field, as it follows from Eq. (2.2), should suit. At the same time, a number of technical reasons are known, such as a breakdown, which limit this field.

Therefore, in the mid-1960s it was decided to focus the attention on the use of helical beams in the devices operating at waves that propagate with small axial wavenumbers, $k_z << \omega/c$, because the Doppler broadening in such devices can be small enough. Through this way, the gyrotron was invented (Gaponov et al. 1967).

The gyrotron schematic is shown in Fig. 2.2. Here, on the left, an adiabatic magnetron injection gun is shown. Note that, although in the figure the anode and cavity are shown separately, they can be at the same potential. The latter case is often called a *single-anode* or *diode-type* electron gun, in contrast to the case shown in the figure, which is a *triode-type* system, where the anode and cavity have different potentials.

At the cathode, under the action of the electric field caused by the voltage applied between the cathode and the anode, electrons acquire both orbital and axial velocities. Then, they are accelerated toward the interaction region and, because in the region between the gun and the cavity they move in the increasing magnetic field, their orbital momentum increases, in accordance with (2.1), and the electron flow undergoes magnetic compression. In the region of the uniform magnetic field, the electrons pass through an open resonator where, under the cyclotron resonance condition, they excite one of the eigenmodes with high enough quality factors. As will be discussed later in more detail, such a cavity is made in a form of a slightly irregular open waveguide. The waves propagating along this waveguide with large axial wavenumbers have large diffractive losses, so their Q-factors are small, while the waves excited near cutoff have small axial wavenumbers, and therefore their diffractive losses are small. A superposition of such waves propagating with small axial wavenumbers forms high-Q modes of resonators open in the axial direction. After interaction, the spent beam propagates in the region of magnetic decompression and reaches a collector. In the simplest gyrotron configuration

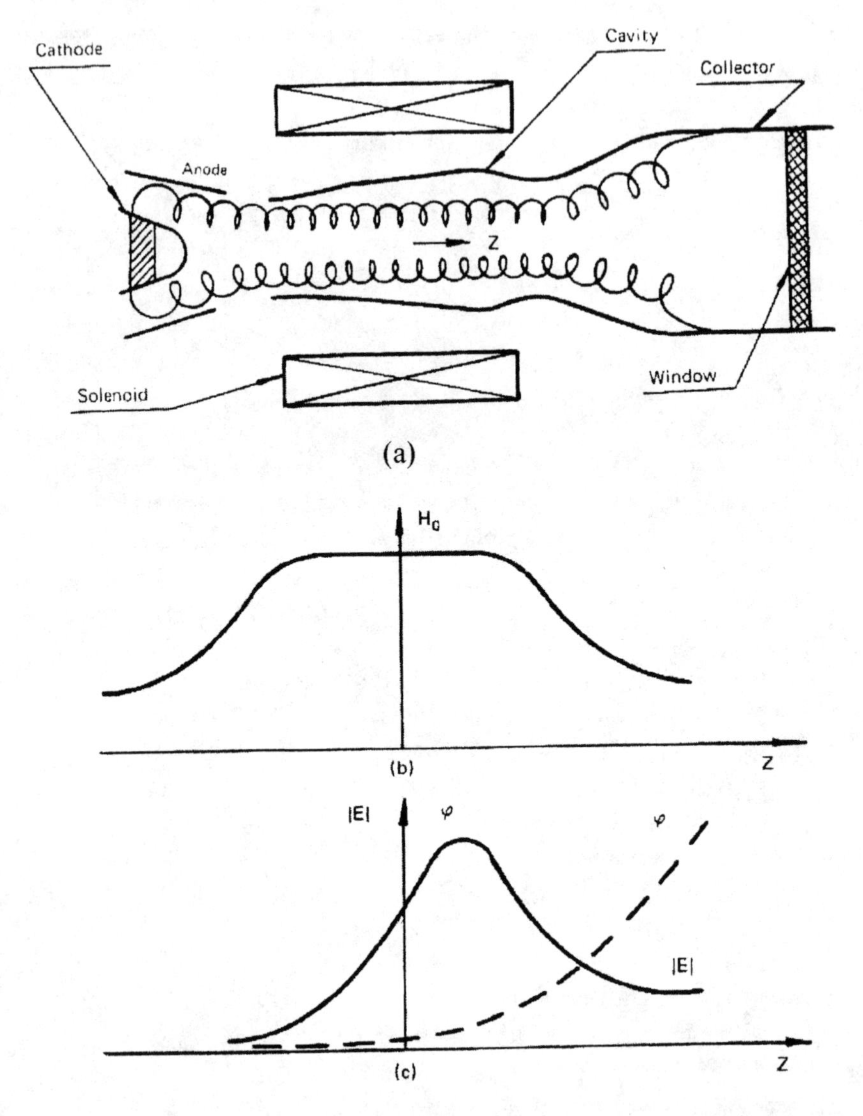

Fig. 2.2. (a) Gyrotron schematic; (b) axial profile of static magnetic field $\vec{H}_0$; (c) axial distribution of the amplitude and phase of the microwave electric field.

shown in Fig. 2.2 the spent beam and the EM radiation propagate together in this region, and a part of the output waveguide is used for collecting the beam. The up-tapering of the output waveguide, which is shown in Fig. 2.2, can significantly increase the collector area and thus reduce the power dissipation density at the collector surface, which is very important for operation at high average power levels. However, in order to avoid the conversion of the outgoing radiation into parasitic modes, this up-tapering should be smooth enough. This contradiction causes some constrains in gyrotron designs. Possible solutions to this problem will be discussed later.

After passing the output waveguide, the radiation is extracted through the output window shown on the right. In addition to the main solenoid shown in Fig. 2.2, some extra coils may be used in the gun region, as well as in the collector region. These additional coils allow for better control of the beam position and other beam parameters in these parts of the tube.

2.2 Electron Optics. Magnetron Injection Guns

The basic principles of operation of such guns were just briefly described above. In this section we will present a very simple adiabatic theory of such guns. In our presentation we will follow Petelin and Goldenberg (1973), where this theory was developed, or, more exactly, Flyagin, Goldenberg, and Nusinovich (1984), where the most important parts of this theory were described.

The adiabatic theory is based on the assumption that the variations of external electric and magnetic fields are small in the vicinity of one electron orbit. This allows one to represent the electron motion as a superposition of the fast gyration with a Larmor radius, a, around a guiding center and a slow motion of this center, which drifts gradually in the fields. When these fields are weakly inhomogeneous, the electron orbital momentum obeys the conservation law given by (2.1), and the drift velocity of electrons can be determined by (2.2). More exactly, the perpendicular motion of the guiding center with respect to the magnetic force line is determined by (2.2), while the parallel motion is determined by the similar equation with the scalar product of $\vec{E}$ and $\vec{H}$ in the numerator instead of the vector product. A more general case of such a motion has been analyzed elsewhere (see, e.g., Nortrop 1963).

Typically, electrons are emitted from the cathode surface with a very small initial velocity. Assuming that this initial velocity is equal to zero, we can conclude that electron initial orbital and transverse drift-velocity components are equal in magnitude and opposite in direction. This allows us to determine the initial orbital velocity of electrons as

$$v_{\perp,c} = cE_{\perp,c}/H_c, \tag{2.3}$$

where the subscript c designates the cathode. Note that this simple consideration is valid for emitters with a smooth enough surface only. In the case of finite roughness of the emitter, as shown by Tsimring (1974) and Y. Y. Lau (1987), the treatment is a little more complicated.

Let us call the readers' attention to the fact that this initial orbital velocity is determined by the component of the electric field perpendicular to the magnetic force line. In principle, the cathode and anode are positioned with respect to the magnetic force lines in such a way that the electric field at the cathode has both perpendicular and parallel components. So, emitted electrons first move azimuthally along cycloidal trajectories over the cathode surface. This motion is quite similar to the electron motion in conventional magnetrons. Then, under the action of the electric field component, which is parallel to the magnetic force line, they become accelerated axially and thus leave the gun region.

In the process of electrons moving from the gun region toward the cavity region, the orbital velocity of electrons varies in accordance with (2.1), which yields the following formula for the orbital velocity in the interaction region:

$$v_{\perp,0} = \alpha_B^{3/2} c(E_{\perp,c}/\gamma_0 H_0). \tag{2.4}$$

Here α_B is the ratio of the magnetic field in the interaction region, H_0, to the magnetic field in the cathode region, H_c. This ratio is often called the magnetic compression. Corresponding axial velocities of electrons can be determined from the general relation between electron energy and momentum given by (1.19). Note that, although typically electrons are accelerated by the electric field produced by applied voltage at shorter distances than their orbital momentum increases significantly in the region of a growing magnetic field, we should always be concerned about possible reflection (mirroring) of some electrons in this magnetic bottle. This means that in the process of electron motion in the transition region from the gun to the cavity, the magnetic compression should not stop and send back the electrons, which acquired significant orbital velocities in the gun region.

In principle, the transverse motion of guiding centers also includes their azimuthal motion. However, in axially symmetric systems, this azimuthal motion does not play a role. At the same time, the radial position of guiding centers changes in accordance with Busch's theorem (see, e.g., Pierce 1954), viz. the beam undergoes the adiabatic compression. Correspondingly, the guiding center radius of any electron orbit can be determined as

$$R_g = R_e[H_c/H(R,z)]^{1/2}. \tag{2.5}$$

Here R_e is the radial coordinate of the cathode point where a given electron was emitted (see Goldenberg and Petelin 1973). Let us also emphasize that the magnetic field $H(R, z)$ in the right-hand side of this equation depends on both R and z coordinates.

As was already discussed above, one of the most important issues in gyrotron operation is the electron velocity spread. There are many reasons that cause velocity spread in the beams formed by magnetron injection guns. Some of the reasons are the space charge, initial thermal velocity spread, emitter roughness, and inhomogeneity of external fields in the region of emitters having a finite width. Here we will not go into details of these quite complicated issues, which were analyzed in many papers (see, e.g., review chapter by Piosczyk 1993 for references). Let us only mention that, when electron guns are well designed, it is possible to generate high-power electron beams with a rather high orbital-to-axial velocity ratio (about 1.5 and more) and an acceptable spread in velocities (several percents in either orbital or axial directions). The spread in electron energies is typically rather small and hence can be neglected. Let us also mention that, depending on the initial slant angle, as shown in Fig. 2.3, the electron trajectories in the gun region can either intersect each other (nonlaminar flow shown in Figs. 2.3a and b) or be quasi-laminar, as shown in Fig. 2.3c. The latter case can be realized when the angle between the cathode surface and the magnetic force line is equal or larger than 25° (Manuilov and Tsimring 1978). Note that with the increase of the slant angle the electron beam gets thicker in the interaction region, which is often undesirable.

In principle, such a simple "zero-order"approach allows one to formulate a set of adiabatic trade-off equations for quick estimates of possible parameters and constrains in the gun design. Tsimring (1974) formulated such trade-off equations for weakly relativistic electron beams; later Lawson with co-workers (Baird and Lawson 1986, Lawson and Specht 1993) generalized them to the case of relativistic beams. In addition to these simple equations, it is also necessary to take into account some serious limitations. The most important of them is the restriction on the ratio of the operating current to the space charge limited current. This ratio should not be too large for mitigating the negative effect of the space charge on the electron velocity spread. Typically in gyrotron electron guns, thermionic cathodes are used, and these cathodes operate in the regime of temperature-limited emission. The current density in such guns primarily depends on the cathode temperature, and this dependence, as is well known (see, e.g., Chodorow and Susskind 1964), is determined by the Richardson-Dushman equation. Numerous simulations and experiments indicate that the beam quality is satisfactory when the current density in these guns does not exceed 0.2–0.3 of the space-charge-limited current density.

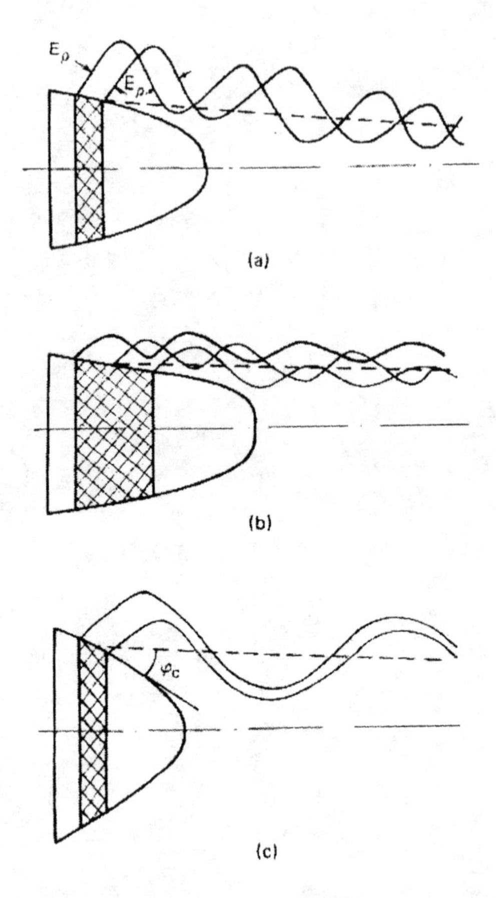

Fig. 2.3. Types of gyrotron electron guns: (a) gun with a narrow emitter; (b) gun with a wide emitter; (c) gun forming a quasi-laminar electron flow.

Voltage Depression

In spite of the gun operation in the temperature-limited regime, the beam space charge effects may cause the voltage depression in the interaction region as well as in the region of the magnetic compression of the beam. In linear beams, this effect was studied a long time ago (see, e.g., Pierce 1954). When electrons have not only axial but also orbital velocities, the voltage depression effect slightly differs from this effect in linear beams (Drobot and Kim 1981, Ganguly and Chu 1984). Therefore, we will briefly discuss this issue below.

To evaluate the voltage depression, one can use the Poisson equation, which we will apply to the case of the beam propagation in a cylindrical pipe of a constant radius. Let us call the region of the beam and the region outside beam Regions I and II, respectively. In Region I the potential obeys the

Poisson equation

$$\frac{1}{r}\frac{\partial}{\partial r}\left(r\frac{\partial \Phi}{\partial r}\right) = -4\pi\frac{|j_z|}{v_z}.$$

Integrating this equation yields for the potential derivative at the outer border of the beam $\left.\frac{\partial \Phi}{\partial r}\right|_{R_{b,\mathrm{out}}} = -2I_b/v_z R_{b,\mathrm{out}}$. Then, solving the Poisson equation for Region II, where there is no current, with a given boundary condition for the potential derivative, yields $\partial\Phi/\partial r = -2I_b/v_z r$. Correspondingly, the potential depression between the cylinder wall of a radius R_w and the beam of a radius R_b is equal to

$$\Delta\Phi = \frac{2I_b}{v_z}\ln\left(\frac{R_w}{R_b}\right). \quad (2.6)$$

In the presence of this depression, the normalized energy of the beam electrons is equal to

$$\gamma = \frac{1}{\left(1-\beta_\perp^2-\beta_z^2\right)^{1/2}} = 1 + \frac{eV-\Delta\Phi}{mc^2}, \quad (2.7)$$

where V is the accelerating voltage applied between the cathode and this pipe. So, the normalized electron energy changes due to the voltage depression by $\Delta\gamma = \Delta\Phi/mc^2$. The normalized orbital velocity of electrons determined by (2.4) also depends on this effect. Denoting the normalized electron energy in the absence of the voltage depression by γ_0, one can rewrite (2.4) as $\beta_\perp = \beta_{\perp 0}(\gamma_0/\gamma)$. Correspondingly, the normalized axial velocity can be determined, in accordance with (2.7), as

$$\beta_z = \left[\beta_{z0}^2 - \left(1-\beta_{z0}^2\right)\left(\gamma_0^2/\gamma^2 - 1\right)\right]^{1/2}.$$

So, the orbital-to-axial velocity ratio varies due to the voltage depression as

$$\alpha = \alpha_0\frac{1+\delta\gamma}{\left[1-2\delta\gamma\left(1-\beta_{z0}^2\right)/\beta_{z0}^2\right]^{1/2}}. \quad (2.8)$$

Here we denoted $\Delta\gamma/\gamma_0$ by $\delta\gamma$. For small enough $\delta\gamma$'s (2.8) reduces to $\alpha = \alpha_0(1+\delta\gamma/\beta_{z0}^2)$.

In conclusion to this brief overview, let us mention that these magnetron injection guns (MIGs) can be designed by two methods: analysis and synthesis. Typically, designers use numerous codes for analyzing the beam properties. The most known among these codes is the EGUN code developed by W. Herrmannsfeldt (Herrmannsfeldt 1979). There were some attempts to develop methods of synthesis of such guns, which are applicable to the guns forming quasi-laminar electron flows (Manuilov and Tsimring 1978, Fliflet et al. 1982). However, so far these methods are not actively used in practice.

2.3 Microwave Structures (Cavities and Waveguides)

In this section we will consider only fast-wave microwave structures, which are used nowadays for excitation of EM fields by electron beams in gyrodevices.

Waveguides are extremely simple because in the case of operation at fast waves we do not need any slow-wave structure. So, it could be any waveguide, in which, as is known (see, e.g., Jackson 1962), a fast wave propagates along the waveguide axis with the phase $v_{ph} = \omega/k_z$, and group $v_{gr} = d\omega/dk_z$, velocities related as

$$v_{ph}v_{gr} = c^2. \tag{2.9}$$

(So, the refractive index $n = \beta_{ph}^{-1}$ introduced above in our case is equal to $\beta_{gr} = v_{gr}/c$.)

Most complicated parts of these waveguides are couplers, which couple the excited wave to the output part of the device. Also, in the case of traveling-wave amplifiers, the input signal should excite the waveguide at the entrance. There is no one specific type of the input coupler that is widely recognized as the best for all waveguides. So, depending on the operating mode and wavelength, various input couplers are in use. (Since this is often considered as a standard microwave technique, these couplers are barely mentioned in papers describing experiments with gyroamplifiers. Note, however, that development of input couplers for excitation of high-order modes can be a very complicated business.)

Outgoing microwave radiation propagates from the interaction region to the output window through an up-tapered waveguide, as is shown in Fig. 2.2. In this waveguide, as the wave cutoff frequency decreases with up-tapering, the axial wavenumber, k_z, in accordance with (1.16), increases. Since the group velocity, $v_{gr} = c^2/v_{ph} = c^2k_z/\omega$, increases proportionally to k_z, this facilitates extraction of the microwave power from the interaction region. The main concern in the design of such up-tapers is minimization of the conversion of the operating wave into other waveguide modes in the tapers of a finite length (see Solymar 1959). Presently, optimization of these tapers is often made with the use of Dolph-Chebyshev distributions (see, e.g., Lawson 1990). Recently, the research in this field was overviewed by Mobius and Thumm (1993).

Resonators, which are typically used in gyrotrons, can also be considered as irregular waveguides (Vlasov et al. 1969). Almost all of these resonators have a cylindrical symmetry, and their radius varies slowly along the axis. As mentioned in Sec. 2.1, in gyrotrons *open* resonators are used. The fact that these

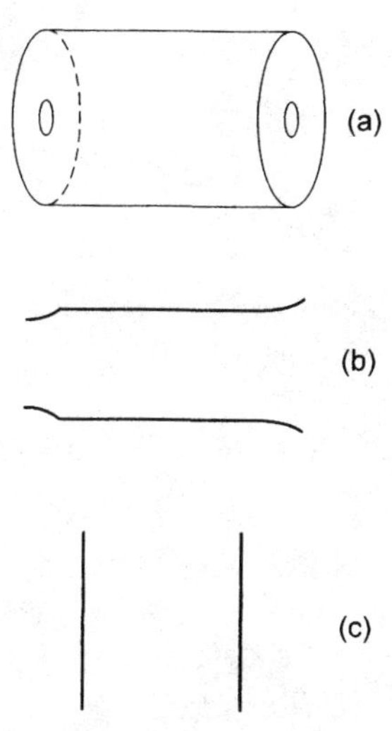

Fig. 2.4. Schematics of (a) closed cavity; (b) axially open resonator; (c) Fabry-Perot resonator.

resonators are open provides their selectivity, which is extremely important for selective excitation of the desired mode in a volume with dimensions much larger than the wavelength. To illustrate this statement, recall that in closed cavities, as in the one shown in Fig. 2.4a, the spectrum density of modes is given by the Rayleigh-Jeans formula

$$\Delta N \sim \frac{V}{\lambda^3}\frac{\Delta\omega}{\omega}.$$

Here V is the resonator volume and ΔN is a number of modes with frequencies in the $\Delta\omega$ range. When the resonator is open in the axial direction (Fig. 2.4b), the density of high-Q modes is equal to

$$\Delta N \sim \frac{S}{\lambda^2}\frac{\Delta\omega}{\omega},$$

where S is the resonator cross section, because the modes with only one axial variation have smaller diffractive losses than others. Further rarefaction of the mode spectrum can be achieved in Fabry-Perot resonators (Fig. 2.4c) used in lasers. Recall that a Fabry-Perot resonator is formed by a pair of mirrors. Therefore, the high-Q modes here are those modes that have only one field variation at the mirror surface. The spectral density of such modes depends on the distance L between mirrors,

$$\Delta N \sim \frac{L}{\lambda}\frac{\Delta\omega}{\omega}.$$

The idea to use such resonators in gyrodevices has led to the development of quasi-optical gyrotrons, which will be discussed later. Now we will come back to the resonators open in the axial direction.

It was stated in Sec. 2.1 that these resonators provide selectivity among modes having different axial structure. This statement can easily be understood, once we introduce, ignoring, for the sake of simplicity, reflections from the ends, the diffractive Q-factor for such a resonator of a length L:

$$Q_{\text{dif}} \simeq \frac{\omega L}{v_{\text{gr}}}. \tag{2.10}$$

This is a minimum diffractive Q-factor, which characterizes the time required for the field stored in such a resonator to be radiated through open ends. In (2.10) the group velocity, as we already discussed, is proportional to the axial wavenumber. Therefore, the modes with a large number of axial variations, i.e., modes that have large axial wavenumbers, have lower diffractive Q's. The axial wavenumber of the mode with one axial variation at the resonator length L can be estimated as $k_z \sim \pi/L$. Correspondingly, the group velocity for this mode is on the order of $v_{\text{gr}} = c^2 k_z/\omega \sim c\lambda/2L$, and hence the minimum diffractive Q is on the order of

$$Q_{\text{dif,min}} \approx 4\pi\,(L/\lambda)^2\,. \tag{2.11}$$

The EM field in such a resonator can be considered as a superposition of waves. To describe these wave fields in irregular waveguides in general is quite complicated. Usually (see, e.g., Vlasov et al. 1969), these fields are represented as the waves of the waveguide of comparison, which is a regular waveguide whose radius is equal to the radius of our irregular waveguide in a given cross section. However, as it was discussed by Kuraev (1979), the fields of azimuthally rotating waves of the comparison waveguide, strictly speaking, do not obey the boundary conditions at the irregular waveguide wall. (One of possible methods to correctly solve this problem is based on the transformation of the coordinate system suggested by Ilyinskiy and Sveshnikov (1968). Their method implies introducing a new normalized radial coordinate $\rho = r/R_w(z)$, where $R_w(z)$ is the axially variable radius of the waveguide wall.)

The fields in irregular waveguides can be represented in terms of the waves of the comparison waveguide under certain conditions only. According to Vlasov, Orlova, and Petelin (1981), this condition can be formulated as

$$L^2/\lambda D >> 1. \tag{2.12}$$

Here D is the waveguide diameter. Note that a wave propagating almost perpendicular to the waveguide axis can be considered between two successive reflections from the wall as the wave radiated at the point of the first reflection, which then experiences a certain diffraction spread before reaching the wall the next time. This process is illustrated by Fig. 2.5. Here an azimuthally rotating $TE_{m,p}$-wave (m and p are the azimuthal and radial indices, respectively, for rotating modes m is a nonzero index) is represented as a set of rays, which in the process of propagation form a caustic with the radius $R_c = (m/\nu_{m,p})R_w$. Here $\nu_{m,p}$ is the pth root of the equation $J'_m(\nu) = 0$, which is the boundary condition for the $TE_{m,p}$-mode at the waveguide wall of the radius $R_w(z)$. Such diffraction, as is shown in Fig. 2.5, is similar to the diffraction of the wave radiated from a hole of the radius $R = L/2$ in the screen, at the distance, which corresponds to the distance between reflections. In our case this distance is

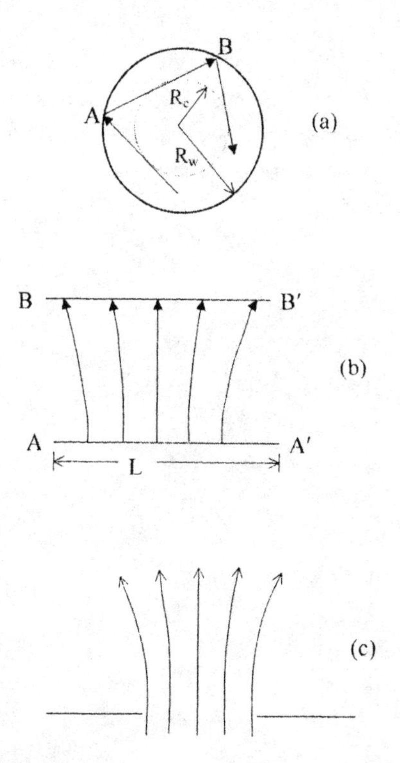

Fig. 2.5. (a) Ray propagation in the case of a rotating wave in a circular waveguide; (b) ray propagation (with the account for diffraction) between two consecutive reflections at AA' and BB' [A and B are shown in figure (a)]; (c) ray propagation in the case of an incident plane wave passing through a hole in a screen.

on the order of D, or, more exactly, $D_{\text{eff}} = D\sqrt{1 - (m^2/\nu^2_{m,p})}$. Thus, the ratio $L^2/\lambda D_{\text{eff}}$ plays a role of the Fresnel parameter. Correspondingly, Eq. (2.12), which should be more accurately rewritten as $L^2/\lambda D_{\text{eff}} >> 1$, can be treated as the condition of applicability of the geometric optics approach, in which the diffraction effects are negligibly small. (For accurate description of diffraction effects in various open resonators and waveguides see Weinstein 1969.)

By using this approximation, one can assume that the transverse structure of any mode in our resonator is the same as the transverse structure of corresponding wave in the comparison waveguide. Then, from Maxwell equations one can readily derive for the axial structure of this mode the nonuniform string equation

$$\frac{d^2 f}{dz^2} + k_z^2(\omega, z) f = 0\,, \tag{2.13}$$

where $k_z^2 = (\omega/c)^2 - k_\perp^2(z)$. In the case of TE-modes, which we will focus on later, it is convenient to use this function f for describing the axial component of the magnetic field of the mode. The transverse wavenumber, $k_\perp$, is determined by the boundary conditions at the wall: for a TE-wave in a cylindrical irregular waveguide it is equal to $\nu_{m,p}/R_w(z)$. Note that (2.13) has the form of the stationary Schrodinger equation for a particle in a one-dimensional potential well with the potential $U = k_\perp^2(z)$. As is known (see, e.g., A. Messiah, n.d.),

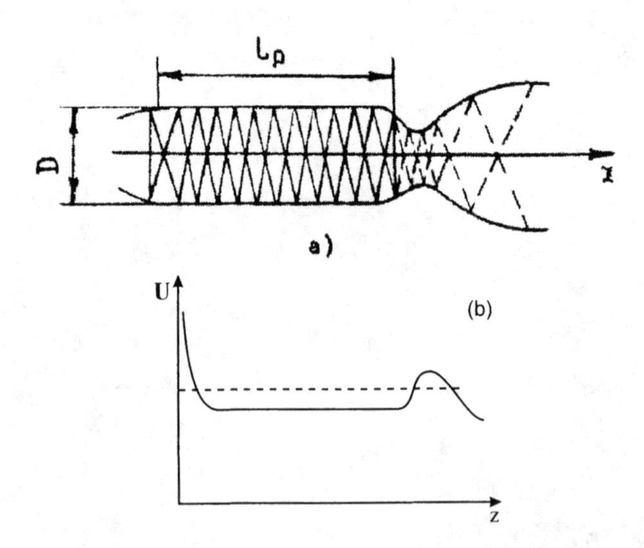

Fig. 2.6. (a) Gyrotron resonator with rays forming the field of an operating mode; (b) potential well corresponding to the resonator profile shown in (a). Horizontal dashed line in (b) corresponds to $k_z^2 = 0$.

the solutions of this equation depend on the well shape. In our case, this well shape is determined by the profile of the irregular waveguide. Typically, the irregular waveguides used as open resonators in gyrotrons have a strong cutoff narrowing at the entrance, which protects the gun side of the tube from penetration of microwave radiation. At the exit, these waveguides may have a small narrowing neck. Such a profile and a corresponding potential well are shown in Fig. 2.6. The neck at the exit has a finite width and depth, which are important for diffractive losses of the mode. This effect is illustrated with Fig. 2.7 reproduced from Vlasov et al. (1969). This figure shows the resonator profile (a) and the corresponding dependence of diffractive losses on the parameter of output up-tapering, $\xi = \nu_{m,p}^2 (L/D)^3 \theta_t$ (here θ_t is the angle of the up-taper) for several configurations of the output neck (b). (In our consideration these

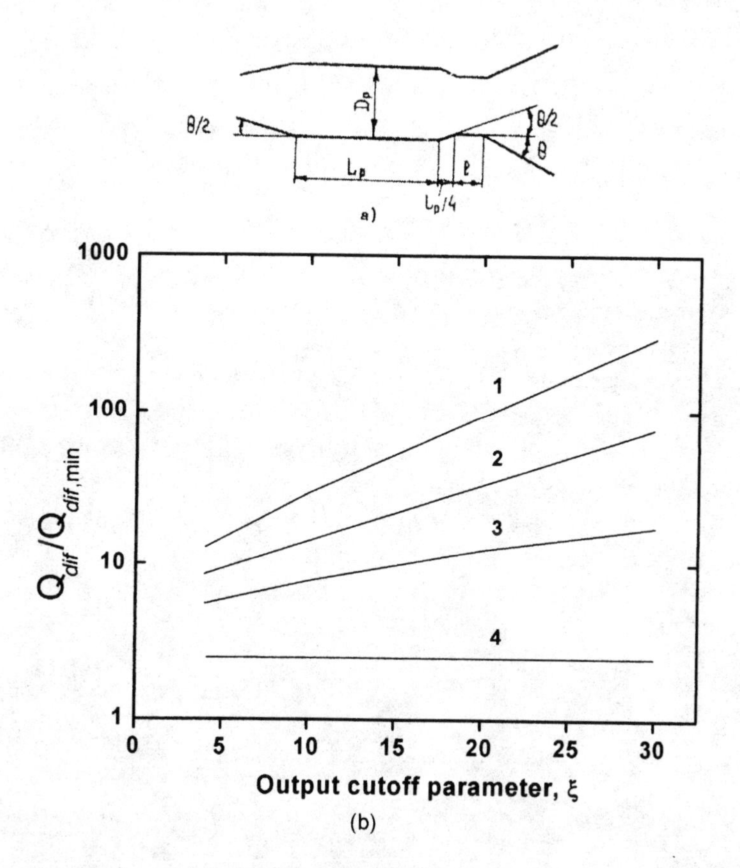

Fig. 2.7. (a) Resonator profile, (b) ratio $Q_{dif}/Q_{dif,min}$ as the function of the output taper parameter ξ for several configurations of the output neck: 1, $l/L = 0.23$; 2, $l/L = 0.115$; 3, $l/L = 0$; 4, no neck (a regular part is directly connected to the uptaper).

losses are quite similar to the effect of tunneling radiation through a finite barrier in the potential well.) Note that, very often, in high-power gyrotrons the open resonators without such necks at the output are used. This allows one to reduce Q-factors of operating modes, and thus, in accordance with (1.30), to operate with the maximum efficiency at higher powers.

Coming back to the nonuniform string equation (2.13), let us note that, of course, (2.13) should be solved together with the corresponding boundary conditions. For resonators like the one shown in Figs. 2.6 and 2.7, they are the condition of the cavity field decay at the entrance, $\frac{df}{dz}|_o = |k_z|\, f(0)$ and the condition $df/dz|_L = -ik_z f(L)$ for the outgoing radiation at the exit. (Here the radiated field is proportional to $\exp\{i(\omega t - k_z z)\}$.)

2.4 Types of Gyrodevices

The mechanism of the orbital bunching based on the relativistic dependence of electron cyclotron frequency on its energy can be used for developing a large variety of devices. As it was stated by A. V. Gaponov in 1966 at the Microwave Electronics Conference, Saratov, USSR (see also available references in Flyagin et al. 1977 and Andronov et al. 1978), a gyro counterpart can be found for any microwave tube driven by a linear electron beam. Key members of this large family are shown in Fig. 2.8. Here, in the first column one can see the gyromonotron, which we described already. Next is the gyroklystron, which is shown in its simplest, two-cavity configuration. Certainly, the number of cavities can be larger. The two subsequent devices are those based on the electron interaction with traveling waves: the gyro-traveling-wave tube (gyro-TWT) and the gyro-backward-wave oscillator (gyro-BWO). These devices are much simpler than their linear beam counterparts, because gyrodevices do not need any slow-wave structure. Also, in the linear backward-wave oscillator

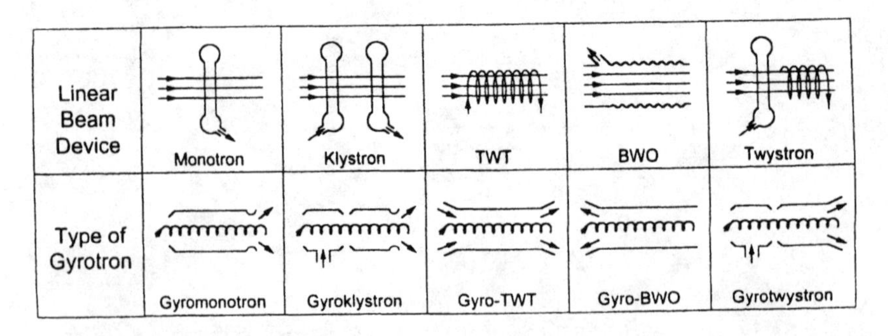

Fig. 2.8. Schematics of linear beam devices and corresponding gyrodevices.

the backward wave is a wave whose phase and group velocities have opposite signs: the phase velocity of the synchronous space harmonic is in Cherenkov synchronism with electrons, while the microwave energy propagates toward the cathode. On the contrary, the wave used in the gyro-BWO is simply the wave whose phase and group velocities have the same sign [these velocities obey (2.9)], so this wave propagates in a smooth-wall waveguide in the direction opposite to the beam. In the last column, a hybrid device, gyrotwystron, is shown. Like a linear-beam twystron, this device is a hybrid of two devices: the gyroklystron and the gyro-TWT. Therefore, it can combine the merits of both: efficient electron bunching and, hence, potential for high gain due to the energy modulation in the input cavity, which is typical for klystrons, and large bandwidth due to the use of the output waveguide, which is typical for TWTs. There are also some other types of gyrodevices, which will be discussed later.

2.5 Magnets and Solenoids

As follows from the cyclotron resonance condition (1.11), the magnetic field required for generating a radiation with the wavelength λ is equal to

$$B_0(T) = \frac{1.07\gamma_0}{s\lambda(\mathrm{cm})}(1 - n\beta_{z0}). \tag{2.14}$$

This field should be produced in large enough volumes because, first, the cross section of the interaction space is much larger than λ^2; and second, the walls of the microwave structure as well as other parts of a tube should be cooled, which requires cooling channels, etc.

Permanent magnets are capable of producing in such volumes the fields below 1T. Widely used Sm-Co permanent magnets produce up to 0.5–0.6 T. New materials (such as, for example, Nd-Fe-B) allow to slightly increase the field. (We are not discussing here some specific configurations like "magic spheres"(Leupold, Potenziani, and Tilak 1992), in which higher fields can be achieved, but configurations that can be utilized in gyrotrons.) This means that such magnets can, in principle, be used for generating at the fundamental cyclotron resonance only centimeter and longer wavelengths' radiation. Operation at the second cyclotron harmonic allows one to generate 8-mm wavelength (Ka-band) radiation, which is important for many applications.

It is necessary to note that gyrotrons require quite specific configurations of the magnetic field. Thus, design of reasonably lightweight magnet systems for gyrotrons is rather complicated. For a number of reasons permanent magnets

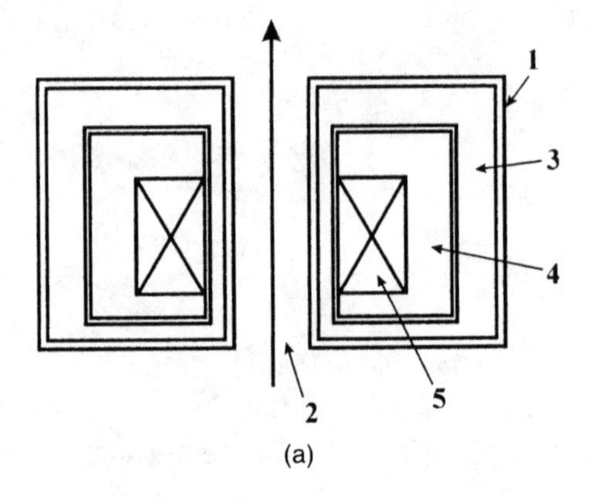

(a)

(b)

Fig. 2.9.

(c)

Fig. 2.9. (a) Schematic of a cryostat with a superconducting solenoid: 1, body; 2, inner bore; 3, liquid nitrogen vessel; 4, liquid helium vessel; 5, solenoid; (b) cryomagnet producing a nominal magnetic field of 5.5 T; (c) a 140 GHz, 100 kW CW gyrotron in the cryomagnet shown in (b). [Figures (b) and (c) courtesy of CPI]

are not widely used in gyrotrons so far. However, there were several experiments in which permanent magnets were used in combination with small solenoids, which allow one to correct the magnetic field distribution for optimizing the device performance.

Water-cooled solenoids are often used in experiments with second-harmonic Ka-band gyrotrons. The power consumption of these solenoids increases with the magnetic field and the volume. Therefore, typically, such solenoids are used for producing magnetic fields up to 0.6–0.7 T. In the case of operation at the second harmonic it allows one to generate the Ka-band radiation; shorter wavelengths can be generated at higher harmonics. As an example, let us mention a 100 kW, 5.5 mm wavelength, third-harmonic gyrotron, which operated in a water-cooled solenoid (S. Malygin 1986).

As was mentioned above, in addition to the main solenoid/magnet, which creates a uniform magnetic field in the interaction region, some solenoids are also often used in the electron gun and collector regions. The gun coil allows one to optimize the quality of an electron beam. This typically means simultaneous maximizing of the electron orbital-to-axial velocity ratio without causing reflected electrons and minimizing the velocity spread. Additional coils in the collector region allow one to optimize the beam power deposition at the collector surface.

Superconducting solenoids are the solenoids that are routinely used for producing strong magnetic fields in millimeter- and submillimeter-wave gyrotrons. Solenoids manufactured from Nb-Ti wires can produce magnetic fields up to 8–10 T in large enough volumes (bore size larger or on the order of 100 mm). Achievable magnetic fields can be increased up to 12–15 T when multifilament Nb_3Sn wires are used. In principle, these fields can even reach 20 T and above. However, such solenoids become extremely expensive. (For recent advances in the development of superconducting solenoids, see van Sciver and Marken 2002.)

A superconducting solenoid should be placed in a cryostat. Schematic of such a cryostat is shown in Fig. 2.9a. Here a solenoid (5) is placed into a vessel (4) filled with liquid helium. This vessel, in turn, is placed into a vessel (3) containing liquid nitrogen. These two vessels are isolated by a vacuum space from each other and from the cryostat body 1. All elements allowing for the solenoid fixation and alignment, as well as current conductors used for feeding the solenoid, are not shown here. A real cryomagnet and this cryomagnet with a gyrotron in it are shown in Figs. 2.9b and 2.9c, respectively.

There are many technical issues to be addressed in the development of such cryomagnets. Among them are such important issues as reducing the consumption of liquid helium and alignment and fixation of solenoids in cryostats. Of course, in principle, the axes of solenoid and cryostat should coincide. Restrictions on their misalignment are quite severe: typically a linear misalignment should be smaller than 1 mm (often, even less than 0.1 mm), and angular misalignment should be smaller than a half of degree (again, sometimes it can be even more stringent). While it is relatively easy to align a gyrotron with respect to the cryostat, it is much more complicated to align it with respect to the solenoid. Since there can be some additional solenoids, which were mentioned above, located outside of the cryomagnet, the effects of their magnetic fields should not result in any displacement of a superconducting solenoid. Thus, this main solenoid should be well fixed. Note that in recent years a significant progress in the development of liquid helium-free cryomagnets has been demonstrated (see, e.g., Yokoo et al. 1996).

Regarding the solenoids located in the collector region, one more comment should be made. In gyrotrons operating at high average-power levels, in order to reduce the beam power density deposition at the collector, very often some additional solenoids are used, which are fed by an alternating current. The use of such current results in a periodic scanning of an electron beam back and forth along the collector area. The frequency of this scanning typically ranges from 0.1 Hz to a few Hz, because too low frequency may cause local overheating of the collector, while too high frequency may result in the skin-effect of a variable magnetic field by a metallic body of a gyrotron tube. The latter issue is especially important in the case of using pulse solenoids in gyrotrons.

Pulse solenoids can produce magnetic fields much higher than superconducting ones. Therefore they attracted the attention of gyrotron developers from the very beginning as the means for generating very short wavelength radiation. Bott (Bott 1964) was the first to generate a submillimeter-wave (up to $\lambda \approx 0.95$ mm) radiation with the use of pulse solenoids. However, the power in his experiments was at a milliwatt level only. There were several more attempts during the 1960s and 1970s. Finally, in the early 1980s there was a series of experiments (Flyagin, Luchinin, and Nusinovich 1983), in which a 100 kW level radiation was generated at frequencies up to 500 GHz. (At the highest frequency achieved, 650 GHz, the power level exceeded 40 kW in about 50 microsecond pulses.)

In the course of these experiments it was necessary to meet quite contradictory requirements. On the one hand, the conductivity of metallic parts of the tube had to be low enough and the pulse duration of the magnetic field should be large enough to allow this variable field for penetrating into the tube. Therefore a tube body was made of a thin stainless steel and the pulse duration was about 5–10 milliseconds. At the same time, an inner surface of the resonator should have high conductivity for reducing ohmic losses of the microwave power. To solve this problem, an inner surface of a resonator was covered by a thin (10–20 microns) layer of copper. As a result, the efficiency of submillimeter-wave gyrotrons at the wavelengths above 0.7 mm exceeded 10% and then gradually decreased to several percentages at the wavelengths shorter than 0.5 mm.

2.6 Problems and Solutions

Problems

1. Assume that in the region between an electron gun and an interaction space the electrons have the same kinetic energy at the entrance to the resonator and in the cross section, where the magnetic field is two times smaller than in the resonator. Find the difference between orbital-to-axial velocity ratios in these two cross sections.
2. Evaluate the effect of the voltage depression on the orbital-to-axial velocity ratio, α, for the case of an 80 kV, 40 A electron beam, whose radius is two times smaller than the waveguide radius. Assume that in the absence of this depression $\alpha = \alpha_0 = 1.5$.
3. To provide a more or less equal coupling of all electrons in a thin annular beam to the electromagnetic field of a waveguide or resonator, the spread of the guiding center radii should be small enough. Estimate the ratio of the radial thickness of a 40 A electron beam to the wavelength in a 100 GHz gyrotron. (The "radial thickness" means here the spread in the radii of electron guiding centers.) Consider an electron gun, which has the cathode loading (current density at the emitter) 3 A/cm^2, the radius of a thin emitter ring equal to 4 cm, and an angle of the emitter surface with respect to magnetic force line equal to 25°. Also assume that the magnetic compression ratio $\alpha_B = 25$.

Solutions

1. From (2.1) it follows that in the cross section $z = z_1$, where $B(z_1) = B_0/2$, the electron orbital momentum is $\sqrt{2}$ times smaller than in the resonator. Denote the orbital-to-axial velocity ratio at $z = z_1$ by α_0. Then, the normalized axial momentum in the resonator p'_z, in accordance with (1.19), can be expressed via its value p'_{z0} at $z = z_1$ as $p'_z = p'_{z0}\sqrt{1 - \alpha_0^2}$. This equation

shows that there will be no electron mirroring in the process of magnetic compression only when $\alpha_0 < 1$. Then, the orbital-to-axial velocity ratio in the resonator can be determined as

$$\alpha = p'_{\perp}/p'_z = \alpha_0 \left[2/\left(1-\alpha_0^2\right)\right]^{1/2}.$$

2. As follows from (2.6), the normalized voltage depression in (2.8) is equal to $\delta\gamma = 0.011676$. Substituting this number together with other given values for the parameters into (2.8) yields $\alpha \approx 2$.
3. The width of the emitter, as follows from given values of the total beam current, current density, and the radius of a thin emitter, is approximately equal to 0.53 cm. In the cross section perpendicular to the magnetic force lines the projection of this width is close to 0.2244 cm. Then, after magnetic compression, the beam thickness in the resonator is equal to 0.045 cm that for a 110 GHz operating frequency results in the ratio of the beam thickness to the wavelength close to 0.15.

Part II

Theory of Gyrotron Oscillators and Amplifiers

■ CHAPTER 3 ■

Linear Theory of the Gyromonotron

3.1 Transversely Homogeneous Model

An electromagnetic wave in any waveguide or resonator has a certain spatial structure. The effect of this structure on the interaction with gyrating electrons will be analyzed in detail later. Before doing this, it makes sense to clarify the physics of interaction in CRMs by considering its simplest model, which implies the interaction of gyrating electrons with a transversely homogeneous wave. (In describing this model we will follow Bratman et al. 1981.)

Let a beam of electrons gyrating in a constant external magnetic field $\vec{H}_0 = H_0\vec{z}$ pass through an interaction region localized in the axial direction ($0 \leq z \leq L$). Let us also represent an EM field in this space as a monochromatic, circularly polarized, plane EM wave with constant amplitude. Assume that, due to the presence of some medium, the wave propagates in the axial direction with the phase velocity, $v_{ph} = \omega/k_z$, which can be different from the vacuum speed of light c. The vector potential of such a wave $\vec{A} = A_x\vec{x}_0 + A_y\vec{y}_0$ can be presented in the complex form as

$$A_+ = A_x + iA_y = Ae^{i(\omega t - k_z z)}, \tag{3.1}$$

where the complex amplitude A in our case is constant.

The momentum of a relativistic electron obeys the standard equation of motion

$$\frac{d\vec{p}}{dt} = -e\left\{\vec{E}_\sim + \frac{1}{c}[\vec{v} \times (\vec{H}_0 + \vec{H}_\sim)]\right\}. \tag{3.2}$$

Here $\vec{E}_\sim$ and $\vec{H}_\sim$ are, respectively, the electric and magnetic fields of the wave, which can be expressed via the vector potential $\vec{A}$ as

$$\vec{H}_\sim = \nabla \times \vec{A}, \quad \vec{E}_\sim = -\frac{1}{c}\frac{\partial \vec{A}}{\partial t}. \tag{3.3}$$

Using the vector potential in the form given by (3.1) and representing the electron transverse momentum in a similar fashion, $p_+ = p_x + ip_y$, one can readily derive from the general equation of motion (3.2) the following pair of equations:

$$\frac{dp_+}{dt} = i\Omega p_+ + i\frac{e\omega}{c}\left(1 - \frac{v_z}{v_{ph}}\right)A_+, \tag{3.4}$$

$$\frac{dp_z}{dt} = -e\mathrm{Re}\left(p_+^* \frac{\partial A_+}{\partial z}\right). \tag{3.5}$$

Also, the general equation for electron energy (1.2) can be rewritten in these variables as

$$\frac{d\mathcal{E}}{dt} = e\mathrm{Re}\left(p_+^* \frac{\partial A_+}{\partial t}\right). \tag{3.6}$$

Since in our simple case, as it follows from Eq. (3.1), $\partial A_+/\partial z = -ik_z A_+$ and $\partial A_+/\partial t = i\omega A_+$, Eqs. (3.5) and (3.6) yield the same integral of motion

$$\mathcal{E} - v_{ph}\, p_z = \text{const.}, \tag{3.7}$$

which was derived earlier, in Sec. 1.3, from simple arguments based on quantum mechanics. This integral, in combination with the general relation between electron energy and momentum given by (1.19), allows us to reduce equations (3.4)–(3.6) to only two equations for the normalized energy $u = (\mathcal{E}_0 - \mathcal{E})/\mathcal{E}_0$ and the phase of the wave with respect to the gyrophase, $\vartheta = \omega t - k_z z - \arg(p_+)$. (Here the complex transverse momentum, p_+, is represented as $|p_+|\exp\{i[\arg(p_+)]\}$.) Introducing the normalized amplitude $\alpha = eA/mc^2\gamma_0$ and the axial coordinate $Z = \omega z/v_{z0}$, one can write these equations as

$$\frac{du}{dZ} = \alpha\beta_{z0}\frac{p'_\perp}{p'_z}\sin\vartheta, \tag{3.8}$$

$$\frac{d\vartheta}{dZ} = \frac{\beta_{z0}}{p'_z}\{\delta - (1-n^2)u - \frac{\alpha}{p'_\perp}[1 - n\beta_{z0} - (1-n^2)u]\cos\vartheta\}. \tag{3.9}$$

Here the primed momenta designate normalization of these components to $mc\gamma_0$; also $\delta = 1 - n\beta_{z0} - \Omega_0/\omega$ is the initial cyclotron resonance mismatch. As it follows from Eqs. (3.7) and (1.19), these components of the momentum can be represented as

$$p'_z = \beta_{z0} - nu, \tag{3.10}$$

$$(p'_\perp)^2 = \beta_{\perp 0}^2 - 2(1 - n\beta_{z0})u + (1-n^2)u^2. \tag{3.11}$$

At the entrance, $Z = 0$: $u = 0$, $p'_{z0} = \beta_{z0}$, $(p'_{\perp 0})^2 = \beta_{\perp 0}^2$, and the initial phase ϑ_0 is homogeneously distributed from 0 to 2π.

In order to ease the derivation of (3.9), let us note that, first, in accordance with the definition of θ, in the figure brackets of this equation we got the variable detuning of the cyclotron resonance, $\omega - k_z v_z - \Omega$. Then, with the use of (3.10) and obvious relation $p_z = m\gamma v_z$, this detuning was represented as $[\omega - k_z v_{z0} - \Omega_0 - \omega(1-n^2)u]/(1-u)$. Then, using this representation and taking into account for $p_z' = \beta_z(1-u)$, one can readily get two first terms in the right-hand side of (3.9).

Integrating Eqs. (3.8)–(3.11) allows one to calculate the interaction efficiency,

$$\eta = \frac{1}{1-\gamma_0^{-1}} < u(Z_{out}) > . \tag{3.12}$$

Here Z_{out} is the normalized interaction length and the angular brackets mean averaging over the initial phases.

When the field amplitude α is large, this calculation can only be done numerically. (This will be discussed later.) Now we will present the analytical results for the case of small field amplitudes, when Eqs. (3.8) and (3.9) can be solved by the method of successive approximations.

In the zero-order approximation ($\alpha = 0$), Eqs. (3.8) and (3.9) yield

$$u_{(0)} = 0, \qquad \vartheta_{(0)} = \vartheta_0 + \delta Z. \tag{3.13}$$

The first-order approximation results in

$$u_{(1)} = \alpha \frac{\beta_{\perp 0}}{\delta}(\cos\vartheta_0 - \cos\vartheta_{(0)}), \tag{3.14}$$

$$\vartheta_{(1)} = \frac{\alpha}{\delta}\left\{\beta_{\perp 0}\left[\delta\frac{n}{\beta_{z0}} - (1-n^2)\right] z\cos\vartheta_0 \right. \tag{3.15}$$

$$\left. - (\sin\vartheta_{(0)} - \sin\vartheta_0)\left[\frac{1-n\beta_{z0}}{\beta_{\perp 0}} + n\frac{\beta_{\perp 0}}{\beta_{z0}} - \frac{1-n^2}{\delta}\beta_{\perp 0}\right]\right\}.$$

In this approximation, the averaging of $u_{(1)}$ over the entrance phases in (3.12) yields zero. So, to calculate the efficiency, it is necessary to determine the second-order term $u_{(2)}$, which obeys the following equation:

$$\frac{du_{(2)}}{dZ} = \alpha\beta_{\perp 0}\left\{\vartheta_{(1)}\cos\vartheta_{(0)} + \left(\frac{n}{\beta_{z0}} - \frac{1-n\beta_{z0}}{\beta_{\perp 0}^2}\right)u_{(1)}\sin\vartheta_{(0)}\right\}. \tag{3.16}$$

Here, in deriving the last term in figure brackets, we used the fact that the first-order perturbations in the axial and transverse normalized momenta, as it follows from Eqs. (3.10) and (3.11), are equal to $-nu_{(1)}$ and $-(1-n\beta_{z0})u_{(1)}/\beta_{\perp 0}$, respectively.

Integrating (3.16) [with the account for Eqs. (3.14) and (3.15)] and averaging over initial gyrophases yield the following expression for the linearized efficiency:

$$\eta_{\text{lin}} = \frac{\alpha^2 Z_{\text{out}}^2}{1-\gamma_0^{-1}} \left\{ -(1-n\beta_{z0})\Phi + \frac{n\beta_{\perp 0}^2}{2\beta_{z0}}(\Phi + \Theta\Phi') - \frac{\beta_{\perp 0}^2(1-n^2)}{2} Z_{\text{out}}\Phi' \right\}. \tag{3.17}$$

Here $\Theta = \delta Z_{\text{out}}$ is the transit angle of electrons; the function $\Phi = (1-\cos\Theta)/\Theta^2$ describes the spectrum of the Lorentz force acting on electrons in the case of the constant amplitude wave and the prime denotes its derivative over Θ. The expression given here in figure brackets was discussed by Bratman et al. (1981) for numerous partial cases, so we will not discuss all of them here. Let us only note that in the case of the operation near cutoff ($n << 1$) this expression reduces to

$$\Re = -\Phi - \mu\Phi', \tag{3.18}$$

where $\mu = (\beta_{\perp 0}^2/2)Z_{\text{out}} = \pi(\beta_{\perp 0}^2/\beta_{z0})(L/\lambda)$ is the parameter, which characterizes the role of relativistic dependence of the cyclotron frequency on electron energy. (As one can easily verify, once we neglect the term proportional to the normalized energy u in (3.9) for the phase, Eq. (3.18) will be reduced to $-\Phi$.) Also note that the condition of net stimulated emission derived by J. Schneider (1959), which was given in Sec. 1.2, in our notations means $\mu > 1$ (this issue is discussed in more detail by Gaponov, Petelin, and Yulpatov 1967).

The plots of the function $\Re$ as the function of the transit angle Θ are shown in Fig. 3.1 for the cases of (a) zero and (b) large values of the parameter μ. Other than the difference in notations, this figure essentially reproduces Fig. 1.4 obtained with the use of the quantum theory. This figure shows that the coherent radiation of gyrating electrons into the waves propagating almost perpendicular to the external magnetic field is possible only due to the relativistic dependence of Ω on $\mathcal{E}$. Note that Eq. (3.17), being combined with the balance equation (1.30), allows one to determine the self-excitation condition of an oscillator. Recall that in Eq. (1.30) the microwave energy W depends on the field intensity, A^2, so, in the framework of the linear theory, this intensity ($\alpha^2 \propto A^2$) is canceled in (1.30).

The function $\Re$ can also be expressed in terms of the electron beam susceptibility χ, conductivity σ, or dielectric constant ε, which are related as

$$\varepsilon = 1 + 4\pi\chi = 1 - i4\pi\sigma/\omega. \tag{3.19}$$

This statement can be illustrated by an example considered by Gaponov, Petelin, and Yulpatov (1967). There, a cavity filled with an electron beam

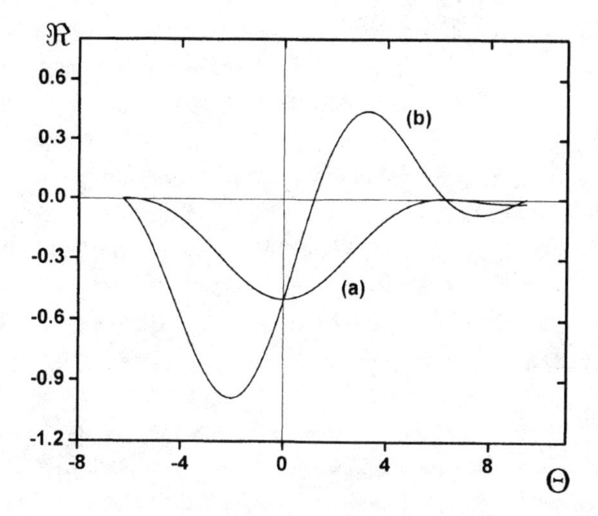

Fig. 3.1. Dependence of the gain function on the transit angle for two values of the parameter μ characterizing the role of cyclotron maser instability: (a) $\mu = 0$; (b) $\mu = 5$.

was replaced by an ordinary LC-circuit with lumped inductance L, resistance R, and capacitance C. Treating the beam as a dielectric medium filling the capacitor yields the standard expression $C = \varepsilon C_0$ for the capacitance, where C_0 is the capacitance of an empty capacitor. For low enough resistance, the equation, which determines the normal frequencies of such a lumped circuit, has a form

$$\omega - \omega_s - i\nu = 2\pi i\sigma. \tag{3.20}$$

Here $\omega_s = c(LC)^{-1/2}$ and $\nu = c^2 R/2L$ are the frequency and the damping rate of the cavity in the absence of the beam, respectively. The imaginary part of this equation is equivalent to the balance equation (1.30). So, the real part of the conductivity should be negative in order to make it possible for a beam to excite oscillations. Thus, $-\sigma'$ plays a role of the function $\mathfrak{R}$ here.

3.2 Gyro-Averaged Equations of Electron Motion: General Approach

The method described in this section is a general method, which is valid for gyrodevices of any geometry. It is based on the representation of the EM field acting upon any gyrating electron as a superposition of angular harmonics of the waves rotating around an electron guiding center. The cyclotron resonance condition allows one to distinguish among all harmonics the only one, which is resonant with electrons, while all others vanish after averaging the equation of motion over fast gyrations. The averaged (or reduced) equations can be written in a quite compact and general form. They contain a reasonably small number of parameters, which greatly simplifies their analysis.

Initially, A. V. Gaponov developed this method for analyzing the linear theory of cyclotron resonance masers (Gaponov 1959a,b, 1961). A little later, V. K. Yulpatov used it for developing the nonlinear theory (Yulpatov 1960, Gaponov, Petelin and Yulpatov 1967). Since in this chapter we consider the gyromonotron, we will describe this method here in application to the excitation of the resonator. (Interaction in waveguides will be considered later.) This general theory of the gyromonotron was described in detail only in the lectures (Yulpatov 1974), which are unavailable for Western readers. Therefore, we will make an attempt to reproduce these results here in good detail.

We will focus on the operation in open cavities at the frequencies close to cutoff. In such a case, the axial momentum, as it follows from Eq. (3.7), remains practically unchanged in the process of interaction (see also Problem 2 in this chapter and Problem 1 in Chapter 7). The transverse momentum, as it follows from Eq. (3.2), in Lagrangian coordinates can be determined by

$$\frac{d\vec{p}_\perp}{dz} + h_H[\vec{p}_\perp \times \vec{z}_0] = -\frac{e}{v_z}\mathrm{Re}\{C_s\vec{G}e^{i\omega t}\}. \tag{3.21}$$

Here $h_H = eH_0/cp_z$, also an obvious relation $dz/dt = v_z$ has been used. The right-hand side of (3.21) represents the RF Lorentz force acting on electrons. The expression for this force is based on the standard representation of electric and magnetic fields in the resonator

$$\vec{E}(\vec{r},t) = \mathrm{Re}\{C_s\vec{E}_s(\vec{r})e^{i\omega t}\}, \qquad \vec{H}(\vec{r},t) = \mathrm{Re}\{C_s\vec{H}_s(\vec{r})e^{i\omega t}\}, \tag{3.22}$$

where ω is the oscillation frequency, s designates the mode under consideration, C_s is the mode amplitude, and the functions $\vec{E}_s(\vec{r})$ and $\vec{H}_s(\vec{r})$ describe the spatial structures of this field. Correspondingly, in (3.21) $\vec{G} = \vec{E}_s + [\frac{\vec{v}_\perp}{c} \times \vec{H}_s]$.

Below we will restrict our consideration by transverse-electric (TE-) modes, which, as it will be shown later, can be excited in gyrotrons easier than transverse-magnetic (TM-) modes. The functions $\vec{E}_s(\vec{r})$ and $\vec{H}_s(\vec{r})$ for these modes in the case of operation near cutoff can be expressed via the scalar Hertz potential H_s as

$$\vec{H}_s = iH_s\vec{z}_0, \qquad \vec{E}_s = (c/\omega)[\nabla_\perp H_s \times \vec{z}_0]. \tag{3.23}$$

As is seen from Eq. (3.23), the Hertz potential H_s describes the spatial structure of the RF magnetic field. In axially open resonators this potential can be represented as $H_s = \Psi_s(\vec{r}_\perp)f(z)$, where the membrane function $\Psi_s(\vec{r}_\perp)$ obeys the equation $\Delta_\perp \Psi_s + k_\perp^2 \Psi_s = 0$ with corresponding boundary conditions and the function $f(z)$ determined by (2.13) describes the axial structure of the field.

The procedure of derivation of gyro-averaged equations is described in detail in Appendix 1. It is shown there that these equations can be written in a quasi-Hamiltonian form as

$$\frac{d\mathcal{E}}{dz} = e\mathrm{Re}\left\{C_s \frac{\partial \mathcal{H}}{\partial \vartheta}\right\},$$

$$\frac{d\vartheta}{dz} + \frac{\omega}{v_z} - sh_H = -e\mathrm{Re}\left\{C_s \frac{\partial \mathcal{H}}{\partial \mathcal{E}}\right\},$$

$$\frac{dX}{dz} = -\frac{1}{k_\perp}\mathrm{Re}\left\{\frac{C_s}{H_0}\frac{\partial \mathcal{H}}{\partial Y}\right\},$$

$$\frac{dY}{dz} = \frac{1}{k_\perp}\mathrm{Re}\left\{\frac{C_s}{H_0}\frac{\partial \mathcal{H}}{\partial X}\right\}. \tag{3.24}$$

Here a role of the Hamiltonian is played by the function $\mathcal{H} = i\frac{p_\perp}{p_z}f(z)\frac{dJ_s(\xi)}{d\xi}L_s(X, Y, z)e^{-i\vartheta}$. In (3.24) and in this Hamiltonian, we use the normalized gyroradius $\xi = \omega a/c$, the slowly variable gyrophase $\vartheta = s\theta - \omega t$, and the operator L_s. This operator describes a transverse structure of the RF Lorentz force acting upon electrons with transverse coordinates of the guiding center X and Y [L_s is determined by Eqs. (A1.10) and (A1.13) for arbitrary and cylindrical geometries, respectively]. The first two equations have the form typical for nonisochronous oscillators in the EM field (Gaponov, Petelin, and Yulpatov 1967). Our system is, however, more complicated because the center of oscillations (guiding center) also experiences a certain transverse drift, which obeys two last equations. The latter equations can be presented in the vector form as

$$\frac{d\vec{R}}{dz} = -\frac{c}{\omega}\mathrm{Re}\left\{\frac{C_s}{H_0}[\nabla_\perp \mathcal{H} \times \vec{z}_0]\right\}.$$

It is, possibly, even more useful to rewrite these equations in polar coordinates that yields

$$\frac{1}{2}\frac{dR^2}{dz} = -\frac{1}{k_\perp}\mathrm{Re}\left\{\frac{C_s}{H_0}\frac{\partial \mathcal{H}}{\partial \psi}\right\}, \tag{3.25}$$

$$R\frac{d\psi}{dz} = \frac{1}{k_\perp}\mathrm{Re}\left\{\frac{C_s}{H_0}\frac{\partial \mathcal{H}}{\partial R}\right\}. \tag{3.26}$$

For widely used cylindrical resonators, the operator L_s is derived in Appendix 1; correspondingly, the Hamiltonian for those resonators is equal to

$$\mathcal{H} = i\frac{p_\perp}{p_z} f(z) \frac{dJ_s(\xi)}{d\xi} J_{m\mp s}(k_\perp R) e^{i[(s\mp m)\psi - \vartheta]}. \tag{3.27}$$

Since in Eq. (3.27) the phases ψ and ϑ are present in one combination, the equation for electron energy in (3.24), being combined with (3.25), yields the following integral of motion:

$$2(s \mp m)(\mathcal{E} - \mathcal{E}_0) = eH_0 k_\perp (R^2 - R_0^2). \tag{3.28}$$

As follows from (3.28), when the cyclotron resonance harmonic number is equal to the azimuthal index of the mode, $s = m$ (that is always the case in large-orbit gyrotrons considered later in Sec. 13.6), there is no radial drift of electron guiding centers. In general, Eq. (3.28) relates the radial drift to the changes in electron energy. As follows from this equation, the derivative $\partial/\partial R$ in (3.26) can be expressed via the derivative in $\mathcal{E}$ and, therefore, this equation can be rewritten as

$$-(s \mp m)\frac{d\psi}{dz} = -e\mathrm{Re}\left\{C_s \frac{\partial \mathcal{H}}{\partial \mathcal{E}}\right\}.$$

This equation has the same form as the equation for phase in (3.24). Thus, these two phases can be combined into one phase by introducing $\vartheta' = \vartheta - (s \mp m)\psi$. Therefore, instead of four equations given in (3.24), one can use just two equations for the energy $\mathcal{E}$ and phase ϑ'.

Let us now estimate the radial drift. As it follows from (1.24), in gyrotrons ($n = ck_z/\omega << 1$) the changes in electron energy are restricted by the condition

$$\frac{|\Delta\mathcal{E}|}{\mathcal{E}_0} \leq \frac{1}{sN}. \tag{3.29}$$

In accordance with Eq. (3.28), these changes in energy and in the radial position of guiding centers are related as

$$\frac{|\Delta\mathcal{E}|}{\mathcal{E}_0} = \Omega_0 \frac{k_\perp R_0}{c} \frac{1}{|m \mp s|} |\Delta R|.$$

Substituting this formula into (3.29) and accounting for the cyclotron resonance condition, $\Omega_0 \simeq \omega/s = 2\pi c/s\lambda$, result in

$$\frac{|\Delta R|}{\lambda} \leq \frac{|m \mp s|}{2\pi N k_\perp R_0}. \tag{3.30}$$

In the case of symmetric modes ($m = 0$), the beam position typically corresponds to $k_\perp R_0 \geq 1$. In the case of whispering gallery modes, which are the modes with $m >> p$, as well as other modes with large azimuthal indices, the beam is located in the region of the caustic, where $k_\perp R_0$ is close to m (this will be discussed later in more detail). So, in any case, as it follows from (3.30), when $N >> 1$, the radial drift is negligibly small, and thus will later be ignored.

Eq. (3.29) allows us to further simplify Eqs. (3.24). Indeed, since in our case $p_z \approx$ const and the changes in total energy are small, from (1.19) it follows that the changes in energy and orbital momentum are related as

$$2\mathcal{E}_0 \Delta\mathcal{E} = c^2 \left(p_{\perp 0}^2 - p_{\perp}^2\right). \qquad (3.31)$$

Correspondingly, the ratio $\Delta\mathcal{E}/(\mathcal{E}_0 - mc^2)$, which, after averaging over initial distributions [cf. (3.12)], determines the interaction efficiency, can be represented as

$$\frac{\beta_{\perp 0}^2}{2(1-\gamma_0^{-1})}\left(1 - \frac{p_{\perp}^2}{p_{\perp 0}^2}\right). \qquad (3.32)$$

Here the first term, $\beta_{\perp 0}^2/2(1-\gamma_0^{-1})$, is the single-particle efficiency, which also follows from (1.27) for the case $n << 1$. Now, when $p_z \approx$ const, this ratio determines the fraction of rotational energy in the electron kinetic energy. (This rotational energy is in our case the only energy available for withdrawing by the RF field.) Note that, in principle, the separation of the energy of a relativistic electron into two parts associated with the axial and transverse motion, respectively, as follows from (1.19), is not so obvious. Therefore, it is more accurate to discuss the changes in the orbital and axial components of electron momentum.

The second term in (3.32), which is given in parentheses, determines the changes in the orbital momentum of electrons. After averaging, this term determines what is often called the orbital (or perpendicular) efficiency. This efficiency characterizes the orbital bunching of electrons, which was preliminarily discussed in Sec. 1.1.

In order to describe the interaction process, it makes sense to introduce a new variable $w = p_{\perp}^2/p_{\perp 0}^2$, which, in accordance with (3.31), is equal to

$$w = \frac{p_{\perp}^2}{p_{\perp 0}^2} = 1 - \frac{2}{\beta_{\perp 0}^2}\frac{\mathcal{E}_0 - \mathcal{E}}{\mathcal{E}_0}. \qquad (3.33)$$

[This variable w relates to the variable u used in Sec. 3.1 as $w = 1 - (2/\beta_{\perp 0}^2)u$.] The argument ξ in the Bessel function $J_s(\xi)$ of the Hamiltonian can be represented as $\xi = \omega a/c = s\beta_{\perp 0}\sqrt{w}$. Taking into account that the Bessel function can be reasonably well approximated by polynomials $(1/s!)(\xi/2)^s$ up to $\xi \leq s$, one can represent $dJ_s(\xi)/d\xi$ in the Hamiltonian as $(\frac{s^s}{2^s s!})(\beta_{\perp 0}\sqrt{w})^{s-1}$. Then, introducing normalized axial coordinate

$$\zeta = \left(\beta_{\perp 0}^2/2\beta_{z0}\right)(\omega z/c),$$

normalized RF field amplitude

$$F = i\frac{eC_s}{mc\omega\gamma_0}\beta_{\perp 0}^{s-4}\frac{s^s}{2^{s-1}s!}L_s,$$

and normalized cyclotron resonance mismatch

$$\Delta = \frac{2(\omega - sh_H v_{z0})}{\beta_{\perp 0}^2\omega} = \frac{2}{\beta_{\perp 0}^2}\frac{\omega - s\Omega_0}{\omega},$$

one can represent equations for the normalized variable w and the phase ϑ in the following compact form

$$\frac{dw}{d\zeta} = 2\mathrm{Im}\{Ff(\zeta)w^{s/2}e^{-i\vartheta}\}, \tag{3.34}$$

$$\frac{d\vartheta}{d\zeta} + \Delta + w - 1 = -s\mathrm{Re}\{Ff(\zeta)w^{(s/2)-1}e^{-i\vartheta}\}.$$

Here, in the equation for the phase we used the relation

$$\omega/v_z - sh_H = (\omega\mathcal{E}/p_{z0}c^2) - sh_H = (\omega - sh_H v_{z0})/v_{z0} + \omega(\gamma - \gamma_0)/v_{z0}\gamma_0$$

$$= \frac{\omega}{v_{z0}}\left[\frac{\omega - s\Omega_0}{\omega} + \frac{\beta_{\perp 0}^2}{2}(w-1)\right].$$

Note that we chose above the operating frequency ω to be a carrier frequency. In some cases, such as multimode gyrotrons, in which an electron beam can simultaneously interact with several modes at different cyclotron harmonics, or frequency-multiplying, multistage gyroamplifiers in which different stages operate at different harmonics, it is more convenient to use the initial electron cyclotron frequency, Ω_0, as the carrier frequency. In this case, one can introduce a new set of normalized variables: $\varsigma' = (\beta_{\perp 0}^2/2\beta_{z0})(\Omega_0 z/c) \simeq \varsigma/s$, $\vartheta' = \theta - (\omega/s)t = \vartheta/s$ and $F' = sF$. In these variables, which will be used below omitting primes, Eqs. (3.34) have the following form:

$$\frac{dw}{d\zeta} = 2\mathrm{Im}\{Ff(\zeta)w^{s/2}e^{-is\vartheta}\}, \tag{3.35}$$

$$\frac{d\vartheta}{d\zeta} + \Delta + w - 1 = -\mathrm{Re}\{Ff(\zeta)w^{(s/2)-1}e^{-is\vartheta}\}.$$

In the absence of electron spread in velocities and radii of guiding centers, the orbital efficiency can be determined as

$$\eta_\perp = < 1 - w(\mu) > . \tag{3.36}$$

Here angular brackets denote averaging over entrance phases ($0 \le \vartheta_0 < 2\pi$) and $\mu = (\beta_{\perp 0}^2/2\beta_{z0})(\omega L/c)$ is the normalized interaction length, which we already introduced in Sec. 3.1 after Eq. (3.18). Correspondingly, the total interaction efficiency [cf. (3.32)] is equal to

$$\eta = \frac{\beta_{\perp 0}^2}{2(1-\gamma_0^{-1})}\eta_\perp. \tag{3.37}$$

Note that in gyrotrons with a small beam current the amplitude of oscillations can be rather small. Thus, it is expedient to increase the interaction length in order to improve, at least, the electron bunching. In such a case, Eqs. (3.34) can be greatly simplified: the RHS in the equation for the phase can be neglected, since the main contribution to the phase bunching will originate from modulation in the electron energies. Differentiating this simplified equation one more time and using the equation for electron energy yield for the phase, the following nonlinear pendulum equation

$$\frac{d^2\vartheta}{d\hat{\zeta}^2} = -\mathrm{Im}\{f(\hat{\zeta})e^{-i\vartheta}\}. \tag{3.38}$$

This equation should be supplemented with the boundary condition for the phase (it should be randomly distributed from 0 to 2π at the entrance) and for its derivative $d\vartheta/d\hat{\zeta}\big|_{\zeta=0} = -\hat{\Delta}$. Here $\hat{\zeta} = \sqrt{F}\zeta$ and $\hat{\Delta} = \Delta/\sqrt{F}$ are the axial coordinate and cyclotron resonance mismatch normalized to the field amplitude, respectively. With this normalization the equation for the energy in (3.34) shows that the changes in the energy are also proportional to $\sqrt{F}$. Correspondingly, the orbital efficiency determined above by the equation (3.36) can now be expressed as $\eta_\perp = \sqrt{F}\hat{\eta}$, where the new normalized efficiency is equal to

$$\hat{\eta} = < \frac{d\vartheta}{d\hat{\zeta}}\bigg|_{\hat{\zeta}=\hat{\mu}} + \hat{\Delta} > . \tag{3.39}$$

So, if, simultaneously with the amplitude decrease, the interaction length increases, the efficiency decreases as $\eta_\perp = \sqrt{F}\hat{\eta}$, while in the case of a device with the constant interaction length, as it will be shown below in Sec. 3.4, the efficiency at small amplitudes is proportional to F^2. Note that a formalism similar to that given by (3.38) and (3.39) is also adopted in the nonlinear theory of resonant TWTs and FELs.

3.3 Excitation of Resonators: General Form

In principle, excitation of resonators at one mode, whose fields are given by (3.22), has been described in numerous textbooks (see, e.g., J. Slater 1950). Nevertheless, for completeness, we will reproduce the derivation of these equations here.

In order to be able to discuss later some nonstationary processes, we will assume that the mode amplitude can slowly vary in time, $|dC_S/dt| << \omega C_S$. We will also use the standard assumption that the functions $\vec{E}_s(\vec{r})$ and $\vec{H}_s(\vec{r})$ obey Maxwell equations for an empty resonator:

$$\nabla \times \vec{H}_s = i\frac{\omega_S}{c}\vec{E}_S, \qquad \nabla \times \vec{E}_s = -i\frac{\omega_S}{c}\vec{H}_S. \tag{3.40}$$

Here $\omega_s = \omega_s' + i\frac{\omega_s'}{2Q_s}$ is the cold-cavity frequency. It contains the real part, which determines the mode eigenfrequency, and the imaginary part describing some losses. Typically, for gyrotron open resonators, two sorts of losses—ohmic and diffraction—can be important. Thus, Q_s should be determined as

$$\frac{1}{Q_s} = \frac{1}{Q_{\text{dif}}} + \frac{1}{Q_{\text{ohm}}}. \tag{3.41}$$

The minimum diffractive Q-factor was determined above by (2.11). In the presence of some reflections at the ends of an open resonator, this Q-factor can be approximated by

$$Q_{\text{dif}} = \frac{Q_{\text{dif,min}}}{1 - |R_1||R_2|}. \tag{3.42}$$

Here $R_{1,2}$ are reflection coefficients from both ends. Typically, at the entrance $|R_1| \approx 1$. The dependence of Q_{dif} on parameters of the output neck was shown above in Fig. 2.7b.

The ohmic Q-factor, by definition, is the ratio of the EM energy stored in the resonator to the energy of the magnetic field stored in the skin layer. For open resonators, this ratio is on the order of the ratio of the resonator radius (or any other typical transverse size, when noncylindrical resonators are under study) to the skin depth

$$\delta_{\text{sk}} = \frac{1}{2\pi}\sqrt{\frac{\lambda c}{\sigma}}. \tag{3.43}$$

Here σ is the wall conductivity, which for copper is approximately equal to $5{\cdot}10^{17}\text{sec}^{-1}$ (in cgs units) or $5.8 \cdot 10^7/\Omega\text{m}$ (in SI). Note that Eq. (3.43) is a theoretical estimate, which follows from the boundary conditions at the metal surface analyzed by M. A. Leontovich. In practice, however, due to the finite roughness of the wall surface, the skin depth is about 1.5–2 times larger than that given by (3.43). Also, for cylindrical cavities, in the case of rotating modes, which have a caustic of the radius $R_c = (m/\nu_{m,p})R_w$ and whose rays are not normally incident on the wall, a more accurate estimate of Q_{ohm} yields (see, e.g., Vlasov, Zagryadskaya, and Orlova 1976)

$$Q_{ohm} = \frac{R_w}{\delta_{sk}}\left(1 - \frac{m^2}{\nu_{m,p}^2}\right). \tag{3.44}$$

In the process of derivation, we will take into account that, strictly speaking, the amplitudes of the electric and magnetic fields are slightly different. Therefore, instead of Eqs. (3.22), the following representation for these fields will be used:

$$\vec{E} = \mathrm{Re}\{A\vec{E}_s(\vec{r})e^{i\omega t}\}, \qquad \vec{H} = \mathrm{Re}\{B\vec{H}_s(\vec{r})e^{i\omega t}\}. \tag{3.45}$$

Substituting the fields given by (3.45) into Maxwell equations

$$\nabla \times \vec{H} = \frac{1}{c}\frac{\partial \vec{E}}{\partial t} + \frac{4\pi}{c}\vec{j}, \qquad \nabla \times \vec{E} = -\frac{1}{c}\frac{\partial \vec{H}}{\partial t}, \tag{3.46}$$

and making use of (3.40) for the functions $\vec{E}_s(\vec{r})$, $\vec{H}_s(\vec{r})$, results in the following equations

$$A = \frac{1}{\omega_s}\left(\omega B - i\frac{dB}{dt}\right), \tag{3.47}$$

$$i\omega_s B\vec{E}_s = \left(i\omega A + \frac{dA}{dt}\right)\vec{E}_s + 4\pi\vec{j}_\omega. \tag{3.48}$$

In (3.48) we represented the electron current density $\vec{j}$ as $\mathrm{Re}(\vec{j}_\omega e^{i\omega t})$, because here we consider the excitation of the resonator by a given current at the frequency ω. [Correspondingly, $\vec{j}_\omega = \frac{1}{\pi}\int_0^{2\pi}\vec{j}e^{-i\omega t}d(\omega t)$.] Then, let us multiply (3.48) by $\vec{E}_s^*$, integrate it over the resonator volume, introduce the norm of the operating mode

$$N_s = \frac{1}{4\pi}\int_V |\vec{E}_s|^2 dv = \frac{1}{4\pi}\int_V |\vec{H}_s|^2 dv \tag{3.49}$$

(here V is the resonator volume), and use (3.47) for expressing A via B. With the account for the condition $|dB/dt| << \omega B$, this results in the equation describing the field excitation:

$$\frac{dB}{dt} + i(\omega - \omega_s)B = -\frac{1}{2N}\int_V \vec{j}_\omega \cdot \vec{E}_s^* dv. \tag{3.50}$$

The equation, which describes nonstationary excitation of the resonator field by an electron beam in such a form, was, possibly, first derived by L. A.

Weinstein (Weinstein and Solntsev 1973). In the regime of stationary oscillations at the frequency ω, real and imaginary parts of this equation can be rewritten as

$$\frac{\omega_s'}{2Q_s}|B| = \text{Re}\left\{-\frac{1}{2N_s}e^{-i\varphi}\int_V \vec{j}_\omega \vec{E}_s^* dv\right\}, \tag{3.51}$$

$$\omega - \omega_s' = \frac{1}{|B|}\text{Im}\left\{-\frac{1}{2N_s}e^{-i\varphi}\int_V \vec{j}_\omega \vec{E}_s^* dv\right\}. \tag{3.52}$$

Here we represented the complex amplitude B as $|B|e^{i\varphi}$. Eq. (3.51) is essentially the same balance equation that was given above in a general form by (1.30). Note that in our present notations the EM energy stored in the resonator, W, is equal to $\frac{1}{2}|B|^2 N_s$. Eq. (3.52) determines a so-called frequency pulling effect, i.e., the shift of the oscillation frequency ω with respect to the cold-cavity frequency. L. A. Weinstein also called these two equations [(3.51) and (3.52)] the balance equations for the active and reactive powers, respectively, bearing in mind that the right-hand side of (3.50) is proportional to the complex power of interaction between electrons and the resonator field. To establish this proportionality, it also makes sense to rewrite (3.51) as

$$\frac{\omega}{2Q_s}|B|^2 N_s = -\frac{1}{2}\text{Re}\{B^* \int_V \vec{j}_\omega \vec{E}_s^* dv\}. \tag{3.53}$$

Now the left-hand side is exactly equal to the left-hand side of (1.30), and thus the right-hand side can be treated as the power withdrawn by the EM field from the beam.

Gyrotron Form

So far, we have been involved in the derivation of general equations of excitation. Now we should express these equations in terms of the parameters and variables, which were introduced above (in Sec. 3.2) in the equations for electron motion in gyrotrons. First, let us note that for TE-modes, in the right-hand side of (3.50)–(3.53)

$$\vec{j}_\omega \vec{E}_s^* = \vec{j}_{\omega\perp} \vec{E}_s^* = \frac{1}{\pi}\int_0^{2\pi} \vec{j}_\perp \vec{E}_s^* e^{-i\omega t} d(\omega t).$$

As follows from the charge conservation law for one elementary electron beam (filament), which propagates through the interaction region without loss of electrons, a charge crossing the entrance at the time interval dt_0 is conserved, i.e., at any time interval dt

$$dq = j_z dt = j_{z0} dt_0 = dq_0.$$

Therefore, in the integral given above $\vec{j}_\perp d(\omega t) = \frac{\vec{j}_\perp}{j_z} j_{z0} d(\omega t_0)$, and hence, the right-hand side of the excitation equation (3.50) can be rewritten as

$$-\frac{1}{N}\int\limits_{S_\perp} j_{z0} W(\vec{R}_{\perp 0})\left[\frac{1}{2\pi}\int\limits_0^{2\pi}\left(\int\limits_0^L \frac{\vec{p}_\perp}{p_z}\vec{E}_s^* e^{-i\omega t} dz\right) d(\omega t_0)\right] ds_\perp. \qquad (3.54)$$

Here we introduced the distribution function $W(\vec{R}_{\perp 0})$ of electrons over coordinates of guiding centers at the entrance. This function obeys the normalization condition $I_b = -\int_{S_\perp} j_{z0} W(\vec{R}_\perp) ds_\perp$. Here and above, $S_\perp$ is the cross-sectional area of the interaction region, and I_b is the absolute value of the beam current. Also, since we consider the operation in TE-modes near cutoff, the axial momentum of electrons in (3.54) is constant.

For electrons gyrating in θ-direction $\vec{p}_\perp = p_\perp \vec{\theta}_0$. The θ-component of the electric field, in accordance with (3.23), is equal to $-\frac{c}{\omega} f(z) \frac{\partial \Psi_s}{\partial r}$. Expanding the membrane function in the Fourier series of θ-harmonics and using representation of these harmonics, which is given by Eq. (A1.9), result in

$$E_{s,\theta}^* = -\frac{c}{\omega} f^*(z) \sum_k \frac{dJ_k(k_\perp a)}{da} L_k^* e^{ik\theta}.$$

Here we replaced $\partial/\partial r$ by d/da because we consider electrons gyrating with the radius $r = a$. Therefore, for the cyclotron resonance harmonic, in (3.54)

$$\vec{p}_\perp \vec{E}_s^* e^{-i\omega t} = p_{\perp 0}\sqrt{w}\left(-\frac{c}{\omega}\right) f^*(z) k_\perp \frac{dJ_s(\xi)}{d\xi} L_s^* e^{i\vartheta}.$$

Then, using the same normalized amplitude F, coordinate ς and polynomial approximation of the Bessel function $J_s(\xi)$ as in Sec. 3.2, one can readily transform (3.54) into

$$-\frac{1}{N}\frac{\beta_{\perp 0}^s}{\beta_{z0}}\frac{s^s}{2^s s!}\int\limits_{S_\perp} j_{z0} W(\vec{R}_{\perp 0}) L_s^* \left\{\frac{1}{2\pi}\int\limits_0^{2\pi}\left[\int\limits_0^L w^{s/2} e^{i\vartheta} f^*(z) dz\right] d\vartheta_0\right\} ds_\perp. \qquad (3.55)$$

Now we can come back to the excitation equation (3.50) and rewrite it in normalized variables. At this step, let us assume that we consider a thin annular electron beam in a cylindrical cavity, so all electrons have the same coupling to the cavity mode, i.e., the operator L_s is constant. Then we can rewrite (3.50) for the normalized amplitude F determined in the text after (3.33) that yields

$$\frac{dF}{dt} = F\left\{I_s \Phi - \frac{1}{2Q} - i\frac{\omega - \omega_s'}{\omega}\right\}. \qquad (3.56)$$

Here we normalized time to ω, separated the real and imaginary parts of the cold-cavity frequency, and introduced the normalized beam current parameter

$$I_s = 8\frac{eI_b}{mc^3}G\left(\frac{s^s}{2^s s!}\right)^2 \frac{\beta_{\perp 0}^{2(s-2)}}{\gamma_0 \beta_{z0}} \frac{1}{\int_0^\mu |f(\zeta)|^2 d\zeta}, \tag{3.57}$$

and the gain function

$$\Phi = -\frac{i}{F}\frac{1}{2\pi}\int_0^{2\pi}\left\{\int_{\varsigma_{in}}^{\varsigma_{out}} w^{s/2}e^{i\vartheta} f^*(\varsigma)d\varsigma\right\}d\vartheta_0. \tag{3.58}$$

In (3.57) we, in turn, introduced the parameter

$$G = \frac{J_{m-s}^2(\nu_{m,p}R_0/R_w)}{\left(\nu_{m,p}^2 - m^2\right)J_m^2(\nu_{m,p})}, \tag{3.59}$$

which describes the coupling of a beam having the radius of electron guiding centers, R_0 to the field of a cylindrical resonator. In deriving (3.57) we used the fact that for cylindrical resonators

$$N_s = \frac{\lambda^2}{16\pi^2}\left(\nu_{m,p}^2 - m^2\right)J_m^2(\nu_{m,p})\int_0^L |f(z)|^2 dz. \tag{3.60}$$

(Recall that we consider operation near cutoff, so $k_\perp \approx \omega/c$.) Note that (3.56) has the same form as excitation equations in the theory of multimode gyrotrons (Moiseev and Nusinovich 1974).

The gain function Φ can be expressed via the susceptibility of an electron beam with respect to the resonator field, χ, which was used in (3.19). As it follows from comparison of (3.56) with Eq. (1) in Ergakov and Moiseev (1975), these two functions are related as

$$I_s\Phi = -2\pi i\chi = -(i/2)(I_0/Q)\hat{\chi}. \tag{3.61}$$

In (3.61) our normalized beam current parameter I_s relates to the normalized current parameter I_0 used by Ergakov and Moiseev as $I_s = I_0/Q$, so $\Phi = -(i/2)\hat{\chi}$.

Stationary oscillations were determined above in a general form by balance equations (3.51) and (3.52). In new notations, the balance of active powers, as follows from (3.56), can be determined by the balance equation

$$2I_s Q\Phi' = 1. \tag{3.62}$$

The real part of the gain function present in (3.62), as it follows from the equation for w in (3.34) and the definition of the orbital efficiency (3.36), can be represented as $\Phi' = \eta_\perp/2\,|F|^2$. Correspondingly, the balance equation (3.62) can be rewritten as

$$|F|^2 = I_0\eta_\perp. \tag{3.63}$$

Since very often the axial structure of the field in open resonators is approximated by the Gaussian function $f(\zeta) = \exp\{-(\frac{2\zeta}{\mu} - 1)^2\}$, for which

$$\int_0^\mu |f(\zeta)|^2 d\zeta \simeq \int_{-\infty}^{\infty} |f(\zeta)|^2 d\zeta = \frac{1}{2}\sqrt{\frac{\pi}{2}}\mu, \tag{3.64}$$

it also makes sense to rewrite Eq. (3.57) for $I_0 = I_s Q$ and the axial Gaussian distribution. This yields

$$I_0 = 0.238 I_b(A) Q 10^{-3} G \frac{\lambda}{L}\left(\frac{s^s}{2^s s!}\right)^2 \frac{\beta_{\perp 0}^{2(s-3)}}{\gamma_0}. \tag{3.65}$$

This is the normalized beam current parameter used by Nusinovich and Erm (1972) and by Danly and Temkin (1986).

3.4 Self-Excitation Conditions

As mentioned at the end of Sec. 3.1, the self-excitation conditions can be found from the balance equations, in which the efficiency (or the analogous interaction power) is calculated by the method of successive iterations assuming the smallness of the field amplitude. In Sec. 3.1, this linearized efficiency was calculated for the simple, transversely homogeneous model. Here, we will derive a similar expression for the gyrotron with a transversely inhomogeneous field. Such a derivation for the case of the constant external magnetic field, $\vec{H}_0$, was given by Yulpatov (1974). Later, Nusinovich (1988) reproduced these derivations for a gyrotron with a weakly tapered magnetic field. Since for enhancing the gyrotron efficiency this field is often weakly tapered while the derivation procedure is essentially the same, we will present the derivation, which will include possible tapering of $\vec{H}_0$. Assuming that the tapering is weak enough, we will take this deviation into account only in the cyclotron resonance mismatch Δ, which will be considered below as a function of the axial coordinate ς. (Conditions, which allow one to neglect the effect of this tapering on other parameters, are discussed by Nusinovich 1988.)

Let us apply now the method of successive iterations, which was used in Sec. 3.1, to Eqs. (3.34), assuming that the normalized amplitude F is small and the detuning Δ depends on ς. The zero-order approximation yields

$$w_{(0)} = 1, \vartheta_{(0)} = \vartheta_0 - \int_0^{\zeta} \Delta d\zeta'. \tag{3.66}$$

Then, the first-order approach results in

$$w_{(1)} = 2\mathrm{Im}\left\{ F \int_0^{\zeta} f(\zeta') e^{-i\vartheta_{(0)}} d\zeta' \right\}, \tag{3.67}$$

$$\vartheta_{(1)} = -\int_0^{\zeta} w_{(1)}(\zeta') d\zeta' - s\mathrm{Re}\left\{ F \int_0^{\zeta} f(\zeta') e^{-i\vartheta_{(0)}} d\zeta' \right\}. \tag{3.68}$$

Again, since these terms being averaged over entrance phases yield zero, we should calculate the second-order perturbation in the electron energy $w_{(2)}$. As it follows from Eq. (3.34), this perturbation obeys equation

$$\frac{dw_{(2)}}{d\zeta} = 2\mathrm{Im}\left\{ F f(\varsigma) e^{-i\vartheta_{(0)}} \left(\frac{s}{2} w_{(1)} - i\vartheta_{(1)} \right) \right\}. \tag{3.69}$$

After averaging over ϑ_0, this equation, with the use of (3.67) and (3.68), reduces to

$$\frac{d\langle w_{(2)} \rangle}{d\zeta} = 2|F|^2 \,\mathrm{Im}\left\{ f' \left[is \int_0^{\zeta} f'^* d\zeta' - \int_0^{\zeta}\int_0^{\zeta'} f'^* d\zeta'' d\zeta' \right] \right\}. \tag{3.70}$$

Here we introduced $f'(\zeta) = f(\zeta) \exp(i \int_0^{\varsigma} \Delta d\varsigma')$. Integrating this equation yields a rather long formula, which can be simplified by using the relation

$$\int_0^{\mu} f' \left(\int_0^{\zeta} f'^* d\zeta' \right) d\zeta + \int_0^{\mu} f'^* \left(\int_0^{\zeta} f' d\zeta' \right) d\zeta = \left| \int_0^{\mu} f' d\zeta \right|^2.$$

It also makes sense to represent the cyclotron resonance mismatch Δ as $\Delta_0 + \Delta'(\varsigma)$, where Δ_0 is its constant part (it can be determined, for instance, by the magnetic field in the middle of the resonator), and use the relation $\frac{\partial f'}{\partial \Delta_0} = i\zeta f'$. The use of these formulas results in the following compact expression for the linearized orbital efficiency

$$\eta_{\perp,\mathrm{lin}} = -\langle w_{(2)}(\mu) \rangle = -|F|^2 \left(s + \frac{\partial}{\partial \Delta_0} \right) \left| \int_0^{\mu} f' d\zeta \right|^2. \tag{3.71}$$

This expression is valid for gyrotrons operating at arbitrary cyclotron harmonics, in a TE-mode with arbitrary transverse and axial structures, and for an arbitrary, weakly tapered magnetic field. For a widely used Gaussian axial distribution of the field in open resonators and a constant external magnetic field, (3.71) yields (Nusinovich and Erm 1972)

$$\eta_{\perp,\mathrm{lin}} = -|F|^2 \frac{\pi}{4}\mu^2 \left(s + \frac{\partial}{\partial \Delta}\right) \exp\left\{-\frac{(\Delta\mu)^2}{8}\right\}. \quad \textbf{(3.72)}$$

Combining this equation with the balance equation (3.63), in which the normalized beam current is determined by (3.65) [in a general case, (3.71) and (3.57) should be used in (3.63)], one can easily find the starting current for gyrotron oscillations. Let us emphasize that the balance equation determines the starting value of the normalized beam current. Some examples of the dependence of this current on the cyclotron resonance mismatch for different values of the normalized length are shown in Fig. 3.2. Then, the starting value of the real current can be determined, which, as it follows from (3.65), is inversely proportional to the Q-factor of the cavity, coupling parameter G and $\beta_{\perp 0}^{2(s-3)}$. So, as the cyclotron harmonic number increases, the starting current increases as well, which corresponds to the weakening of the intensity

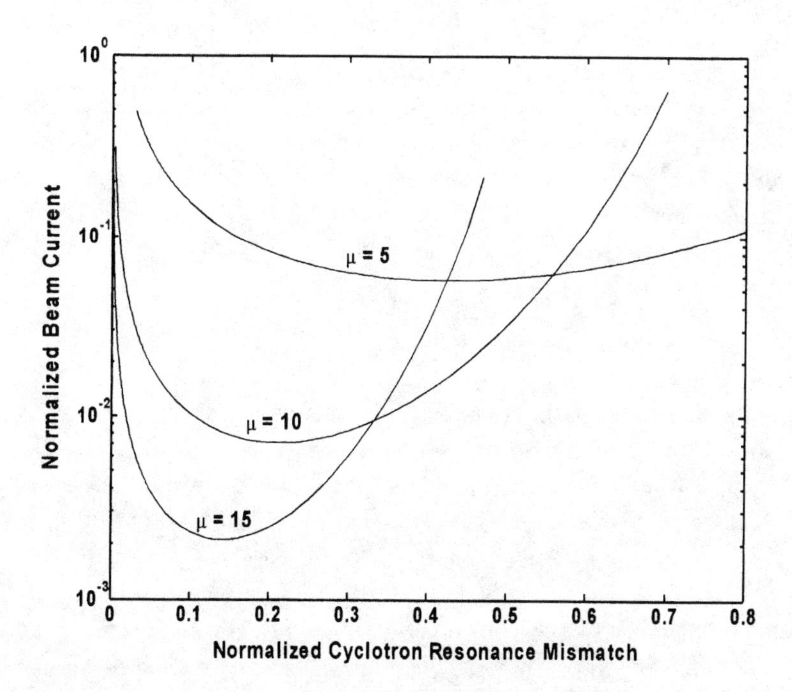

Fig. 3.2. Normalized starting current as the function of cyclotron resonance mismatch Δ for three values of the normalized length μ ($\mu = 5$, 10, and 15) in a gyrotron with Gaussian axial distribution of the resonator field. The curves are shown for the interaction at the fundamental cyclotron resonance ($s = 1$).

of cyclotron radiation at harmonics given by the Schott formula (see, e.g., Landau and Lifshitz 1975).

3.5 Mode Selection

As the gyrotron power and operating frequency increases, the interaction space (in terms of the wavelength) should increase as well. Since the increase in power and frequency is one of the main goals in the gyrotron development, the issue of mode selection becomes more and more important. Some aspects of mode selection were already discussed in Sec. 2.3. Here we will discuss this issue in more detail.

The axial mode selection in axially open resonators, as was discussed in Sec. 2.3, can easily be realized. The mode selection in transverse indices is more complicated. If we restrict our consideration by conventional cylindrical cavities and thin annular electron beams, then it immediately becomes clear that such a system does not provide any mode selection in azimuthal indices in the sense that Q-factors and coupling impedances of these modes are practically the same. However, for modes with different radial indices, there is a possibility of mode selection by a proper positioning of a thin beam.

Since in gyrotrons only the modes, which are in cyclotron resonance with electrons, can be excited, this allows one to provide the mode selection by a proper choice of the external magnetic field. Whispering gallery modes, which are the modes with azimuthal indices much larger than the radial ones ($m >> p$), have frequencies whose separation can be estimated as

$$\frac{\Delta\omega}{\omega} \sim \frac{1}{m}.$$

So, when the azimuthal index m is not too large, the frequency separation of such modes is larger than the cyclotron resonance band, in a part of which the modes can be self-excited. Thus, in the case of such separation it is possible to selectively excite only one of these modes. Typically the width of the self-excitation region, which is inversely proportional to the electron transit time, $T = L/v_z$, does not exceed 1–2%, so this selection can be quite effective up to very large azimuthal indices ($m \leq 40 - 50$).

Modes with different radial indices have different dependencies of the coupling coefficient determined by (3.59) on the beam radius. Even more, in cylindrical waveguides and resonators the modes with different directions of azimuthal rotation have a polarization degeneracy. This means that both modes, whose fields are described by the membrane function $\Psi_s = J_m(k_\perp r)e^{-im\psi}$ and $\Psi_s = J_m(k_\perp r)e^{im\psi}$, have the same frequency. A beam of gyrating electrons, as

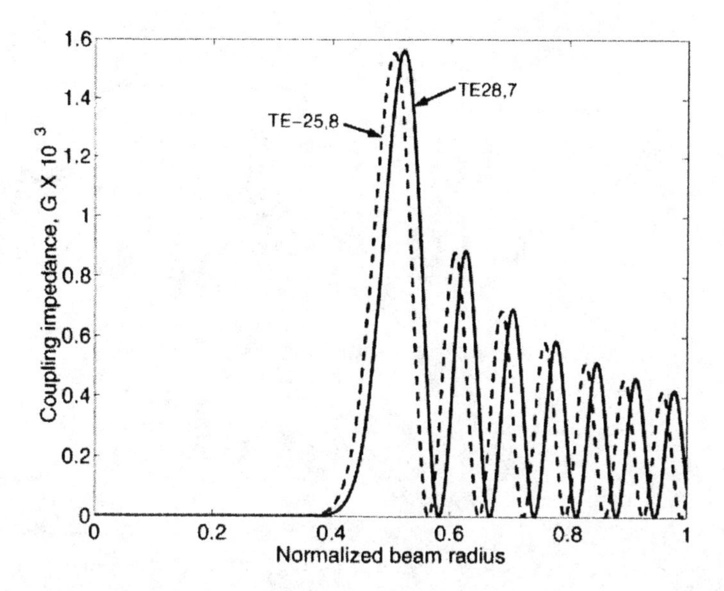

Fig. 3.3. Coupling impedances G for two modes with close frequencies as functions of the electron guiding center radius normalized to the resonator wall radius. Curves are shown for the fundamental cyclotron resonance $s = 1$. Minus sign indicates that the mode rotates in the direction opposite to electron gyration.

a gyrotropic medium, violates this degeneracy. Correspondingly, the coupling coefficients for these modes are different, since the order of Bessel function in the numerator of Eq. (3.59) is $m \mp s$. Here signs "minus" and "plus" correspond, respectively, to the co- and counterrotating waves with respect to electron gyration.

A typical example of the dependencies of G on R_0 for high-order modes with close frequencies is shown in Fig. 3.3. Here the solid and dashed lines correspond to co- and counterrotating waves, respectively. It is obvious that the inner peak has the largest maximum. Therefore, it is preferable to position a thin beam just on this peak. Then, the modes with a smaller number of radial variations will be very weakly coupled to the beam, and the only concern will be about modes with larger radial indices.

Regarding these modes we, first, should point out that among them the most dangerous are the modes with $(p + 1)$-radial index (p is the radial index of the operating mode) and opposite polarization, because their largest peaks are located closer to the beam as shown in Fig. 3.3. Second, these modes, once they have frequencies close to the operating mode frequency, occupy a larger area in the resonator cross section, since their caustic radius is smaller than that for the operating mode. Correspondingly, in Eq. (3.59) the denominator

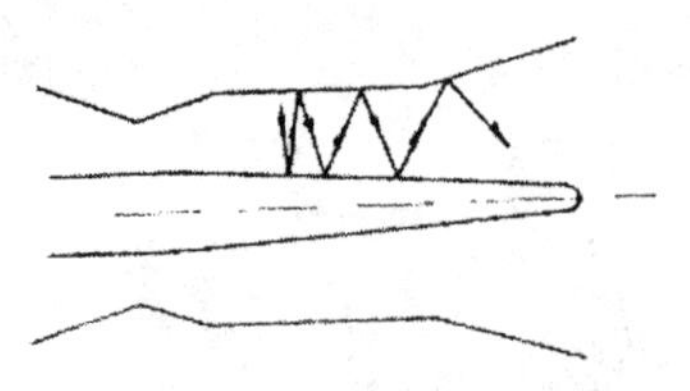

Fig. 3.4. Ray representation of the propagation of parasitic modes in a coaxial resonator.

for them is larger, and hence, their coupling coefficient G is smaller. This difference can be significant in the case of a small number of radial variations. However, when the radial index is large, the coupling coefficient for these modes is almost the same as for the operating one.

To provide mode selection in the latter case, one can use a cylindrical coaxial resonator (Vlasov, Zagryadskaya, and Orlova 1976). When an inner conductor is properly tapered, as is shown in Fig. 3.4, and its radius is a little smaller than the caustic radius of the operating mode, there can be quite effective mode selection. Indeed, the rays representing the field of the operating mode will propagate in the resonator without interception with the inner coax, while the rays of the parasite with a larger radial index will be pushed by the tapered coax out of the resonator. In terms of the diffractive Q-factor, it means lowering the Q-factor of the parasite, while keeping the Q-factor of the operating mode unchanged.

So far we have discussed the mode selection in sufficiently long resonators whose modes can be characterized by large values of the Fresnel parameter, $L^2/\lambda D_{\text{eff}} >> 1$ (see Sec. 2.3). Here $D_{\text{eff}} = D\sqrt{1-(m^2/\nu)^2}$ is the distance between two successive reflections of the wave from the wall. When the value of this parameter is close to 1 for the operating mode, there are some possibilities of the selection of modes with larger radial indices, because for them D_{eff} is larger, and thus the ratio $L^2/\lambda D_{\text{eff}}$ is smaller. Such modes should experience a larger diffraction spread, which can lower their diffraction Q-factor. This method was discussed elsewhere (Vlasov, Orlova, and Petelin 1981).

3.6 Problems and Solutions

Problems

1. In Sec. 3.1, the linearized efficiency (3.17) was derived for the model, in which a plane EM wave propagates along the external magnetic field $\vec{H}_0$.

When such a wave propagates at a certain angle to $\vec{H}_0$, the electron motion can be described by (3.24) with the Hamiltonian $\mathcal{H} = i\frac{p_\perp}{p_z}\frac{dJ_s(\xi)}{d\xi}e^{-i\vartheta}$. Consider this model: assume that the drift of electron guiding centers can be neglected, derive the expression for the linearized efficiency, compare it with (3.17), and make conclusions.

2. In Sec. 3.2, equations for electron motion in a gyrotron operating at one of TE-modes were derived. Then, self-excitation conditions for such a gyrotron were considered in Section 3.4 with the account for formalism presented in preceding sections. Modify this treatment for the case of operation at one of TM-modes and analyze the difference in starting currents. For simplicity, assume that the axial structure of the resonator field is smooth enough, and therefore, the effect of its non-uniformity on electron motion can be neglected.

Solutions

1. Since we assumed that the drift of guiding centers can be neglected, we can start from Eqs. (3.8)–(3.9), in which the account for a new propagation angle results in multiplying the normalized field amplitude by the factor $J_s'(\xi)$. The argument of this Bessel function is the normalized Larmor radius $\xi = k_\perp a/c = k_\perp p_\perp/eH_0$. In accordance with (1.26) we can introduce $u = [2(1-n\beta_{z0})/\beta_{\perp 0}^2](\mathcal{E}_0 - \mathcal{E})/\mathcal{E}_0$ and rewrite $p_\perp$ as $p_{\perp 0}\sqrt{1-u}$ and p_z as $p_{z0}(1-bu)$. Here $b = n\beta_{\perp 0}^2/2\beta_{z0}(1-n\beta_{z0})$ is the recoil parameter, which characterizes the changes in the electron axial momentum in the process of radiation/absorption. In these notations, equations for the electron normalized energy u and phase ϑ can be rewritten in the form similar to (3.8)–(3.9):

$$\frac{du}{dZ} = \alpha\frac{\sqrt{1-u}}{1-bu}J_s'(\xi_0\sqrt{1-u})\sin\vartheta,$$

$$\frac{d\vartheta}{dZ} = \frac{1}{1-bu}\left\{\delta - \mu w - \frac{\alpha}{2\sqrt{1-u}}J_s''(\xi_0\sqrt{1-u})\cos\vartheta\right\}.$$

Here $Z = \omega z/c$, and normalization of other parameters can readily be derived from (3.24). Applying the same method, which was used for deriving (3.17), results now in the following expression for the linearized efficiency

$$\eta_{\text{lin}} = \frac{\beta_{\perp 0}^2}{2(1-n\beta_{z0})\left(1-\gamma_0^{-1}\right)}\frac{\alpha^2 Z_{\text{out}}^2}{2}[J_s'(\xi_0)]^2\left\{-\left(1+\xi_0\frac{J_s''}{J_s'}\right)\Phi\right.$$
$$\left.+\, b(\Phi + \Theta\Phi') - \mu\Phi'\right\}.$$

Here $\mu = [\beta_{\perp 0}^2(1-n^2)/2\beta_{z0}(1-n\beta_{z0})]Z_{out}$ and Θ and Φ have the same meaning as in (3.17). When the normalized Larmor radius is small, $1 + \xi_0 J_s''/J_s' \simeq s$. Also, when the normalized axial wavenumber is small, the second term in figure brackets can be neglected because $b << 1$. Then, the expression in figure brackets has the same structure as one in the definition of the gyrotron linearized orbital efficiency (3.71). Note that at large values of ξ_0 the term $1 + \xi_0 J_s''/J_s'$ can be negative. This indicates that an additional possibility of CRM excitation can originate from the transverse nonuniformity of the RF field.

2. In line with definitions of electric and magnetic fields of the TE-mode given by (3.23), let us determine the fields of the TM-mode by

$$\vec{E}_s = i\Psi_s f(z)\vec{z}_0,\ \vec{H}_s = -(c/\omega)[\nabla_\perp \Psi_s \times \vec{z}_0] f(z).$$

The RF Lorentz force acting upon electrons is $\vec{G} = \vec{E}_s + [\frac{\vec{v}}{c} \times \vec{H}_s]$. So, using the same formalism as was developed above for TE-modes results in the following expressions for resonant components of this force in the case of TM-modes:

$$G_{s,r} = -f(z)\beta_z L_s J_s'(\xi),\ G_{s,\theta} = if(z)(\beta_z/\beta_\perp)L_s J_s(\xi)$$

(cf. Eqs. (A1.13) for TE-modes). Note that the resonant term in the axial component of the RF Lorentz force is equal to zero because the effect of the axial electric field is compensated by the axial component of the RF Lorentz force in the case of wave propagation near cutoff. Once we assume that the normalized Larmor radius of electrons is small enough ($\xi = \omega a/c = s\beta_\perp < 1$), and correspondingly $J_s(\xi) \approx (1/s!)(\xi/2)^s$, Eqs. (A1.13) for TE-modes are reduced to

$$G_{s,r}^{TE} = -if(z)L_s(s/2)^s(\beta_\perp^{s-1}/s!),\ G_{s,\theta}^{TE} = -f(z)L_s(s/2)^s(\beta_\perp^{s-1}/s!).$$

Similarly, our expressions for these components of TM-modes are reduced to

$$G_{s,r}^{TM} = -f(z)\beta_z L_s(s/2)^s(\beta_\perp^{s-1}/s!),\ G_{s,\theta}^{TM} = if(z)\beta_z L_s(s/2)^s(\beta_\perp^{s-1}/s!).$$

So, these components are now related as $G_{s,r}^{TM} = -i\beta_z G_{s,r}^{TE}$ and $G_{s,\theta}^{TM} = -i\beta_z G_{s,\theta}^{TE}$. From this relationship it follows that one can introduce instead the function $f(z)$ describing the axial structure of the RF field a new function $g(z) = -i\beta_z f(z)$ and apply the formalism that was developed in Chapter 3 to consideration of TM-modes replacing everywhere $f(z)$ by $g(z)$. (Of course, different boundary conditions for membrane functions in both cases result in different expressions for norms of the modes and their coupling impedances.) From this relationship it also readily follows that the expression (3.71) for the linearized orbital efficiency now contains an extra term β_z^2 in its right-hand side. Correspondingly, as follows from the balance equation (3.63) for the case of the linear theory, the starting current of TM-modes should be $1/\beta_z^2$ higher than the starting current of similar TE-modes.

■ CHAPTER 4 ■

Nonlinear Theory of the Gyromonotron (Single-Mode Treatment)

4.1 Cold-Cavity Approximation

Cold-cavity approximation is the approximation we used so far. It implies that we neglect the effect of the electron beam on the axial structure of the EM field excited in an open resonator. Correspondingly, this structure is determined by the nonuniform string equation (2.13), which for axially open resonators was analyzed in detail by Vlasov et al. (1969).

The analysis of the gyrotron large-signal operation thus reduces to the integration of equations for electron motion (3.34) or (3.35) with given parameters Δ, μ, and F and calculation of the orbital efficiency determined by (3.36). Then, the balance equation (3.63) allows one to establish correspondence between the beam current and the EM field amplitude and, hence, translate the nonlinear dependence of the efficiency on the amplitude F, which describes the saturation effects, into the nonlinear dependence of the efficiency on the beam current.

Results of first calculations of the efficiency for gyrotrons with the uniform EM field ($f(\zeta) = 1$) and constant external magnetic field ($\Delta = \Delta_0$) were presented by Gaponov, Petelin, and Yulpatov (1967) for the first five cyclotron harmonics ($1 \leq s \leq 5$). It was shown that the maximum orbital efficiency slowly decreases as the harmonic number increases: this maximum efficiency is equal to 0.42, 0.29, 0.22, 0.17, and 0.14 for $s = 1, 2, 3, 4$, and 5, respectively. These numbers were calculated for the case when all beamlets are equally coupled to the resonator EM field. Rapoport, Nemak, and Zhurakhovskiy (1967) have studied the effect of the transverse nonuniformity of this field on the efficiency at the first three harmonics. In their paper the beam interaction with the standing wave formed by a pair of mirrors was considered. (Later, this sort of gyrotron configuration was called "quasi-optical gyrotrons"; these devices will be discussed in Chapter 13.) It was assumed that a beam diameter is larger than a wavelength, so the transverse nonuniformity of the beam coupling to the wave is important. Rapoport, Nemak, and Zhurakhovskiy (1967) showed that this nonuniformity reduces the maximum orbital efficiency at the first three harmonics from the above-mentioned efficiencies for the case of uniform interaction 0.42, 0.29, and 0.22 to 0.31, 0.22, and 0.16, respectively.

Later, due to the reasons discussed above, attention was mostly paid to gyrotrons with the Gaussian axial distribution of the resonator field. First calculations (Moiseev, Rogacheva, and Yulpatov 1968) showed that in gyrotrons with such an axial structure of the resonator field the maximum orbital efficiency is equal to 0.72, 0.71, and 0.55 for $s = 1$, 2, and 3, respectively. Then, a more detailed parameter search was done for the first two harmonics by Nusinovich and Erm (1972) and for the first five harmonics by Danly and Temkin (1986). An example of these results is shown in Fig. 4.1. In all these studies it was assumed that the interaction starts and ends in the cross sections where the field is e^{-3} times smaller than in the middle of resonator. Note also that when the interaction is restricted by cross sections where the field decays by e times only, the maximum orbital efficiency is much smaller (Gryaznova, Koshevaya, and Rapoport 1969): 0.52, 0.42, and 0.38 for $s =$ 1, 2, and 3, respectively. This efficiency degradation can be associated with the fact that in the latter case the electron interaction starts in a relatively strong field, while for preparation of a compact bunch it is preferable to start interaction gently, in the region of a weak field. Such a study of the effect of "tails"was done later (Gaponov et al. 1975). It was shown that the account for the cathode tail of the Gaussian distribution increases the maximum orbital efficiency up to 0.79 and 0.76 for $s = 1$ and 2, respectively, while the account for the collector tail does not play a big role.

It seems quite obvious that not only the tapering of the axial profile of the resonator field, but also the tapering of the external magnetic field can improve the interaction efficiency. First theoretical studies done by Moiseev, Rogacheva, and Yulpatov (1968) showed that for the efficiency enhancement the external magnetic field should be slightly up-tapered. Results of more detailed calculations reported by Gryaznova, Koshevaya, and Rapoport (1969) indicated that the magnetic field tapering can increase the maximum orbital efficiency at the first three cyclotron harmonics from the above-mentioned 0.52, 0.42, and 0.38 to 0.7, 0.67, and 0.53 for $s = 1$, 2 and 3, respectively. These predictions were later confirmed by theoretical studies done by other groups (Kuraev et al. 1970, Sprangle and Smith 1980, Chu, Read, and Ganguly

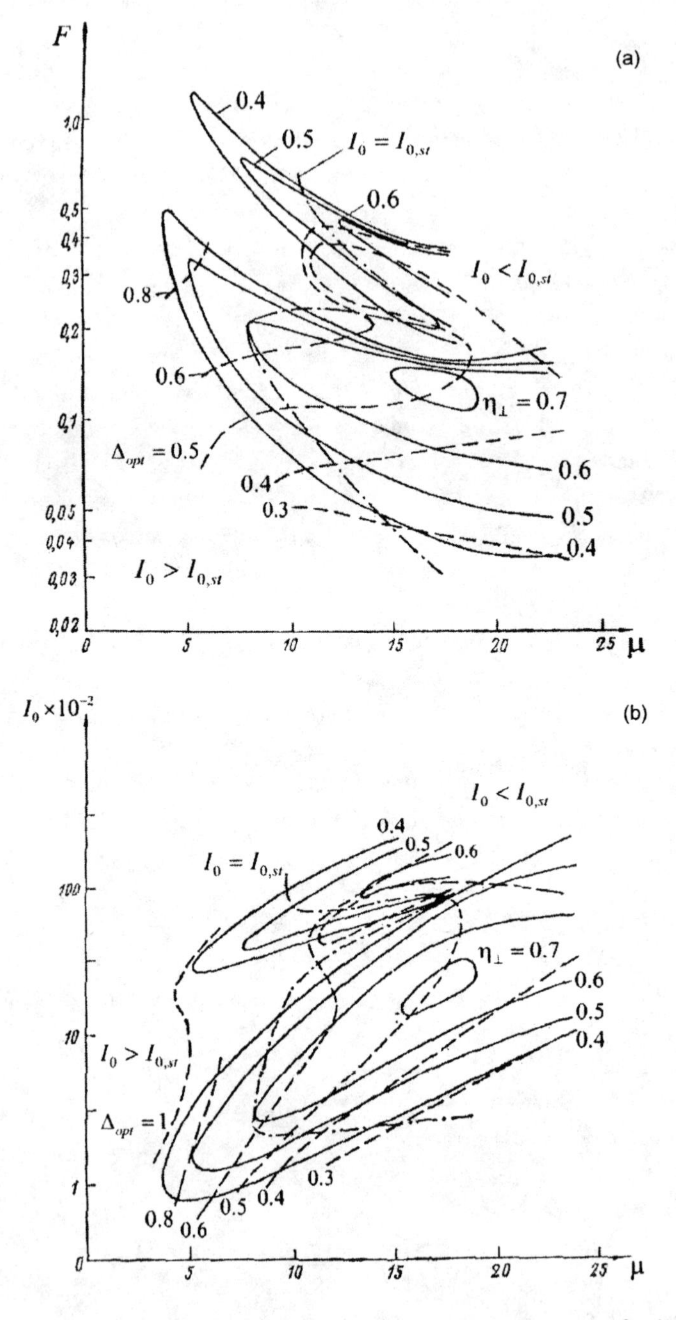

Fig. 4.1. Lines of equal orbital efficiencies in gyrotrons operating at the fundamental cyclotron resonance in the planes of normalized RF field amplitude, F, versus the normalized length, μ (a) and normalized beam current, I_0, versus the normalized length, μ (b). Dashed lines show optimal cyclotron resonance mismatches. Dash-dotted line $I_0 = I_{0,st}$ shows the border between the regions of soft ($I_0 > I_{0,st}$) and hard ($I_0 < I_{0,st}$) self-excitation.

1980), as well as by experiments (Glushenko, Koshevaya, and Prus 1977, Read, Chu, and Dudas 1982).

It was also shown that the tapering of the magnetic field allows one to realize the maximum efficiency in the regime of soft self-excitation (Kuraev 1979), while in the case of constant magnetic field the maximum efficiency, as is shown in Fig. 4.1, corresponds to the hard self-excitation. The difference between these two regimes is in the relationship between the beam current and the starting current. In the soft excitation regime the beam current exceeds the starting value, i.e., the self-excitation conditions are fulfilled (see Fig. 4.2a). So the oscillation amplitude can start to grow from the noise level. On the contrary, in the regime of hard excitation the starting current exceeds the beam current (see Fig. 4.2b). So the oscillations can start to grow only when the initial amplitude exceeds a certain bifurcation threshold. In practice, in order to realize the operation in the hard excitation regime, one should use some kind of start-up scenarios (Nusinovich 1974, Whaley et al. 1994). Those imply that, by varying some gyrotron parameters, we can first pass through the soft excitation region, in which the oscillations will be excited, and then reach the maximum efficiency point of destination located in the

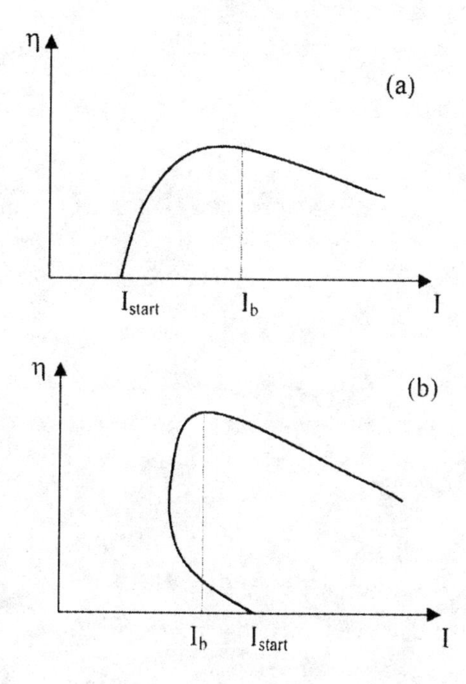

Fig. 4.2. Efficiency as the function of the beam current in regimes of soft (a) and hard (b) self-excitation.

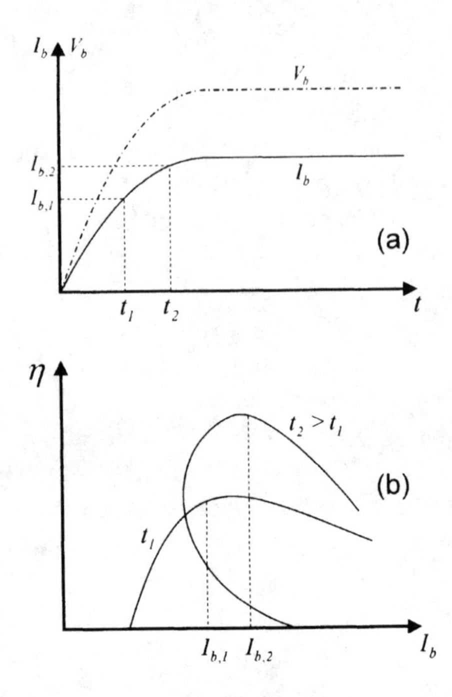

Fig. 4.3. (a) Beam voltage and current rise in the case of pulsed operation; (b) efficiency as the function of the beam current for two instants of time ($t_1 < t_2$) in the case of proper changes in gyrotron parameters.

region of hard excitation. This statement is illustrated with Fig. 4.3. Note that, as the beam voltage increases, the relativistic cyclotron frequency decreases. Correspondingly, the cyclotron resonance mismatch $\omega - \Omega_0$, which, as we discussed in Chapter 1, should be positive for realizing the coherent cyclotron radiation, becomes larger. This makes the self-excitation harder, because only the EM field with large enough amplitude can trap electrons with substantial slippage of the gyrophase with respect to the phase of the EM field. Since there are a number of limitations on variation of gyrotron parameters (especially in pulsed operation regimes), which will be discussed later, it is obvious that it is preferable to operate in the regime of soft excitation.

4.2 Self-Consistent Approach

A self-consistent treatment of interaction between gyrating electrons and EM fields in axially open resonators implies that we include into consideration the modification of an axial structure of the field by the high-frequency component of the electron current density. This modification becomes significant when the diffractive Q-factor is close to its minimum value, or, in other words, when the wave reflections, at least at the output, are negligibly small. (When the reflections at both ends are large enough, the axial structure is fixed, so the beam cannot modify it significantly.)

To describe the field excitation in such a case, one cannot use the field representation, which was used in Chapter 3, because in that representation the field amplitude C_s, determined by the beam current, was considered separately from the function $f(z)$ describing the axial structure of the field. Instead, the electric field should be represented as

$$\vec{E} = \mathrm{Re}\left\{A(z,t)\vec{E}_s(\vec{R}_\perp)e^{i\bar{\omega}_0 t}\right\}. \quad (4.1)$$

Here $\bar{\omega}_0$ is an arbitrary chosen carrier frequency. For instance, one can choose as $\bar{\omega}_0$ the cutoff frequency in the output cross section. Such a representation, which was, in particular, used by Ginzburg, Nusinovich, and Zavolsky (1986), allows one to describe self-consistently not only stationary but also nonstationary processes. Again, assuming that the Fresnel parameter for open resonators under consideration is large enough, one can determine the function $\vec{E}_s(\vec{R}_\perp)$ by the membrane function, which obeys Helmholtz equation (AI.1), with corresponding boundary conditions. Recall (see Sec. 2.3) that in an irregular waveguide the transverse wavenumber, which is present in the Helmholtz equation, depends on z.

Introducing the field representation given by (4.1) into Maxwell equations and making use of the assumption about slow temporal variations in the field amplitude ($|\partial A/\partial t| << \omega|A|$) results in the parabolic equation (see, e.g., Ginzburg, Nusinovich, and Zavolsky 1986)

$$\frac{\partial^2 A}{\partial z^2} + \frac{\bar{\omega}_0^2 - \omega_0^2}{c^2}A - 2i\frac{\bar{\omega}_0}{c^2}\frac{\partial A}{\partial t} = -i\frac{\bar{\omega}_0}{cN_s}\int\limits_{S_\perp}\vec{j}_\omega \cdot \vec{E}_s^* ds_\perp. \quad (4.2)$$

Here $\omega_0 = \omega_0(z) = k_\perp(z)c$ is the cutoff frequency in a given cross section, i.e., the cutoff frequency of the comparison waveguide. In the right-hand side of (4.2) the norm N_s is different from the norm used in Chapter 3 because we consider an irregular waveguide near cutoff. Now the norm is determined as

$$N_s = \frac{c}{4\pi}\int\limits_{S_\perp}\vec{E}_s \cdot \vec{E}_s^* ds_\perp. \quad (4.3)$$

The integral in the right-hand side (RHS) of this equation can be reexpressed in gyrotron normalized variables, which results in the following equation

$$\frac{\partial^2 f}{\partial \zeta^2} - i\frac{\partial f}{\partial \tau} + \delta f = \frac{I_0}{2\pi}\int\limits_0^{2\pi} w^{s/2}e^{i\vartheta}d\vartheta_0. \quad (4.4)$$

Here we introduced the normalized time $\tau = (\beta_{\perp 0}^4/8s^2\beta_{z0}^2)\bar{\omega}_0 t$, the parameter $\delta = (8s^2\beta_{z0}^2/\beta_{\perp 0}^4)[(\bar{\omega}_0 - \omega_0)/\bar{\omega}_0]$ characterizes the deviation of the cutoff frequency in an irregular waveguide. The variables w and ϑ obey the same equations (3.34), in which the product $Ff(\varsigma)$ should be replaced by $-2f(\varsigma, \tau)$. The normalized beam current parameter I_0 in (4.4) is equal to

$$I_0 = 64\frac{eI_b}{mc^3}\frac{\beta_{z0}\beta_{\perp 0}^{2(s-4)}}{\gamma_0}s^3\left(\frac{s^s}{2^s s!}\right)^2 G. \tag{4.5}$$

Eq. (4.5) is quite similar to (3.57). Here G is determined by (3.59).

Certainly, it is necessary to supplement Eq. (4.4) with the initial $f(\tau = 0) = f_0(\zeta)$ and boundary conditions. The boundary condition at the entrance, in the input cutoff cross section, can be given simply by $f(\zeta = 0) = 0$. At the open exit, the boundary condition in nonstationary regimes is more complicated. It can be simplified by assuming that there are no waves entering the interaction space from the output waveguide. Then, one can write for Fourier components of the field $f_\Omega = \int_0^\infty f(\tau')e^{-i\Omega\tau'}d\tau'$ the condition for outgoing radiation $\frac{\partial f_\Omega}{\partial\zeta}|_{\zeta_{out}} = -i\xi f_\Omega(\zeta_{out})$. Here ξ is the normalized axial wavenumber in the output cross section; its normalization corresponds to normalization of z. Making an inverse Fourier transform of this condition results in the integro-differential equation for $f(\zeta_{out})$, which was derived by Ginzburg, Nusinovich, and Zavolsky (1986).

In the stationary regime, the field amplitude $f(\varsigma, \tau)$ can be represented as $f(\zeta)e^{i\Omega\tau}$, which reduces (4.4) to the equation

$$\frac{d^2 f}{d\zeta^2} + \Omega f = \frac{I_0}{2\pi}\int_0^{2\pi} w^{s/2}e^{i\vartheta}\,d\vartheta_0, \tag{4.6}$$

with the boundary conditions $f(0) = 0$, $df/d\zeta|_{\zeta_{out}} = -i\sqrt{\Omega}f(\zeta_{out})$. Here Ω and the normalized axial wavenumber ξ introduced above are related as $\Omega = \xi^2$. Certainly, the neglect of the beam effect on the field structure ($I_0 \to 0$) reduces (4.6) to the nonuniform string equation (2.13).

The stationary self-consistent theory developed by Bratman et al. (1973) (see also Fliflet et al. 1982) allows one to study the modification of the field structure with the variation in gyrotron parameters. It also allows one to determine optimal values of parameters for realizing the maximum efficiency. Note that the optimal parameters found in the self-consistent treatment are not very different from those determined in the cold-cavity approximation. For instance, now the optimal value of the normalized length for the operation at the fundamental cyclotron resonance is equal to 14.5 (Bratman et al. 1973), while for the fixed Gaussian profile of the field it was about 17 (see

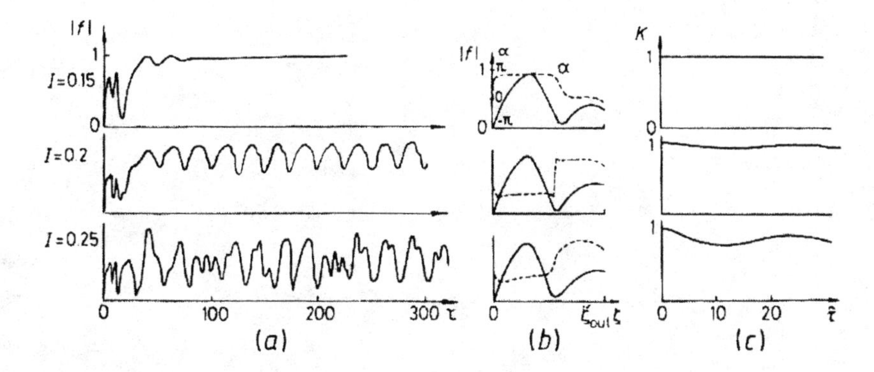

Fig. 4.4. Nonstationary processes at positive cyclotron resonance mismatch $\Delta = 0.3$ (a), and corresponding axial structure of the module and phase of the RF field (b) and the correlation function (c).

Fig. 4.1). However, the maximum efficiency operation is realized now in the soft-excitation regime.

The nonstationary self-consistent theory (Ginzburg, Nusinovich, and Zavolsky 1986) describes the onset of oscillations, which, at relatively low currents, can be stationary oscillations with the constant amplitude. As the current increases, the device may exhibit transition from such oscillations to automodulation, and then to irregular oscillations, which can be interpreted as chaotic ones. An example of such transitions is shown in Fig. 4.4. These

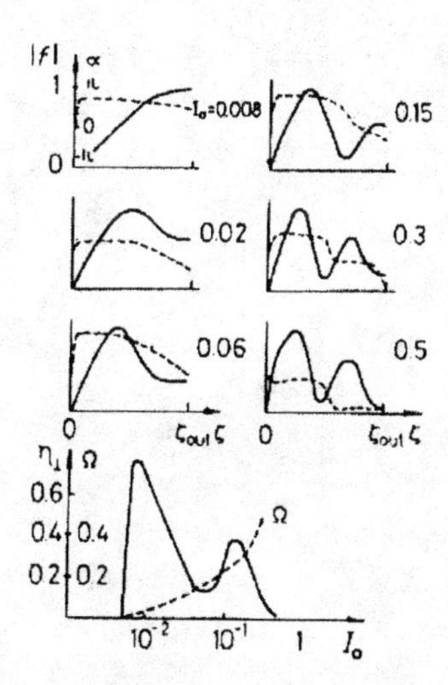

Fig. 4.5. Evolution of the axial structure of the RF field with the beam current increase. Bottom figure shows corresponding dependencies of the electron orbital efficiency (solid line) and operating frequency shift (dashed line).

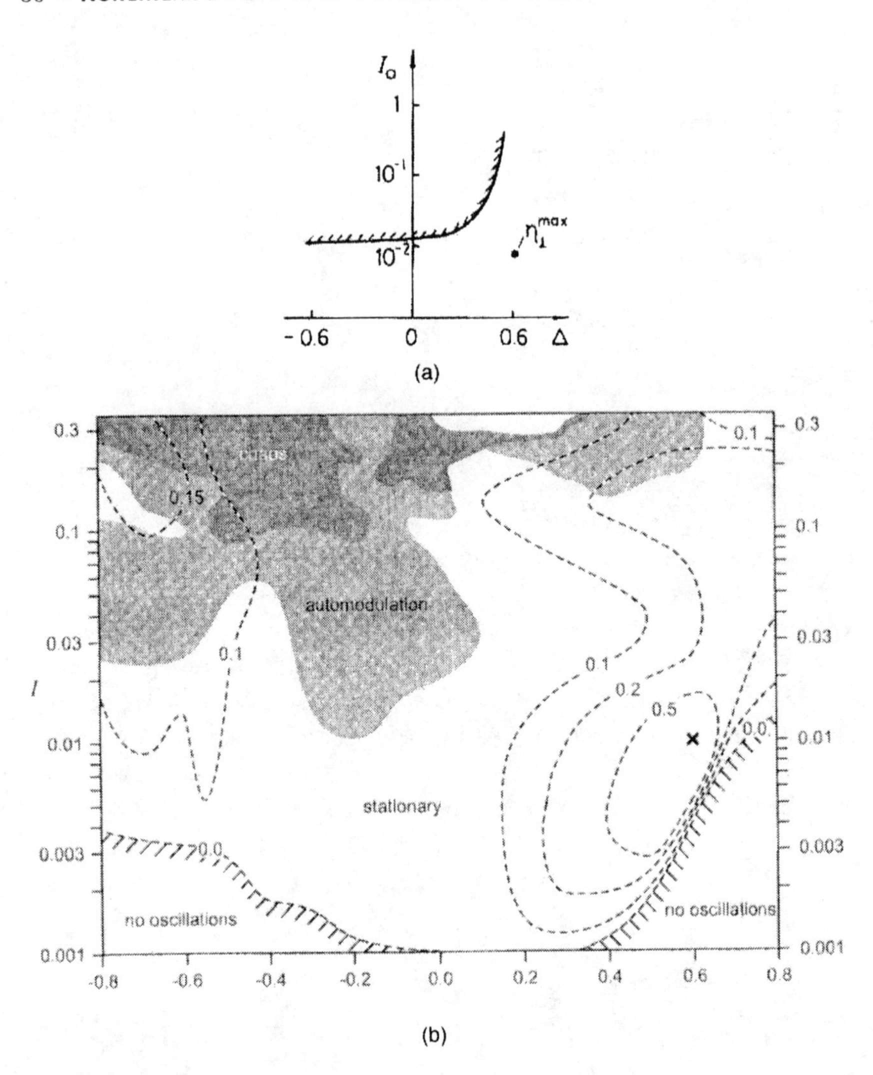

Fig. 4.6. (a) Boundary of the region of stable oscillations with constant amplitude in the plane of gyrotron parameters: normalized beam current versus cyclotron resonance mismatch, (b) a more detailed map of the same plane.

transitions can easily be observed in the region of backward wave operation (negative cyclotron resonance mismatch) and even in the region of small positive mismatches (gyrotron regimes with a low efficiency).

In the region of gyrotron operation with a high efficiency (large positive mismatches), the oscillations with constant amplitude, however, remain, stable even when the current significantly exceeds the optimum value. As is shown in Fig. 4.5, as the current increases, the axial structure of the EM field experiences substantial modification, but this does not change the stability of operation. The resulting boundary, which limits the region of stationary oscillations with constant amplitude, is shown in Fig. 4.6a. Later, Airila et al. (2001) explored this parameter space in more details. The resulting map is presented in Fig. 4.6b.

Presently, such codes as MAGY (Botton et al. 1998) are available and can accurately describe self-consistent, nonstationary processes in gyrotron oscillators. One of the important results of using such codes was identification of possible parasitic excitation of various modes in the region of up-tapering in the output waveguide. In general, in a region that is not too far from the resonator exit, the cutoff frequency is still close to the operating frequency, and the magnetic field is also close enough to its resonant value. So the cyclotron resonance conditions can be fulfilled there and, hence, the interaction can continue.

4.3 Effect of Velocity Spread

As was already discussed in Chapter 2, the spread in electron axial velocities may cause the inhomogeneous Doppler broadening of the cyclotron resonance band. This, in turn, can result in the efficiency deterioration. Even in the case of operation near cutoff the difference in electron axial velocities makes a difference in electron transit times through the interaction region, which spoils the efficiency.

This effect can be illustrated with the use of Eqs. (3.24) and (3.34). As was described in Chapter 3, the equation for the slowly variable phase ϑ in (3.24) contains in its left-hand side (LHS) the combination $\omega/v_z - sh_H$, which can be presented [see comments to (3.34)] as $\frac{\omega}{v_{z0}}\frac{\beta_{\perp 0}^2}{2}(\Delta + w - 1)$.

During the initial stage of electron interaction with a weak EM field near the entrance, we can assume that the energy modulation given by the last term, proportional to $w - 1$, is practically the same for all electron fractions, i.e., here $v_{z0} \approx \bar{v}_{z0}$, where $\bar{v}_{z0}$ is the mean value of the initial axial velocity. However, for the first term, which describes the electron slippage with respect to the EM field, this difference in axial velocities can be significant. So, for different fractions the difference in this term can be estimated as $\frac{\omega}{\bar{v}_{z0}}\frac{\beta_{\perp 0}^2}{2}\Delta|\frac{\Delta v_z}{\bar{v}_{z0}}|$. This means that the effect of energy modulation becomes visible only when this modulation $|w - 1| \sim F|\int_{\zeta_{entr}}^{\zeta} f(\zeta')d\zeta'|$ is larger than $\Delta|\frac{\Delta v_z}{\bar{v}_{z0}}|$. For the field with the Gaussian axial structure, $f(\zeta) = \exp\{-(2\zeta/\mu)^2\}$, this condition can be given as

$$F\mu\frac{\sqrt{\pi}}{4}[1 - \mathrm{erf}(t)] > \Delta\left|\frac{\Delta v_z}{\bar{v}_{z0}}\right|, \tag{4.7}$$

where erf(t) is the error function of the argument $t = 2\zeta/\mu$, which is tabulated elsewhere (Abramovitz and Stegun 1964).

Using the results of efficiency calculations presented in Sec. 4.1, one can readily figure out how the spread in v_z reduces the efficiency. Recall that in that section it was mentioned that, when we take into account all the input tail of the Gaussian distribution, the maximum orbital efficiency is 0.79. Correspondingly, when we assume that the interaction starts in the cross section where the field is e^{-3} times smaller than in the maximum, the maximum orbital efficiency is 0.72, and finally, in the case when it starts at e^{-1} level, the maximum orbital efficiency is 0.52 only. Let us take for parameters F, μ, and Δ their optimal values: 0.14, 17, and 0.5, respectively. Then, the effect of energy modulation becomes visible only when $1 - \text{erf}(t) > 0.474|\frac{\Delta v_z}{\bar{v}_{z0}}|$. So, in the absence of the spread we can achieve the maximum orbital efficiency equal to 0.79. The axial velocity spread of about 3% makes the energy modulation visible only starting from the cross section, where the Gaussian distribution is e^{-3} times smaller than in its maximum. This reduces the maximum orbital efficiency to 0.72. Correspondingly, the effective interaction starts only in the cross section, where the field is e^{-1} times smaller than in the maximum, when the axial spread is close to 33%. Note that in typical electron beams formed by the magnetron injection guns the axial velocity spread is on the order of a few percentages. Therefore, our simple consideration can be used as an explanation to the fact that for many experiments the results of calculations done by Nusinovich and Erm (1972) and Danly and Temkin (1986) were in reasonable agreement with experimental data. (Recall that in those studies it was assumed that the interaction space is limited by the cross sections where the field is at e^{-3} level.)

Needless to say that the velocity spread also restricts the maximum orbital-to-axial velocity ratio, which can be used in gyrotrons. Since typically just the energy of orbital motion is extracted in the process of interaction in gyrotrons, it is desirable to make this ratio as large as possible. However, in the case of significant spread, electrons with initially small axial velocities will be then reflected in the magnetic mirror, formed by the increasing magnetic field. This can lead to the formation of a cloud of trapped electrons in the region between the gun and the resonator.

The effect of velocity spread on the efficiency was studied in a large number of papers. One of the first and quite general studies of this effect has been performed by Ergakov, Moiseev, and Erm (1980). It was assumed there that the axial structure of the resonator field can again be described by the Gaussian distribution and also that all electrons have the same kinetic energy, but a certain spread in orbital-to-axial velocity ratios. The latter was described by the Gaussian distribution in orbital velocities

$$f_e(v_\perp) \sim \exp\left\{-4\left(\frac{v_{\perp 0} - \bar{v}_{\perp 0}}{\Delta v_\perp}\right)^2\right\}. \tag{4.8}$$

Note that the spread $\Delta v_\perp$ at e^{-1} level in (4.8) is $2\sqrt{2}$ times larger than the RMS value. Some special means were undertaken in order to eliminate the electrons with too large orbital velocities from consideration, since such mirrored particles cause singularities in the equations of motion written in Lagrangian coordinates.

Ergakov, Moiseev, and Erm (1980) have optimized the gyrotron parameters for maximizing the total electron efficiency

$$\eta = \frac{1}{2\left(1-\gamma_0^{-1}\right)} \int_0^v v_{z0}\beta_{\perp 0}^2 f_e(v_{\perp 0})\eta_\perp dv_{\perp 0}. \tag{4.9}$$

Before presenting their results, let us make some comments to (4.9). This equation is a generalization of (3.37) for the case of a beam with velocity spread. In such a beam, a number of electrons with orbital velocities in the interval $\Delta v_\perp$ is proportional to $\Delta n \sim f_e(v_{\perp 0})\Delta v_\perp$. The microwave power extracted from this fraction of the beam is $\eta V_b \Delta I_b$. Here the efficiency η is determined by (3.37) and the current is given by $\Delta I_b = v_z \Delta\rho = ev_z\Delta n$. So, this power can be determined as

$$\Delta P = \frac{\beta_{\perp 0}^2}{2\left(1-\gamma_0^{-1}\right)}\eta_\perp e V_b v_z f_e(v_\perp)\Delta v_\perp.$$

Correspondingly, the total power extracted from the beam is

$$P = \frac{eV_b}{2\left(1-\gamma_0^{-1}\right)} \int_{v_\perp} \beta_{\perp 0}^2 \eta_\perp v_{z0} f_e(v_{\perp 0}) dv_{\perp 0}. \tag{4.10}$$

The total efficiency is the ratio of this power to the beam power. In (4.10) the distribution function is normalized as $\int_{v_\perp} f_e'(v_\perp)v_z dv_\perp = 1$. This implies $f_e'(v_\perp) = e/I_b f_e(v_\perp)$. Thus, using (4.10) yields (4.9).

Some results of simulations are presented in Fig. 4.7, which shows the evolution of optimal parameters with the increase in the velocity spread. The results shown look quite optimistic. They predict that even in the case of the 60% spread, which is equivalent to more than 20% RMS spread in orbital velocities, the maximum efficiency can be about 30%. The optimal length gets smaller, as the velocity spread increases. This tendency can be associated with the increasing role of slowly moving electrons, which interact with the field

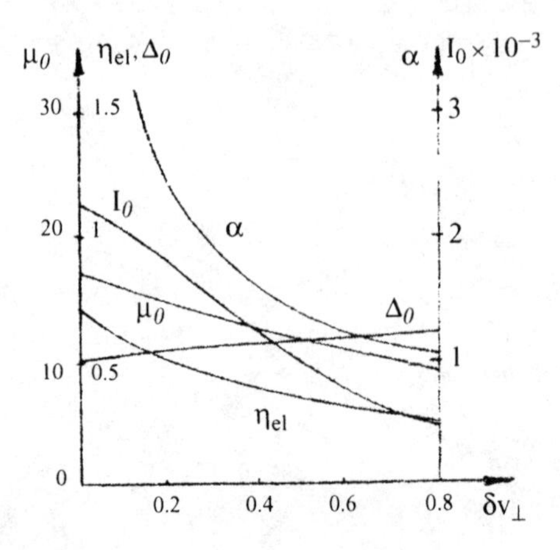

Fig. 4.7. Maximum electron efficiency and optimal cyclotron resonance mismath Δ_o, orbital-to-axial velocity ratio α, and normalized beam current I_o as functions of the orbital velocity spread. (Reproduced from Ergakov, Moiseev, and Erm 1980).

for a rather long time. Correspondingly, as the spread increases, the efficiencies achievable in the regime of soft self-excitation get closer to the maximum efficiency, as is shown by Ergakov, Moiseev, and Erm (1980). An interesting finding from their study was the conclusion that in the range of spreads $0.1 < \delta v_{\perp} < 0.8$ (here $\delta v_{\perp} = \Delta v_{\perp}/\bar{v}_{\perp 0}$) the optimal value of the pitch-ratio $\alpha = \bar{v}_{\perp 0}/\bar{v}_{z0}$ approximately obeys the relation $\alpha^2 \delta v_{\perp} \approx 1$. This makes experimentally achievable α's ($\alpha \leq 2$), in the presence of the velocity spread, closer to their optimal values.

4.4 Space-Charge Effects

As in any microwave source driven by an electron beam, in the gyrotron one can distinguish two kinds of space charge effects, which are caused, respectively, by DC and AC space-charge fields.

The voltage depression caused by the beam DC space charge was considered in Sec. 2.2 (see also Drobot and Kim 1981). In addition to that consideration, let us note that in real devices the vacuum is typically at the level 10^{-7}–10^{-8} Torr. This means that there are always some residual gases, which can be ionized by the beam impact ionization. The presence of these ions can compensate, to some extent, for the DC space charge force. Let us also note that when the beam has a finite thickness, of course, the voltage depression, which varies across the beam, causes a certain spread in electron velocities.

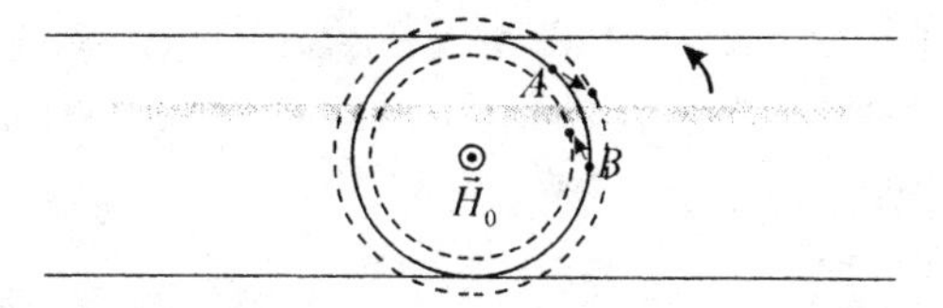

Fig. 4.8. AC space charge effect: enhancement of electron bunching due to the negative mass instability in a layer of gyrating electrons.

The role of AC space charge effects is more specific. Its nature is similar to the negative mass instability (Nielsen and Sessler 1959, Kolomenskiy and A. N. Lebedev 1959), which was briefly mentioned in Chapter 1. These effects can be present not only in the interaction regions, but also in regions free from microwaves (like drift regions of gyroklystrons). Therefore, it makes sense to explain the nature of these effects by considering the electrons in such drift tubes. Such a treatment was, first, done by V. K. Yulpatov (1970), who, however, did not publish his results. Later, Bratman (1976), Symons and Jory (1981), and Charbit, Herscovici, and Mourier (1981) considered the same problem by using slightly different approaches. Since, typically, the beam radius is much larger than the wavelength and the gyroradius of electrons, a thin annular electron beam can be replaced by a thin electron layer, which is schematically depicted in Fig. 4.8. In our explanation of the effect we will follow Bratman (1976) and Symons and Jory (1981).

A gyrating electron experiences the action of the space-charge force, which results from the contributions of all other electrons. Certainly, in an unperturbed beam, this force is zero in the midplane of a beam and has a maximum on the top of the layer and minimum in the bottom. The mean value of this force is equal to zero.

Assume now that for some reason (like a bunching caused by initial modulation of electron energies in the input resonator) there is a region of highest space-charge density, which is shadowed in Fig. 4.8. This layer of the highest density oscillates with the relativistic electron cyclotron frequency. Then, a test particle, A, shown in the figure, which is moving ahead of the bunch, will be outside this shadowed region during its journey up, and thus will be accelerated by the space charge field of the layer. On its way down, this particle will be decelerated. However, particle A will spend a smaller time on its way back because this layer was also moved up during this time interval. Therefore, there will be a net acceleration of this electron. Correspondingly, as a result, it will start to gyrate slower and its orbit increases, as is shown by a small arrow in Fig. 4.8. Likewise, an electron, B, gyrating behind the bunch will be after all decelerated by the bunch, and hence, will gyrate faster and execute smaller orbits, as shown in the figure. So, as a result, the beam space charge increases the density of electrons in the layer, and hence, enhances the electron bunching.

In gyroklystrons, this effect can be used for shortening the drift sections. The AC space charge fields can also be important in the regions between the electron gun and the interaction space, because they may cause some space charge instabilities. As was shown by Liu, Antonsen, and Levush (1996), the growth rate of these instabilities is proportional to ω_p^2/Ω^2, where ω_p is the beam plasma frequency. Since such instabilities, once they occur in the region of beam formation, can induce the spread in electron velocities and energies, they were actively studied for the case of electron beams propagating in adiabatically increasing external magnetic fields (see Liu, Antonsen, and Levush 1996 and references therein).

As we just noted, the role of space-charge effects in gyrotrons is characterized by the ratio ω_p^2/Ω^2. For practical purposes it makes sense to show how this ratio can be expressed in terms of the beam current and beam geometry. Let us start from the standard definition of the electron density, $n = j_z/ev_z$. Since the beam current and current density are related as $I_b = j_z S_b$, where S_b is the beam cross-section area, the electron density can be rewritten as

$$n = \frac{eI_b}{mc^3} \frac{1}{\beta_z r_e S_b}, \tag{4.11}$$

where $r_e = e^2/mc^2 \approx 2.82 \cdot 10^{-13} cm$ is the classical electron radius. For the beam current expressed in Amperes, (4.11) can be rewritten as

$$n(\mathrm{cm}^{-3}) = 0.21 \cdot 10^9 \frac{I_b(A)}{\beta_z S_b(\mathrm{cm}^2)}. \tag{4.12}$$

Correspondingly, the ratio of squared frequencies can be determined as

$$\frac{\omega_p^2}{\omega^2} = \frac{4\pi e^2 n}{m\gamma_0\omega^2} = \frac{eI_b}{mc^3} \frac{1}{\beta_{z0}\gamma_0} \frac{\lambda^2}{\pi S_b}. \tag{4.13}$$

When an electron beam can be treated as a thin annular electron beam with the guiding center radius much larger than the Larmor radius, $S_b \simeq 4\pi R_b v_\perp/\Omega_0$. Then, taking into account the cyclotron resonance condition, (4.13) can be rewritten as

$$\frac{\omega_p^2}{\omega^2} = \frac{eI_b}{mc^3} \frac{1}{s\beta_{z0}\beta_{\perp 0}\gamma_0} \frac{\lambda}{2\pi R_b}. \tag{4.14}$$

The ratio ω_p^2/Ω^2 has a quite similar form: the only difference is in the place of the cyclotron harmonic number s: instead of the denominator in (4.14) it should be placed in the numerator.

The AC space charge effects, as was shown by Bratman and Petelin (1975), slightly decrease the interaction efficiency in gyromonotrons. In their paper it was shown that the dependence of the efficiency on the beam space charge can be approximated by the formula

$$\eta = \eta(S = 0) + S \left. \frac{\partial \eta}{\partial S} \right|_{S=0} . \tag{4.15}$$

Here the space charge parameter S is equal to $(4/\pi\beta_{\perp 0}^2)(\omega_p^2/\Omega^2)$ and the absolute value of the negative derivative $\partial\eta/\partial S$ is on the order of one. A typical example was also studied, for which it was shown that the space-charge effects reduce the efficiency from 42% to 38%.

4.5 Trade-Offs in the Gyrotron Design

Above, we described some methods of optimizing the gyrotron parameters for maximizing the interaction efficiency. However, in the design of gyrotrons not only interaction efficiency but also other operating parameters, which characterize the device performance, are of great interest. Therefore, typically, researchers and designers deal with some trade-offs, part of which will be discussed below. Here, we will discuss the output efficiency and the output power. The power limits in the peak pulse and CW or repetitive-pulse operations will be considered separately.

Output efficiency is the ratio of the power radiated from the resonator to the beam power. Recall that there are two sorts of losses: diffractive losses, which cause the outgoing radiation, and Ohmic losses, which cause the power dissipation in the resonator walls. Correspondingly, the output efficiency, η_{out}, relates to the interaction or electronic efficiency, η_{el}, which we analyzed so far, as

$$\eta_{out} = \left(1 - \frac{Q}{Q_{ohm}}\right)\eta_{el} = \frac{Q_{ohm}}{Q_{ohm} + Q_D}\eta_{el} , \tag{4.16}$$

where the total Q is determined by (3.41). Clearly, when $Q_{ohm} >> Q_D$, the Ohmic losses are negligibly small in comparison with diffractive losses, and $\eta_{out} \approx \eta_{el}$.

As one can easily estimate by using Eqs. (3.42) and (3.44), at wavelengths in the range of centimeters and long millimeters $Q_{ohm} >> Q_D$, so there is no need to distinguish these two efficiencies. However, at short millimeters and especially at submillimeters, the ohmic Q decreases rapidly. Also, in some cases, as the wavelength shortens, the interaction cross section and the beam current get smaller, which makes it necessary to increase the diffractive Q in order to be able to excite oscillations with a reasonable efficiency. In such cases one should optimize the output efficiency instead of the electronic efficiency, or, in other words, to find a trade-off between Ohmic losses and interaction efficiency. This adds one more parameter, the ratio of difractive to ohmic Q-factors, to the set of normalized parameters that characterize the gyrotron efficiency. V. S. Ergakov, M. A. Moiseev, and V. I. Khizhnyak carried out detailed optimization of gyrotrons operating at the first two cyclotron harmonics with the account for ohmic losses. Their unpublished results were presented later in the review paper (Nusinovich and Pankratova 1981). These results demonstrated gradual reduction in the output efficiency with the increase in ohmic losses. Temkin et al. (1979) treated this problem in a similar way.

In the limit of small currents, it makes sense to increase the interaction length in order to get a substantial bunching in a weak EM field. In such a case the optimal value of the ratio Q_D/Q_{ohm}, as is shown by Nusinovich and Pankratova (1981), is ½. A much more favorable scaling for the efficiency occurs in the case of corresponding lengthening of the interaction space, which was discussed in Sec. 3.2. (In more detail, this scaling is discussed by Nusinovich 1984.) Recall that the electron motion and the efficiency in such a case can be described by Eqs. (3.38) and (3.39), respectively.

Peak output power, as follows from the balance equation (1.30), is equal to

$$P_{out} = \eta_{out} V_b I_b = \frac{\omega}{Q_D} W. \tag{4.17}$$

So, to increase this power, one should increase the beam power and try to keep the efficiency unchanged. The latter is not so easy because the orbital efficiency depends on the normalized interaction length and amplitude of the EM field. This length, in turn, determines the diffractive Q, while the amplitude determines the microwave energy stored in the resonator [see RHS of Eq. (4.17)]. Clearly, as the power increases, the field amplitude increases as well. As follows from Fig. 4.1a, electron overbunching in EM fields that are too strong can be to some extent avoided, when the field amplitude increase is accompanied with the shortening of the interaction length. These changes in length will also decrease the diffractive Q. Consistent changes in the field amplitude and interaction length allow designers to combine a substantial power enhancement with a slight efficiency degradation.

Average power is limited by microwave ohmic losses in a cavity. Mean value of the density of these losses can be estimated as

$$\bar{p}_{\text{ohm}} = \frac{Q_D}{Q_{\text{ohm}}} \frac{P_{\text{out}}}{S}, \tag{4.18}$$

where $S = 2\pi RL$ is the wall area in an open resonator. Taking into account definitions of Q_D and Q_{ohm} given above, one can readily show (see, e.g., Flyagin and Nusinovich 1988) that

$$\bar{p}_{\text{ohm}} \propto \frac{P_{\text{out}}}{\lambda^2} \frac{L\delta_{\text{sk}}}{R^2(1 - m^2/\nu^2)}. \tag{4.19}$$

Since this density of ohmic losses is limited by cooling capabilities and typically does not exceed $\bar{p}_{\text{ohm}}^{\text{max}} = 1 - 2kW/\text{cm}^2$, (4.19) sets the limit for the maximum output power in such regimes. Let us point out that this limit can play an important role even when $Q_{\text{ohm}} >> Q_D$. With the account for the dependence $\delta_{\text{sk}} \propto \sqrt{\lambda}$ given by Eq. (3.43), Eq. (4.19) shows that when resonator dimensions scale linearly with the wavelength (L/λ, $R/\lambda = \text{const}$), which implies operation at a given mode in a resonator with a given diffractive Q, the maximum average power scales as $P_{\text{out}}^{\text{max}} \propto \lambda^{5/2}$. This is a standard scaling law for various microwave tubes. Alternatively, Eq. (4.19) shows how to expand the cross section of the interaction region in order to keep the output power constant. It follows from (4.19) that to realize this, when $L/\lambda = \text{const}$, the resonator radius should scale as $R \propto \lambda^{-1/4}$. Certainly, this means the shift of operation into the region of high-order modes with a very dense spectrum, where simultaneous interaction with several modes becomes possible.

4.6 Problems and Solutions

Problems

1. By using assumptions made in Sec. 4.2, derive from the wave equation the parabolic equation (4.2).
2. Determine the mean orbital-to-axial velocity ratio, $\alpha = \bar{v}_{\perp 0}/\bar{v}_{z0}$, at which 1% of electrons in a beam with the Gaussian velocity distribution function given by (4.8), is reflected (mirrored) in the magnetic bottle. Assume the orbital velocity spread to be equal to 10%.
3. Determine the residual gas density at the pressure 10^{-7} Torr and compare it with the electron density in a 40 A, 80 kV beam with the orbital-to-axial velocity ratio 1.5. Assume that the mean value of the beam radius is 1 cm and the spread in guiding center radii can be neglected. Also assume

that the external magnetic field value corresponds to the operation at the 100 GHz frequency at the fundamental cyclotron resonance.

4. Evaluate the space-charge parameter S determined in Sec. 4.4 for the case considered in Problem 3 neglecting the effect of residual gas.
5. Assuming the maximum density of ohmic losses in a copper cavity $\bar{p}_{\text{ohm}}^{\max} = 1\text{kW/cm}^2$ and $L/\lambda = 7$, estimate the maximum CW output power of a gyrotron with the minimum Q_{dif} cavity operating in the $TE_{0,3}$-mode at 100 GHz.

Solutions

1. Let us start from the wave equation written as

$$\Delta\vec{E} - \frac{1}{c^2}\frac{\partial^2\vec{E}}{\partial t^2} = -\frac{4\pi}{c^2}\frac{\partial\vec{j}}{\partial t}.$$

Taking into account that we consider the fields excited near cutoff frequencies, let us represent the electric field and the electron current density, respectively, as

$$\vec{E} = \text{Re}\left\{A(z,t)\vec{E}_s(\vec{r}_\perp)e^{i\bar{\omega}_0 t}\right\}, \vec{j} = \text{Re}\left\{\vec{j}_\omega e^{i\bar{\omega}_0 t}\right\}.$$

Here the field amplitude is a slowly variable function of time $(|\,\partial A/\partial t\,| << \bar{\omega}_0\,|\,A\,|)$, $\bar{\omega}_0$ is an arbitrarily chosen carrier frequency, which can be the cutoff frequency in a certain cross section, and the function describing the transverse distribution of the electric field obeys the known equation $\Delta_\perp\vec{E}_s + k_\perp^2\vec{E}_s = 0$. In this equation, $k_\perp(z) = \omega_0(z)/c$ and $\omega_0(z)$ is the cutoff frequency in a given cross section, i.e., the cutoff frequency of the comparison waveguide. The assumption about slow variation of the field amplitude allows us to ignore the second time derivative in the wave equation, i.e., to write $\frac{\partial^2\vec{E}}{\partial t^2} \simeq \text{Re}\{\vec{E}_s(\vec{r}_\perp)(-\bar{\omega}_0^2 A + 2i\bar{\omega}_0\frac{\partial A}{\partial t})e^{i\bar{\omega}_0 t}\}$. Also, in the right-hand side of this equation, we can represent the time derivative as $\frac{\partial\vec{J}}{\partial t} = \text{Re}\{i\bar{\omega}_0\vec{j}_\omega e^{i\bar{\omega}_0 t}\}$.

Substituting these representations into the wave equation, multiplying both sides of this equation by $\vec{E}_s^*$, and integrating over the cross-section area results in the parabolic equation (4.2). Note that, in general, the electric field can be represented as a superposition of eigenmodes; however, when these modes are orthogonal, the result is the same, because integration over the cross section leaves only one mode of choice from the superposition.

2. As follows from the definition of the velocity distribution function given by (4.8), there will be 1% of reflected particles when the error function

erf(t_0) is equal to 0.99, and hence, the argument $t_0 = 2(v_0 - \bar{v}_{\perp 0})/\Delta v_{\perp 0}$ is equal to 0.01. Here v_0 is the total velocity of all electrons. Thus, when the orbital velocity spread is equal to 10%, this value of the argument t_0 corresponds to the ratio of the total electron velocity to the mean value of the orbital velocity equal to 1.091. Using known expressions for the total and orbital velocities [cf. (1.19) and (1.21)], one can readily find that this ratio corresponds to the mean orbital-to-axial velocity ratio close to 2.29.

3. Using the known relation between the pressure and density, one can easily find that the residual gas pressure 10^{-7} Torr corresponds to the density of the residual gas close to $3.55 \cdot 10^9$ cm^{-3}. Then, to determine the beam density, one should use (4.11). The Larmor radius of electrons for given beam parameters is equal to 0.02 cm, hence the beam area is close to 0.25 cm^2. Substituting this and other given parameters into (4.11) yields for the beam density the value close to $1.2 \cdot 10^{11}$ cm^{-3}. So, the beam density is almost two orders of magnitude higher than the density of a residual gas and, therefore, one should expect a certain voltage depression caused by the beam space charge during the initial stage of a long pulse. (The duration of this stage depends on the ionization cross sections of the gases present as well as on other parameters determining the ionization time.)
4. Using the well-known relation between the electron density and the electron plasma frequency and assuming that the cyclotron frequency is close to the operating frequency (100 GHz), one can easily find that the squared ratio of these frequencies is approximately equal to 10^{-3}. Hence, the space charge parameter S is close to $0.7 \cdot 10^{-2}$.
5. The maximum power can be found from (4.18). As follows from (3.43) and (3.44), the Ohmic Q-factor for the case of an ideal copper is close to $2.3 \cdot 10^4$. (Here we estimated the cavity wall radius assuming that the operating frequency is close to the cutoff frequency, and therefore $R_w \approx$ 0.486 cm.) Then, the minimum diffractive Q-factor, as follows from (2.11), is equal to 616. Hence, the maximum CW power is equal to 237 kW. Note that in the reality, first, the skin-depth is larger than its theoretical value, as we discussed in Chapter 3, and second, the diffractive Q-factor is larger than its minimum value, as shown in Fig. 2.7. Therefore, in real devices, the maximum CW power, which can be achieved in the case of operation at the $TE_{0,3}$-mode at 100 GHz, is lower than this number.

▪ CHAPTER 5 ▪

Mode Interaction in the Gyromonotron

5.1 Preliminary Remarks

As was just stated at the end of Chapter 4, in order to generate high average power at short wavelengths, it is necessary to operate at very high-order modes. As the power and the frequency increase, so does the density of the mode spectrum.

The methods of mode selection discussed above, in Sec. 3.5, are aimed at providing self-excitation conditions for the only operating mode. These methods can be referred to as the methods providing the linear selection of modes. However, as the spectrum gets denser, sooner or later these methods become insufficient. This means that several modes can be excited simultaneously. When these modes are not degenerate, their amplitudes grow independently on each other until they become large enough to cause some saturation effects. At this stage, which can be called a large-signal operation, the modes interact, because they interact with the same electrons. Such a "cross-talk" between modes has been studied in other active systems with several degrees of freedom, such as radio oscillators (Van der Pol 1922, Andronov and Vitt 1934), lasers (Lamb 1965, Yariv 1975), and magnetrons (Weinstein 1969). In such systems numerous effects are possible, depending on the parameters characterizing the device performance. These effects are considered below. (Recently, this problem in gyrotrons was overviewed by Nusinovich 1999.)

5.2 Main Effects in the Mode Interaction

Very often, a mode that is excited first is capable of suppressing other modes that initially have a lower growth rate. This effect is called the *mode suppression*. It can be used for nonlinear mode selection, because it allows one to

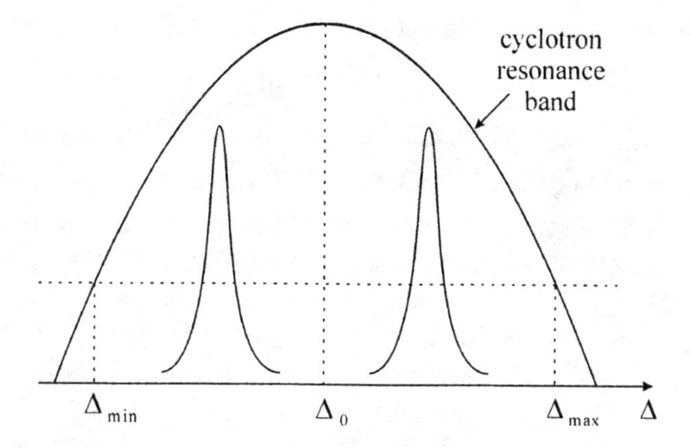

Fig. 5.1. Resonance curves of modes within a cyclotron resonance band. Dotted horizontal line shows the threshold level determined by microwave power losses. The self-excitation condition is fulfilled within the interval where the gain curve is above the threshold line.

realize single-mode oscillations in the cases when the self-excitation conditions are initially fulfilled for several modes. To illustrate this effect, it makes sense to start from considering the interaction between just two modes, for which the self-excitation conditions are initially fulfilled.

Let us consider an instant turn-on of a gyrotron. This means that the beam voltage and current rise time are negligibly small in comparison with the cavity decay time Q/ω. In such a case, we can assume that at the initial instant of time for both modes, the self-excitation conditions are fulfilled in a certain range of cyclotron resonance mismatches, as is illustrated by Fig. 5.1. This figure shows schematically the imaginary part of the susceptibility, χ'' as the function of the cyclotron resonance mismatch Δ. This susceptibility relates to the linearized orbital efficiency given by (3.71) as

$$\eta_{\perp,\text{lin}} = |F|^2 \chi''. \tag{5.1}$$

Correspondingly, the self-excitation conditions, in accordance with (3.63), can be given as

$$I_0 \chi'' = 1. \tag{5.2}$$

The horizontal line in Fig. 5.1 shows $1/I_0$, which is the threshold level for self-excitation. So, the oscillations start to grow in the region where $\chi'' > 1/I_0$. The mismatches $\Delta_{\min}$ and $\Delta_{\max}$ show the boundaries of this self-excitation zone. The mismatch Δ_0 shows the position corresponding to the minimum of the start current. Also in Fig. 5.1 positions of two modes are shown schematically.

The regions of self-excitation of these modes can also be shown (see Fig. 5.2) in the plane of mismatches $\Delta \sim \omega_1 - \Omega_0$ and $\tilde{\Delta} \sim \omega_2 - \omega_1$ (ω_1 and ω_2 are the frequencies of the first and second modes, respectively), as it was done by Zarnitsyna and Nusinovich (1974). In Fig. 5.2a they are shown by dashed lines for the initial instant of time when there are no oscillations yet. Thus, the mode, whose mismatch is closer to Δ_0, which corresponds to the minimum start current, will have a larger growth rate, because this mode has a smaller starting current. The boundaries between these regions are shown in Fig. 5.2a by dashed lines. In regions "1" the first mode has a larger increment, while in regions "2" the second mode has a preference. So, if initially the operating point is located in region "1," where the first mode has a larger growth rate, this mode will grow faster. Correspondingly, the second mode will be to some extent suppressed. In Fig. 5.2a the regions of self-excitation of the second mode in the presence of the first-mode oscillations with constant amplitude are shaded. As one can see, these regions of the second-mode excitation are smaller than in the absence of oscillations of the first mode. This fact demonstrates the effect of the second-mode suppression. Now the self-excitation of the second mode is possible only when the first mode is located close to the boundary of the self-excitation region, while the second mode is positioned much closer to the minimum start current point.

Fig. 5.2a illustrates the case when, for the sake of simplicity, it is assumed that the first-mode amplitude does not depend on the cyclotron resonance mismatch. In real devices, however, it does depend, and for a given current the amplitude of oscillations at large mismatches is larger than at small ones, as is shown in Fig. 5.2b. Therefore, the effect of suppression of the second mode is better pronounced on the right side of the self-excitation region, as is shown in Fig. 5.2c. Here parameter $\hat{q} = Q_2G_2/Q_1G_1$ characterizes the difference in Q-factors and coupling impedances of the modes. When the modes in this sense are equal ($\hat{q} = 1$), the second mode cannot be excited at the right side, and the region of excitation appears at large enough $\hat{q}$'s only.

So far, for the sake of simplicity, we considered the case of an instant turn-on of a gyrotron. In real devices the situation is quite opposite, namely: the rise time of the beam voltage and current is much larger than the cavity fill/decay time, Q/ω. Therefore, one can assume that some modes can be excited and reach their equilibrium state much faster than the voltage and current vary. Thus, the temporal evolution of the voltage and current can be treated as adiabatically slow variations, and hence slow evolution of mode intensities during the voltage and current rise and corresponding start-up scenarios can be considered (see Sec. 5.3).

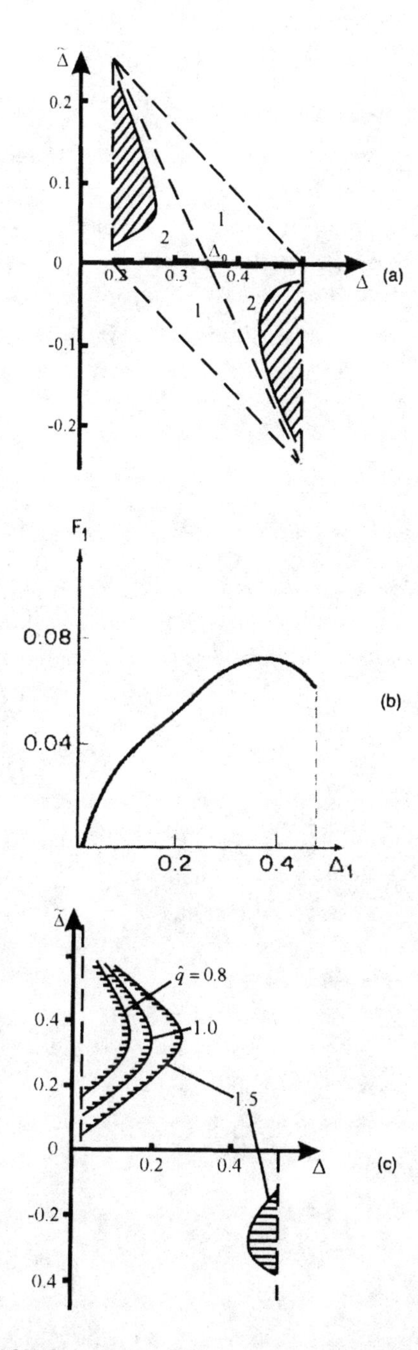

Fig. 5.2. (a) Region of self-excitation of the second mode in the absence of the first mode oscillations (dashed lines) and in the presence of such oscillations with a constant amplitude (shaded areas); (b) amplitude of the first mode oscillations as the function of the cyclotron resonance mismatch; (c) self-excitation regions of the second mode in the presence of the first-mode oscillations whose amplitude depends on the magnetic field (mismatch Δ).

Weak and Strong Coupling between Modes

The effect of mode suppression can also be explained by using a quasi-linear approach, which is based on the expansion of the nonlinear source terms in equations for mode excitation in terms of polynomials of the field intensities, M_s (here the index s designates the mode number). When the mode frequencies are well separated, $|\omega_2 - \omega_1| >> \omega/Q$, or the modes are azimuthally orthogonal, $m_2 \neq m_1$, the phase relations between modes can be eliminated from the equations for mode intensities. Then, corresponding equations can be written as

$$\frac{dM_1}{d\tau} = M_1[\sigma_1 - \beta_1' M_1 - \gamma_{12}' M_2], \quad (5.3)$$
$$\frac{dM_2}{d\tau} = \hat{q} M_2[\sigma_2 - \beta_2' M_2 - \gamma_{21}' M_1].$$

Here $\sigma_{1,2} = \alpha_{1,2}' - 1/2I_{1,2}Q_{1,2}$ are mode increments that show the excess of the beam current over its threshold, coefficients $\beta_{1,2}'$ and γ_{12}' and γ_{21}' describe, respectively, the speed of self- and cross-saturation. To determine these coefficients, one should again use the method of successive iterations used in Sec. 3.4 for analyzing the starting conditions. (Now two more iterations should be made for calculating fourth-order perturbations in the electron motion.) This procedure and the results are described elsewhere (Nusinovich 1981). Note that such a simple polynomial approximation, of course, is valid only in the regime of soft self-excitation, in which the device saturates monotonically with the field intensity increase. In regimes of hard self-excitation the field intensity increase, first, increases the gain function, and only then, at large enough field intensities, saturates it.

Eq. (5.3) is similar to the equations that were derived and analyzed for lasers by W. E. Lamb (1965). He analyzed this simple set of equations and determined the conditions of the stability for single-mode and two-mode oscillations. As one can easily find by analyzing these equations, the stability of single-mode and two-mode oscillations depends on the ratio

$$\Psi = \gamma_{12}'\gamma_{21}'/\beta_1'\beta_2'. \quad (5.4)$$

When this ratio is larger than 1, only single-mode equilibrium states are stable, while a two-mode equilibrium state is unstable and, vice versa, when $\Psi < 1$, two-mode equilibrium is stable, while single-mode equilibria are unstable. Phase portraits (or state spaces) for these two cases are shown in Fig. 5.3. Lamb called these cases "strong"and "weak"coupling between modes, respectively. In the first case, when $\Psi > 1$, both modes interact with the same electrons, and hence, strongly compete. In ecology this result is known as the "Gause

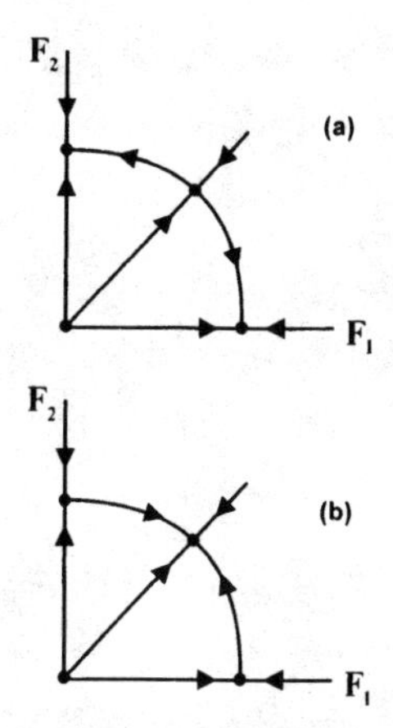

Fig. 5.3. Phase portraits for the cases of strong (a) and weak (b) coupling between two modes.

principle," which states that two species with identical requirements cannot coexist in a habitat (see, e.g., Gause 1934 or Smith 1974). On the contrary, when this ratio is smaller than 1, the modes predominantly interact with different electrons, and hence are weakly coupled. In radio oscillators this ratio is always equal to 4, thus only the case of strong coupling can be observed, at least in the circuits analyzed by Van der Pol (1922) and Andronov and Vitt (1934).

In gyrotrons the coefficients describing the self- and cross-saturation depend on the axial structures of modes and their cyclotron resonance

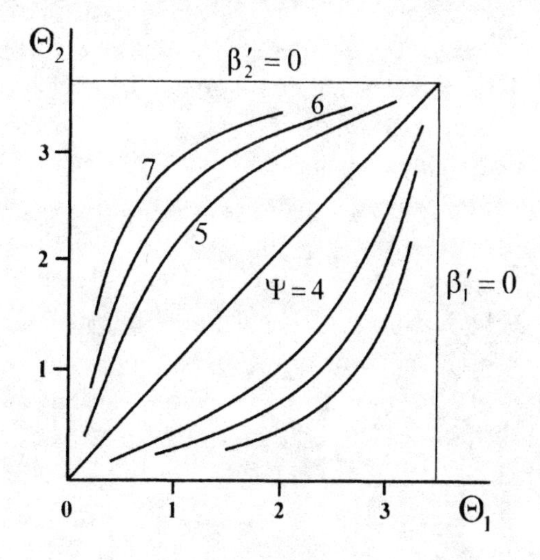

Fig. 5.4. Lines of the equal mode coupling in the plane of transit angles of two modes.

mismatches. These dependencies were analyzed elsewhere. It was shown (Nusinovich 1981) that, when the mode frequency separation is small in comparison with the cyclotron resonance band, the ratio Ψ equals 4. This ratio increases with frequency separation between the modes as shown in Fig. 5.4 reproduced from Zarnitsyna and Nusinovich (1975a). At the same time, such effects as velocity spread and electron beam thickness can weaken the mode coupling.

Nonlinear Excitation

A more detailed analysis of the dependence of saturation coefficients $\beta_{1,2}$ and γ_{12} and γ_{21} on the position of mode frequencies in the cyclotron resonance band shows that the mode suppression is not the only effect in two-mode interaction. These dependencies are shown in Fig. 5.5. Here Fig. 5.5a shows the coefficients responsible for the self-excitation (α') and self-saturation (β'), while Fig. 5.5b shows cross-saturation terms ($\gamma'_{ss'}$). The ratio $K = \Theta_2/\Theta_1 = \Delta_2/\Delta_1$ characterizes the position of competing modes in the cyclotron resonance band. As shown in Fig. 5.5b, at large transit angles $\Theta_s = \Delta_s\mu$ the cross-saturation term becomes negative. This means that, as follows from (5.3), excitation of one mode in this case helps another mode to be excited. Consider, as an example, the case when the second mode, which can be positioned in the region of hard excitation, cannot be initially excited ($\sigma_2 < 0$), while for the first mode the self-excitation conditions are fulfilled ($\sigma_1 > 0$). Then, the first mode intensity starts to grow. However, if for the second mode $\gamma'_{21} < 0$, then this mode can start to get excited when the first mode intensity becomes larger than $|\sigma_2/\gamma'_{21}|$. This effect is called the nonlinear excitation. In order to describe the final result of this nonlinear excitation, one should consider equations more complicated than (5.3).

One of the cases where this nonlinear excitation is especially pronounced is the case of competition between modes resonant with different cyclotron harmonics. Such a competition has been studied analytically and numerically (Zarnitsyna and Nusinovich 1977). A simple analytical model was considered that clearly showed the reasons for the effect of nonlinear excitation. This analytical treatment was based on a quite artificial representation of the axial structure of the resonator field as a sum of two very short sections separated by a region in which the EM field is absent. Such an artificial model, however, makes the physics of the processes involved quite transparent. Note that a similar "point-gap" model is traditionally used for analyzing conventional klystrons driven by linear electron beams. (We will also use it later for analyzing the gyroklystrons in Chapter 8.) The use of this model for analytical

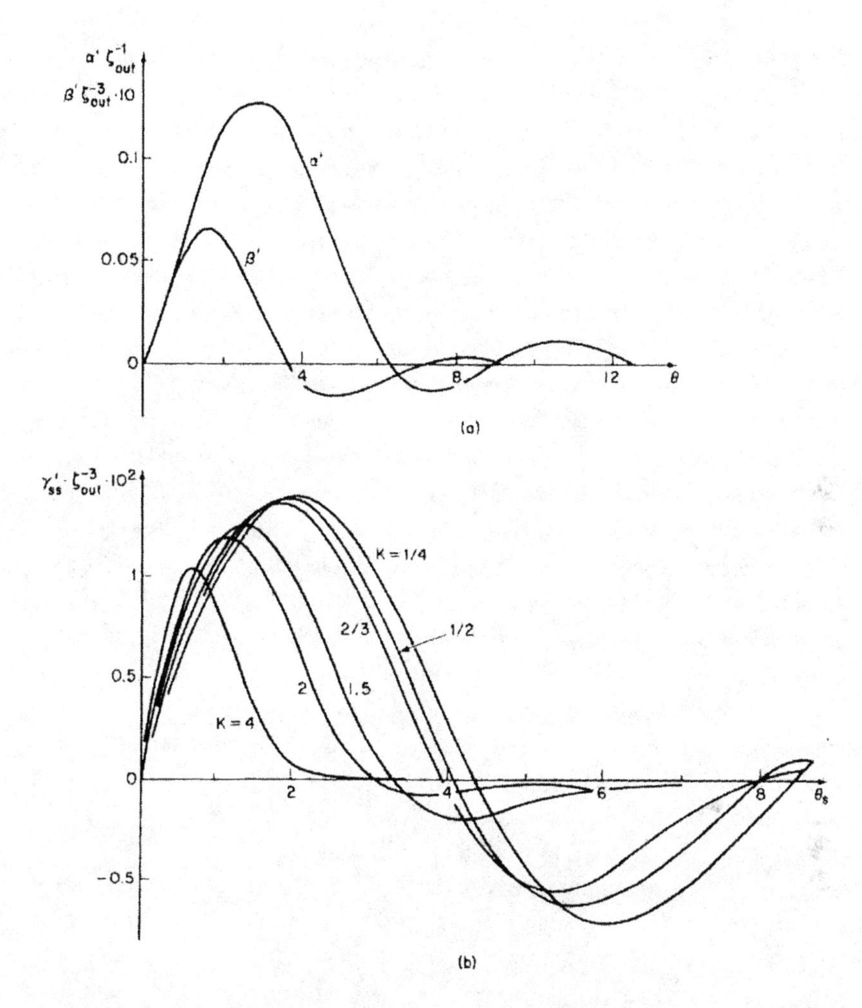

Fig. 5.5. (a) Real parts of coefficients α and β describing, respectively, the mode self-excitation and self-saturation as functions of the transit angle Θ; (b) real parts of coefficients $\gamma_{s,s'}$ describing the cross-saturation effects for different ratios K, which characterize the position of mode frequencies in the cyclotron resonance band.

studies of saturation effects in the gyromonotron was suggested by V. K. Yulpatov (1972).

In the framework of such a model, the axial structure of the resonator field can be represented as $f(\zeta) = (1/2)[\delta(\zeta) + \delta(\zeta - \mu)]$. Here μ is the normalized distance between two short sections. The gain functions for the case of stationary oscillations at one of the two competing modes can than be calculated analytically. As was shown by Zarnitsyna and Nusinovich (1977), when the device operates at a mode resonant with the second cyclotron harmonic, main saturation effects for the operating mode are described by the Bessel functions of the argument $2q$, where $q = F\mu$ and F is the field amplitude. At the same time, the saturation effects for the competing mode resonant with the fundamental gyrofrequency are described by the Bessel functions of the argument q. This difference in arguments of Bessel functions can be explained by the difference in the multipole structure of transverse fields resonant with electrons at different harmonics, which was discussed in Chapter 1. It seems intuitively clear that in the field of a quadrupole structure shown in Fig. 1.2, the saturation effects associated with the orbital electron bunching occur when the phase shift of electrons is on the order of $\pi/2$. However, in the case of the dipole interaction at the fundamental resonance, the electron phase shift should be on the order of π for appearance of the saturation effects. This difference in the speed of saturation of these two modes may result in a very interesting sequence of events. Assume that initially the starting current of the second harmonic mode is smaller than that for the fundamental harmonic mode. Then, as the beam current increases, the second harmonic mode will be excited first and it will suppress the fundamental harmonic mode. However, when the amplitude of this mode, with the increase in the beam current, exceeds a certain critical value, because of the difference in the speed of saturation the second harmonic mode will lose its stability. Instead of this mode, the mode resonant with the fundamental gyrofrequency will be excited. Let us emphasize that this mode at the fundamental resonance can be excited even at beam currents that are smaller than its starting current in the absence of any oscillations. This qualitative analysis was supported by numerical studies done by Zarnitsyna and Nusinovich (1977) for a second harmonic gyrotron with the Gaussian axial structure of the resonator field.

These theoretical results were confirmed in experiments carried out in gyrotrons operating at the fundamental resonance (Kreischer et al. 1984) and at the second cyclotron harmonic (Idehara and Shimizu 1994). In the former case the effect of nonlinear excitation can be associated with the fact that the second mode with a higher frequency was situated in the region of hard excitation, where it cannot be excited from the noise level. At the same time, the first mode can be excited in the regime of the soft excitation. So, just its excitation may help the second mode to get started. (Note that a similar mechanism of hard excitation of the operating mode at the second harmonic with the help of the parasite at the fundamental was theoretically found by Saraph et al. 1993.) In the above-mentioned experiment at the second harmonic (Idehara and Shimizu 1994), it was shown that with the further increase in current the fundamental harmonic mode completely suppresses the second harmonic rival, which helped it to get excited. Detailed analysis of this issue in regard to, at least, one experiment allowed researchers to experimentally avoid the excitation of the fundamental harmonic parasite and thus demonstrate efficient operation at 0.5 MW level (Zapevalov, Malygin, and Tsimring 1993).

Phase-Amplitude Interaction

So far, we have considered the simplest case of interaction between two modes only, assuming that in their interaction only the intensities of modes participate. In some cases, however, not only mode intensities, but also the phase relations between modes play a role. Those are the cases when the modes have overlapping resonance curves, $|\omega_2 - \omega_1| \leq \omega/Q$, and they are not orthogonal azimuthally. A simple example of such a case is the case of two modes having the same transverse structure, but different axial indices. As one can easily find from the definition of the frequencies $\omega_k = \sqrt{\omega_{\text{cut}}^2 + (ck_{z,k})^2}$, where for the first mode with one axial variation $k_{z,1} \approx \pi/L$, while for the second mode with two axial variations $k_{z,2} \approx 2\pi/L$, the frequency separation of these modes, when their frequencies are close to cutoff, is on the order of $\omega(\lambda/L)^2$. So, this separation is of the same order as the width of the resonance curves, ω/Q_{dif}, where Q_{dif} is determined by Eq. (2.11). In the first study of such a case (Zapevalov and Nusinovich 1989) it was shown that now the device is able to exhibit phase-locked oscillations of two modes. Later, this approach was used for explaining some peculiarities in the operation of an inverted gyrotwystron (the device that will be discussed in Chapter 13). It was experimentally demonstrated that such a device exhibits amplification in the frequency band, which significantly increases the width of the resonance curve of a single mode. The explanation for this effect was based on the fact that the resonance curve widens as the axial index increases, and, hence, overlapping of these curves increases as well (see Sec. 2.3). Using this argument, Zhao et al. (2000) showed that the overlapping of resonance curves of modes with more than two axial variations can be the reason for experimentally observed amplification in a large (about 2%) bandwidth.

Certainly, with an increase in the number of modes that can participate in the phase-amplitude interaction, the physics of the mode interaction becomes even more interesting. Some of these effects will be considered below.

Automodulation and Parametric Processes

When more than two modes can be excited, it is not enough to have modes with different azimuthal indices for eliminating the phase relations from the equations for mode intensities. For example, the modes of a whispering gallery with different azimuthal indices ($m >> p$) can form a quasi-equidistant spectrum, and therefore for these modes the conditions of the parametric four-photon interaction can be fulfilled. These conditions can be written as

$$|\omega_1 + \omega_3 - 2\omega_2| \leq \omega/Q, \tag{5.5}$$

$$m_1 + m_3 = 2m_2. \tag{5.6}$$

Eqs. (5.5) and (5.6) can be interpreted as conservation laws in the four-photon decay for energy and angular momentum, respectively. Here index "2" designates the central mode, while indices "1" and "3" designate the low-frequency and high-frequency satellites, respectively.

Stability of the central mode with respect to such symmetric satellites is often called the *automodulation* or *sideband instability*. In the first study of this instability, it was shown (Zapevalov and Nusinovich 1985) that the regime with the maximum efficiency can be stable with respect to the satellites when the resonator length is reasonably short. (The authors studied the gyrotron with the Gaussian axial structure of the resonator field and took the normalized resonator length equal to 10 in units shown in Fig. 4.1.) It was also found that the most dangerous satellites are not the modes located very closely to the operating one, but those that have a certain frequency separation. This conclusion was later confirmed by other authors (Antonsen et al. 1990, Levush and Antonsen 1990). In their studies this instability was analyzed in more detail. Note that there are two conditions of the stability of the central mode oscillations. Antonsen et al. (1990) and Levush and Antonsen (1990) characterized corresponding instabilities as the phase instability and the overbunch instability. The result of the phase instability is the mode jumping from one mode to another, while the result of the overbunch instability can be a multimode equilibrium. Such a phase-locked operation of the gyrotron with a set of modes having a quasi-equidistant spectrum was also described by Nusinovich (1981). It makes sense to call the reader's attention to the fact that such a sideband instability is well studied in FELs; a review of this activity can be found in Freund and Antonsen (1996).

Similar processes may occur in the case of excitation of modes resonant with different cyclotron harmonics. In such a case the conditions for synchronous interaction given above by (5.5) and (5.6) should be replaced in the case of three modes by

$$|\omega_1 + \omega_3 - \omega_2| \leq \omega/Q, \tag{5.7}$$

$$m_1 + m_3 = m_2. \tag{5.8}$$

These conditions can be interpreted in the same way as the energy and angular momentum conservation laws for the three-photon decay processes, in which the energy of one photon is approximately two times larger than the energy of others. One can also envision a degenerate case of such a parametric process, which is the interaction of two modes resonant with different cyclotron harmonics. In this degenerate case Eqs. (5.7) and (5.8) can be reduced to

$$|2\omega_1 - \omega_2| \leq \omega/Q, \tag{5.9}$$

$$2m_1 = m_2. \tag{5.10}$$

Parametric processes in gyrotrons with this kind of interaction were studied in a number of papers (Nusinovich 1992c, Saraph et al. 1995). The analytical study of two-mode and three-mode parametric processes showed (Nusinovich 1992c) that even in a system described by very simple equations not only single-mode oscillations, but also automodulation processes and chaotic behavior are possible. The paper by Saraph et al. (1995) deals primarily with numerical analysis of parametric instabilities in gyrotrons where three modes resonant with different cyclotron harmonics can be excited. The parametric interaction of such modes can take place when the conditions given by Eqs. (5.7) and (5.8) are fulfilled. Now, in these equations one should assume that the second mode is in resonance with the third cyclotron harmonic, while two other modes are in resonance with the beam at the first two harmonics. To satisfy these criteria, three modes were chosen: $TE_{3,1}$, $TE_{1,3}$, and $TE_{4,3}$. It was shown that, depending on the operating parameters, the device is able to exhibit completely different performance, examples of which are shown in Fig. 5.6. Here Fig. 5.6a shows cyclic mode hopping; Fig. 5.6b shows a quite complicated interaction pattern, repeating itself in time. Then, Fig. 5.6c shows a stable coexistence of all three modes; Fig. 5.6d shows the suppression of the modes resonant with the fundamental and third harmonics by the mode resonant at the second harmonic. Finally, the onset of the mode resonant with the third cyclotron harmonic, which was initially excited by the fundamental harmonic mode, is shown in Fig. 5.6e.

5.3 Start-up Scenario

As was mentioned in Sec. 4.1, the maximum interaction efficiency is often realized in the regime of hard self-excitation. Therefore, it is necessary to vary the gyrotron parameters in such a way that, first, the oscillations of the desired mode will be excited in the regime of soft excitation, and then, the variation of parameters will drive this mode to the maximum efficiency point. Certainly, in this way all parasitic modes should be suppressed. (Since, in general, the minimum start current and the maximum efficiency correspond to different operating parameters, a similar variation of parameters in devices with a dense spectrum of competing modes is also desirable in the cases when the maximum efficiency point lies in the region of soft self-excitation.) In other words, to realize the high-efficiency oscillations, especially in gyrotrons with a dense spectrum of competing modes, a proper start-up scenario should be found.

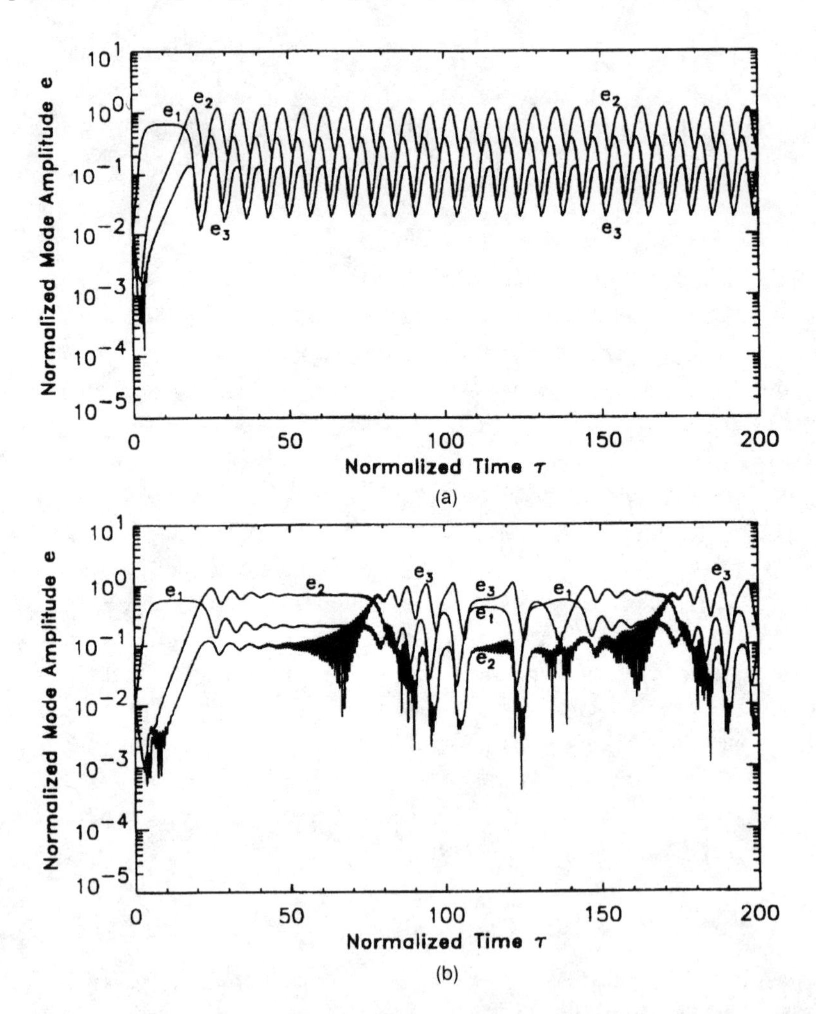

Fig. 5.6. Temporal evolution of amplitudes of parametrically coupled modes resonant with the beam at the first (e_1), second (e_2), and third (e_3) cyclotron harmonics: (a) cyclic mode hopping; (b) complicated periodic beating of modes; (c) stable coexistence of all modes; (d) second harmonic mode suppresses two others; (e) third harmonic mode suppresses two others. (Reproduced from Saraph et al. 1995)

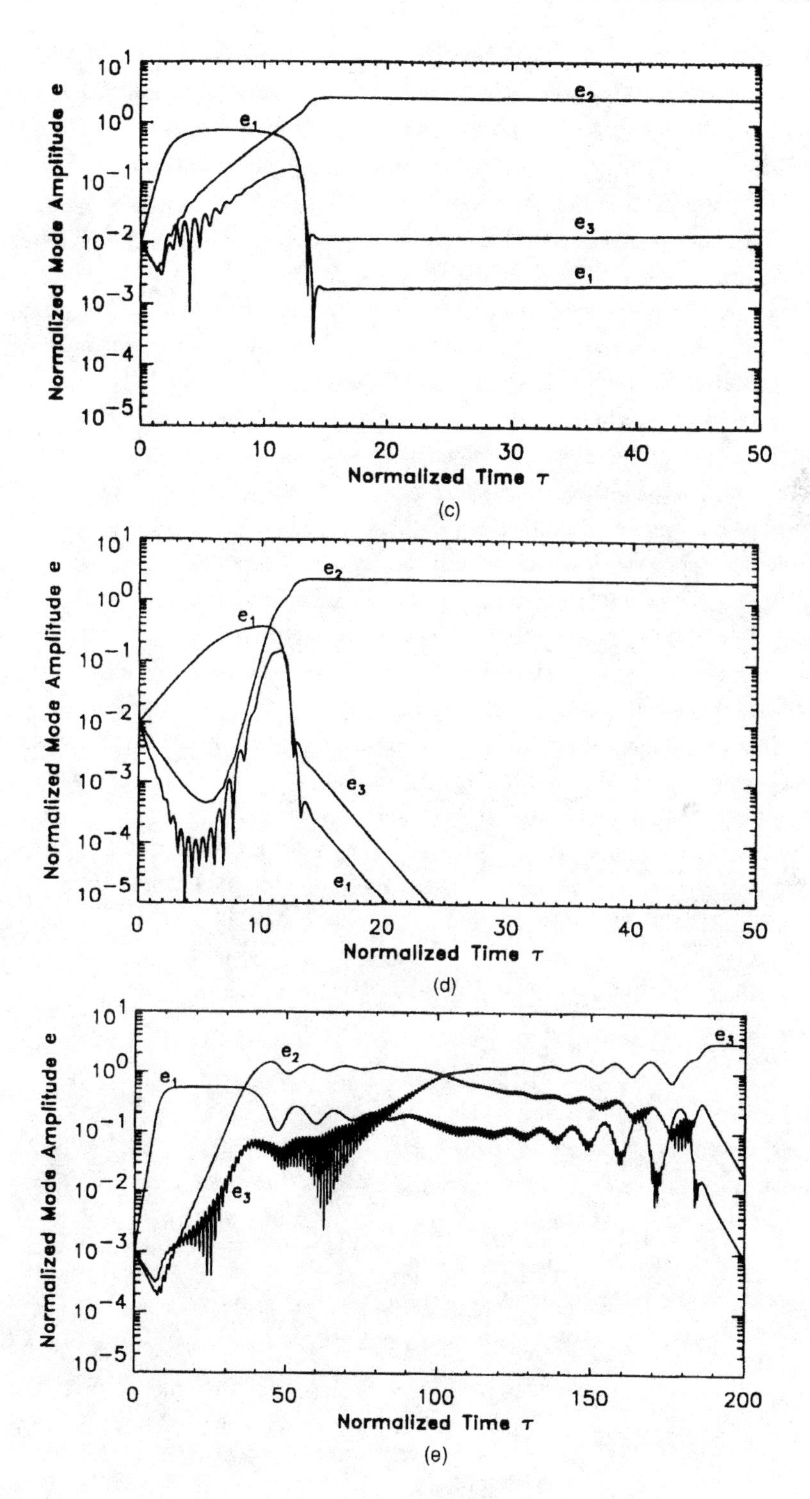

Normalized Mode Amplitude e
Normalized Time τ
e2
e1
e3
e1
(c)
e2
e1
e3
e1
(d)
e3
e2
e1
e3
(e)

It is necessary to emphasize that such a scenario should be realized on a timescale that greatly exceeds the cavity fill time, Q/ω, since all changes made on a shorter timescale can be considered as an instant turn-on. For typical frequencies in the range of 100 GHz and Q-factors on the order of 10^3, this means the variation of parameters with the timescale exceeding 10 ns. Note that typically the beam voltage and current rise times are much longer than microseconds.

In general, it is known that in gyrotrons, as well as in resonant TWTs and FELs, the minimum start current corresponds to smaller mismatches of the resonance than the high-efficiency regime, because the EM field of large amplitude can trap electrons even in the case of large initial mismatches. Since in gyrotrons the cyclotron resonance mismatch is proportional to $\omega - s\Omega_0$, from this statement it readily follows that in the initial excitation phase the relativistic cyclotron frequency should be larger than in the final stage of the efficient operation. So, our scenario should imply either the decrease in the external magnetic field or the increase in the beam voltage. In continuous wave (CW) operation, both of them can be varied. However, in the pulsed operation the possibilities of varying the magnetic field are severely limited, especially in the case of using the superconducting solenoids. Therefore, the studies of start-up scenarios are predominantly focused on the voltage variation. First of all, it is necessary to emphasize that just the beam voltage rise, which is typically slow enough, automatically changes the cyclotron resonance mismatch in the desired way. Triode-type electron guns allow for even more flexibility, because one can control the mod-anode voltage independently on the beam voltage. In such a case, the desired start-up scenario can be realized by providing proper relations between these two voltages during the voltage rise.

Recall that the beam voltage determines the total electron velocity, in accordance with the general relation (1.19):

$$\beta_0^2 = 1 - \frac{1}{(1+V_b')^2}. \tag{5.11}$$

Here the velocity is normalized to the speed of light and the primed voltage is the voltage normalized to 511 kV: $V_b' = eV_b/mc^2$. The mod-anode voltage determines the orbital velocity of electrons. In the framework of the adiabatic theory of magnetron-type electron guns, as follows from (2.4), this dependence can be given by

$$\beta_{\perp,0} = \alpha_B^{3/2} V_m' (c/\Omega_0 d\gamma_0). \tag{5.12}$$

Here the primed mod-anode voltage is normalized in the same way as the beam voltage was normalized above, d is the gap between cathode and anode, $\gamma_0 = 1 + V_b'$.

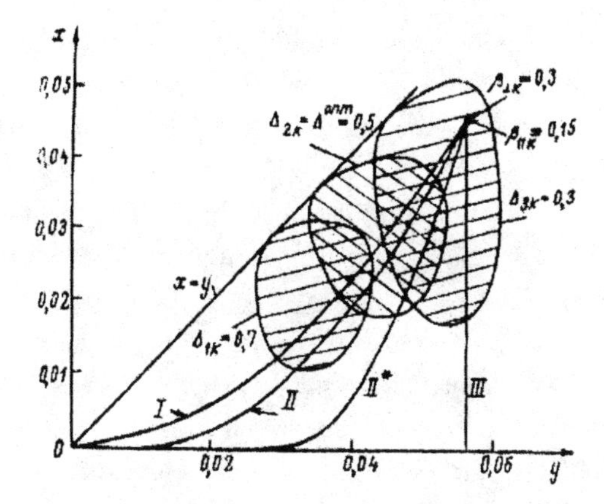

Fig. 5.6. Regions of self-excitation of the operating mode and high- and low-frequency parasitic modes (shaded) and traces showing various start-up scenarios.

This problem of the start-up scenario was first formulated and analyzed a long time ago (Nusinovich 1974). Fig. 5.7 reproduces some of the first results. Here hatched areas indicate the regions of self-excitation of the operating mode (in the middle), high-frequency parasitic mode (on the left), and low-frequency parasite (on the right). The final values of voltages (on the top of the pulses) correspond to the orbital-to-axial velocity ratio equal to 2 and the normalized orbital velocity, $\beta_{\perp,f} = 0.3$. At these voltages the cyclotron resonance mismatch for the operating mode is optimal for the maximum efficiency, $\Delta_{2,\mathrm{opt}} = 0.5$. Coordinates x and y are proportional to the normalized squared orbital and squared total velocities, respectively: $x = \beta_{\perp}^{2}/2$, $y = \beta_{0}^{2}/2 = (\beta_{\perp}^{2} + \beta_{z}^{2})/2$. The straight line $x = y$ corresponds to zero axial velocities, which is the case where the electron mirroring starts.

Traces shown by roman numbers (I, II, II*, and III) designate different scenarios. Trace I corresponds to the case when the mod-anode voltage varies proportionally to the beam voltage. This case can be realized either in the diode-type electron guns or in the triode guns, in which the mod-anode voltage is just a portion of the beam voltage taken with the use of the resistive divider. Traces II correspond to the constant difference between the voltages during the voltage rise: $V_b - V_{\mathrm{mod}}$ =const. Finally, trace III shows the situation when the mod-anode voltage is applied after the beam voltage reached its maximum value.

As is seen in Fig. 5.7, the first scenario, in the presence of parasites, leads to initial excitation of the high-frequency parasite. Then, if the frequency separation is large enough, this mode can lose its stability and be replaced by the operating mode. However, in the case of the dense mode spectrum, it may happen that, at the instant of time when the high-frequency parasite loses its stability, the low-frequency parasite will have a lower start current than the operating mode. In such a case, one will observe a sequence of two excited parasitic modes instead of the desired mode. The third scenario, as seen in Fig. 5.7, leads to the excitation of the low-frequency parasite. So, only the second scenario, when a proper difference between the voltages is provided (trace II*), leads to the single-mode excitation of the desired mode. Note that the regions of excitation shown in Fig. 5.7 correspond to the case when all modes have the same Q-factors and coupling impedance to the beam. Certainly, some methods of mode selection, which we already discussed in Sec. 3.5, can discriminate against parasitic modes. Note also that in the analysis, results of which were presented here, it was assumed that the thermionic cathodes operate in the regime of temperature-limited emission. This implies that the beam current reaches its maximum value at voltages much lower than the final voltage. This is typically the case for long-pulsed gyrotrons. However, it seems obvious from our consideration that sometimes it may be beneficial to use the cathodes that operate in the regime intermediate between temperature-limited and space-charge-limited emission. Indeed, the use of such cathodes allows one to reduce the beam current during the voltage rise, and hence eliminate the excitation of high-frequency parasites.

Let us also illustrate our discussion of a sequence of excited modes in gyrotrons with the relatively rare and dense spectra with Fig. 5.8 reproduced from Nusinovich (1986). Here the cases of the rare and dense spectra are shown in Figs. 5.8a and 5.8b, respectively. Dash-dotted lines show the boundaries of the region of hard self-excitation. Let us assume that an initially excited mode loses its stability just when it crosses the boundary of the hard self-excitation region after passing through the regions of soft and hard excitations. Then, in a gyrotron with a rare enough spectrum and trace I (Fig. 5.8a), the high-frequency parasitic mode will be excited first. Then, it will lose the stability and the operating mode will appear, which will remain stable for the rest of the voltage rise and on the top of the pulse. However, in a gyrotron with a dense spectrum, as is shown in Fig. 5.8b, the situation is quite different. Here, when trace I crosses this boundary for the initially excited high-frequency parasite, the beam current is approximately equal to the start current of the operating mode and, at the same time, is larger than the start current of the low-frequency parasite. So, one should expect that in this case the high-frequency parasite will be followed by the low-frequency parasite. To avoid this parasitic excitation, one should use trace II. However, as the mode spectrum gets denser, even the use of this trace should be more and more precise (cf. Figs. 5.8a,b) to avoid excitation of parasitic modes.

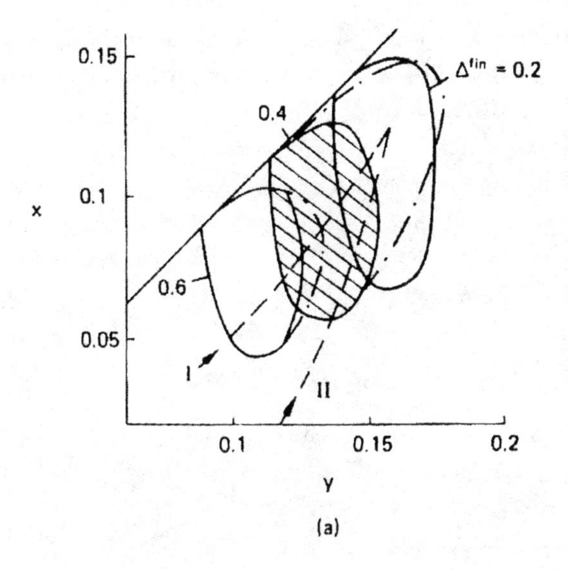

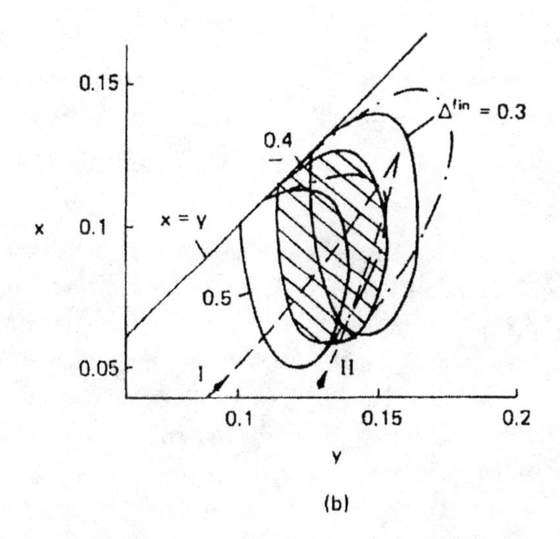

Fig. 5.7. Regions of self-excitation and hard-excitation of the operating and parasitic modes in the cases of rare (a) and dense (b) mode spectrum.

In some cases the excitation of parasitic modes can be avoided by sacrificing the efficiency, i.e., by taking the cyclotron resonance mismatch slightly different from the optimal value. This and many other issues associated with the start-up scenario were actively studied during the 1980s and 1990s (see Borie and Jodicke 1987, Nusinovich, Pavelyev, and Khiznyak 1989, Whaley et al.

1994, and references therein). Soon the analysis of start-up scenarios became a quite routine procedure in the design of high-power gyrotrons. However, the accuracy with which these studies are and should be done is still a serious issue because in gyrotrons with a very dense spectrum of competing modes even minor effects can change the final equilibrium state. Note that this method of analyzing the gyrotron operation can also be used in the studies of frequency step-tunable gyrotrons, in which this frequency step-tunability is provided by the voltage tuning (Dumbrajs and Nusinovich 1992).

5.4 Phase Locking in Multimode Gyrotrons

In this section, by "phase-locking" we will mean the use of an external signal for controlling the oscillations of the desired mode. The use of an external signal for priming the desired mode is an obvious means to help this mode to suppress the competitors. Indeed, while, in the process of excitation, other modes start to grow from the noise level, the desired mode starts from the level determined by the power of a driver, which typically exceeds the noise level by many orders of magnitude. This driving can be realized either by introducing the input signal into the same cavity where the desired mode should be excited, or into a prebunching cavity separated from the main cavity by a drift section as in klystrons. Below we will consider these two methods separately.

Direct Injection

This is a straightforward method of phase locking, although in high-power gyrotrons it may be difficult to realize it in practice, because an input coupler and a driver should be somehow protected from the large amplitude oscillations excited in such a resonator. To qualitatively describe the processes in such an oscillator, in which the drive signal is used for phase locking of the desired mode while the parasite can be self-excited at its own frequency, one can use a polynomial approximation of the nonlinear source term. This approximation is similar to that used for deriving Eqs. (5.3). Assuming that the drive signal can be characterized by the amplitude A_s, frequency ω_s, and phase φ_s, one can describe such an oscillator by the following set of equations (see, e.g., Nusinovich 1975):

$$\frac{dF_1}{dt} = F_1\left(\sigma_1 - \beta_1' F_1^2 - \gamma_{12}' F_2^2\right) + A_s \cos(\varphi_s - \varphi_1), \tag{5.13}$$

$$\frac{d\varphi_1}{dt} = \delta_1 + \alpha_1'' - \beta_1'' F_1^2 - \gamma_{12}'' F_2^2 + \frac{A_s}{F_1}\sin(\varphi_s - \varphi_1),$$

$$\frac{dF_2}{dt} = \hat{q} F_2\left(\sigma_2 - \beta_2' F_2^2 - \gamma_{21}' F_1^2\right).$$

Here primed coefficients β and γ are the same coefficients used in (5.3). Double priming denotes the imaginary parts of these coefficients, which are responsible for nonlinear frequency pulling effects; $\delta_1 = (\omega_1 - \omega_s)/I_1\omega_s$ is the normalized detuning between the signal frequency and the frequency of free-running oscillations of the first mode. (The time variable in (5.14) is also normalized to I_1.) In radio oscillators coefficients β and γ are related as $\gamma_{12} = \gamma_{21} = 2\beta_1 = 2\beta_2 = 2\beta$. The first two equations in (5.13) describe the phase locking of a single-mode oscillator, which is, in general, a well-studied issue. For the radio oscillator with $\beta'' = \gamma'' = 0$ these equations were first studied by van der Pol (1927), while Appleton (1924) analyzed them for the case of nonzero imaginary coefficients. The last equation in (5.13) describes the parasitic mode. This equation clearly shows that the parasitic mode with a positive initial increment ($\sigma_2 > 0$) can be suppressed by the operating mode having large enough amplitude: $F_1^2 > \sigma_2/\gamma'_{21}$. The intensity of the first mode in a steady-state regime is determined by the cubic equation

$$(\delta_0 - r_0u_0)^2 + (1 - u_0)^2 = \xi_0/u_0, \qquad (5.14)$$

in which $\delta_0 = (\delta_1 + \alpha_1'')/\sigma_1$, $r_0 = \beta_1''/\beta_1'$, $\xi_0 = \beta_1' A_s^2/\sigma_1^3$ and $u_0 = \beta_1' F_0^2/\sigma_1$. The stability of these phase-locked oscillations is determined by two equations: (1) $u_0 > 1/2$, which was first obtained by van der Pol, and (2) $(\delta_0 - 2r_0u_0)^2 + (1 - 2u_0)^2 > u_0^2(1 + r_0^2)$, which is similar to one derived by Appleton. The above condition given for the suppression of the second mode can be rewritten in these notations as $u_0 > \beta_1'\sigma_2/\gamma_{21}'\sigma_1$. So, when $\beta_1'\sigma_2/\gamma_{21}'\sigma_1$ is smaller than ½, as the intensity of the phase-locked oscillations of the first mode decreases, first (at $u_0 < 1/2$) the oscillations of this mode at its own frequency of free-running oscillations appear in addition to the field at the signal frequency, and only then can the parasitic mode appear.

The intensity of phase-locked oscillations, u_0, is shown in Fig. 5.9 as the function of the detuning δ_0 for several values of the signal intensity ξ_0. Here Fig. 5.9a corresponds to the system without nonlinear frequency shift, i.e., $r_0 = 0$, while Fig. 5.9b shows the case of the nonzero $r_0 : r_0 = 0.5$. These figures are quite similar to those shown by Walsh et al. (1989). The dashed horizontal line $u_0 = 1/2$ shows, in both of them, the first condition of stability, and the ellipse shown by the dashed line designates the region in which the second condition of stability is not fulfilled. When the boundary $u_0 = \beta_1'\sigma_1/\gamma_{21}'\sigma_2$ for suppressing the parasitic mode is smaller than ½, the presence of the parasitic mode does not limit the locking bandwidth, which, as one can easily find, obeys the Adler relation (Adler 1946). Indeed, the coefficient ξ_0 in the RHS of (5.14) is the normalized intensity of the signal, which is proportional to the

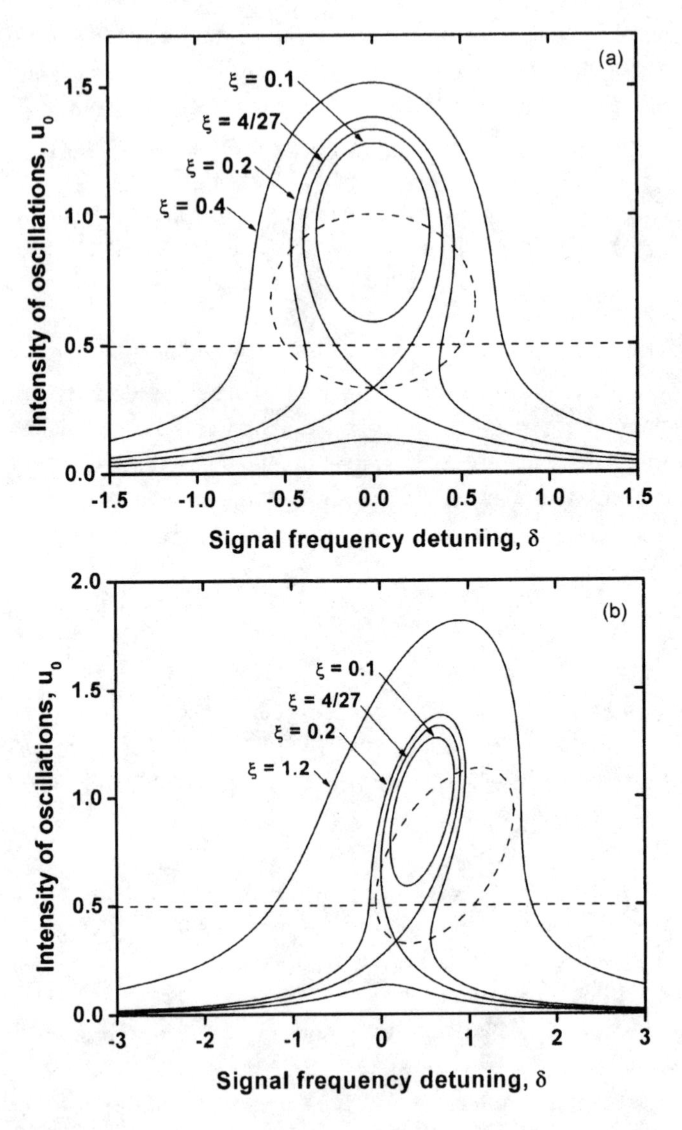

Fig. 5.8. Intensity of the phase-locked oscillations as the function of the signal frequency detuning δ for several values of the normalized signal intensity: (a) in the absence of nonlinear frequency shift, $r_o = 0$; (b) in the presence of this shift, $r_o = 1/2$.

drive power, P_{dr}, while in the LHS the frequency detuning is present as δ_0^2. Thus, the locking bandwidth is proportional to $\sqrt{P_{\mathrm{dr}}}$.

The presence of the parasitic mode limits the locking bandwidth at large signal intensities ($\xi_0 \geq 8/27$) when $u_0 = \beta_1' \sigma_1 / \gamma_{21}' \sigma_2 > 1/2$, and, at small signal intensities ($\xi_0 < 8/27$), when the line $u_0 = \beta_1' \sigma_1 / \gamma_{21}' \sigma_2$ intersects the stationary oscillation curve above the ellipse corresponding to the second condition of stability. Ergakov and Moiseev (1975), Manheimer (1987), Fliflet and Manheimer (1989), and McCurdy, Ganguly, and Armstrong (1989) have analyzed this sort of gyrotron phase-locked operation in more detail.

Phase Locking by a Premodulated Beam

This method of phase locking is similar to the signal amplification in gyroklystrons, the devices briefly mentioned in Chapter 2 and which will be discussed in more detail in Chapter 8. In both cases, an electron beam is modulated by a drive signal in the input cavity. This energy modulation causes the ballistic bunching in the drift section. Then, the prebunched beam excites oscillations in the main, or output, cavity. The only difference is in the fact that in phase-locked oscillators the beam current exceeds the start current of the main cavity, and therefore, the signal is used to phase-lock free-running oscillations. On the contrary, in gyroklystrons the current is below the start current, and, therefore, in the absence of the input signal, the output power equals zero (a so-called zero-drive stable operation of an amplifier).

In the equations describing the operation of the main cavity, the beam prebunching can easily be taken into account in the boundary condition for the electron phase at the cavity entrance (see, e.g., Ergakov and Moiseev 1975, Tran et al. 1986):

$$\vartheta(0) = \vartheta_0 - q \sin \vartheta_0 - (\varphi_2 - \varphi_1), \qquad (5.15)$$

where ϑ_0 is the electron phase at the entrance to the input cavity, $\varphi_2 - \varphi_1$ is the difference in phases of the fields in the main and input cavities with the account for transit time of electrons in the drift section, and $q \propto 2F_1\mu_{dr}$ is the bunching parameter that is proportional to the field amplitude in the input cavity, F_1, and the normalized length of the drift section, μ_{dr}. (A more accurate definition of the bunching parameter can easily be derived from Eqs. (3.34) for electron motion.)

The phase-locked operation of a gyrotron oscillator with the Gaussian axial structure of the main resonator field was studied numerically (Zarnitsyna and Nusinovich 1975b). It was considered the case when both the input and main cavities operate at the fundamental cyclotron resonance and when in the main cavity electrons are also able to interact with the parasitic mode. Some results of these calculations are shown in Figs. 5.10 and 5.11. In Fig. 5.10, the intensity, F^2, and the orbital efficiency of phase-locked oscillations are shown as functions of the normalized detuning between the eigenfrequency of the operating mode, ω_1, and the signal frequency,

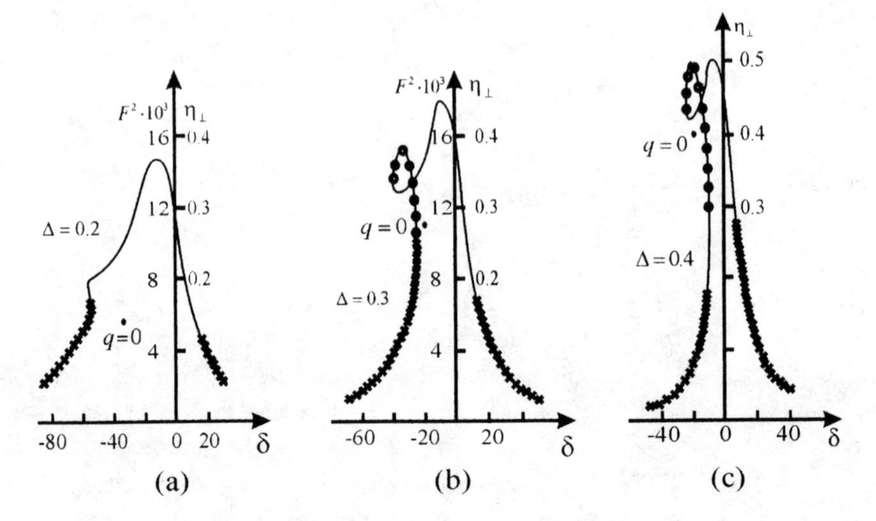

Fig. 5.9. Intensity and orbital efficiency of phase-locked oscillations as the function of the signal frequency detuning δ for several values of the cyclotron resonance mismatch Δ. Other parameters are as follows: bunching parameter is equal to $q = 1_j$; normalized beam current is equal to $I_1 = 0.04$; and the normalized length is equal to $\mu = 10$.

$\delta = (\omega_1 - \omega_{\text{sig}})/I_1\omega_{\text{sig}}$, for several values of the cyclotron resonance mismatch, $\Delta = (2/\beta_{\perp 0}^2)[(\omega_1 - \Omega_0)/\omega_{\text{sig}}]$. Circles and diamonds show the branches of these curves where one of the two conditions for stability of phase-locked oscillations is not fulfilled. As one can see, as the cyclotron resonance mismatch increases, the efficiency becomes larger (see Fig. 5.2b), but the locking bandwidth becomes smaller. Dots indicated by $q = 0$ show the intensity

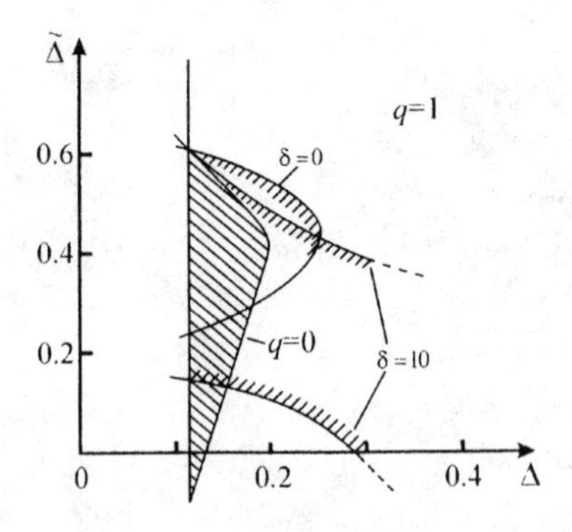

Fig. 5.10. Regions of parasitic mode excitation in the presence of phase-locked oscillations for two values of the signal frequency detuning δ (as in Fig. 5.10, $q = 1$, $I_1 = 0.04$, $\mu = 10$; also $\hat{q} = 1$). The shaded area corresponds to the free-running oscillator ($q = 0$).

and efficiency of free-running oscillations. So, practically in all locking bandwidths, the efficiency is larger than in the free-running regime. Abscissas of these dots show the frequency pulling, i.e., the difference between the eigenfrequency of the operating mode and its oscillation frequency in the free-running regime. Note that these calculations were done for the normalized current parameter equal to $0.04/Q_1$; therefore, the normalized bandwidth, which for $\Delta = 0.4$ is about 30, corresponds to the bandwidth $\Delta f/f$ approximately equal to $1.2/Q_1$. So, while direct injection provides phase locking only in the bandwidth that is much smaller than the width of the resonance curve, ω/Q, the phase locking by the use of a prebunched beam allows one to significantly enlarge the locking bandwidth.

When the parasitic mode has the same Q-factor and the coupling impedance as the operating mode, it can be excited only at small Δ's, as is shown in Fig. 5.11 for two values of the signal frequency detuning. Here, dashed lines for $\delta = 10$ show the region in which phase-locked oscillations of the operating mode become unstable and this mode oscillates at its own frequency, ω_1.

Ballistic bunching in the drift region is, in principle, a nonlinear process. This means that in the process of bunching, not only the signal frequency component, j_{ω_s}, appears in the electron current density but also its harmonics at frequencies $s\omega_s$. Therefore, the main cavity can be excited not only at ω_s, but also in frequency-multiplying regimes at $s\omega_s$. The efficiency of such operation was studied numerically in detail elsewhere (Belousov, Ergakov, and Moiseev 1978). Some results of these authors are presented in Table 5.1, which shows the optimal values of the parameters and corresponding maximum orbital efficiencies for various combinations of cyclotron resonance harmonics in two cavities. For the sake of comparison, Table 5.2 gives similar data for free-running gyrotron oscillators operating at the first five harmonics. These data agree well with the results by Danly and Temkin (1986). So, phase locking

Table 5.1. Maximum Orbital Efficiency and Corresponding Optimal Parameters of Two-cavity, Phase-locked Harmonic Gyrotron Oscillators

s_1	1	1	1	1	1	2	3	4	5
s_2	1	2	3	4	5	2	3	4	5
$\eta_\perp$	0.87	0.8	0.69	0.55	0.46	0.84	0.73	0.6	0.52
I_2	0.026	0.028	0.03	0.028	0.025	0.028	0.029	0.026	0.022
μ_2	26	28	30	32	35	25	30	32	35
θ_2	14	14	14	14	14	14	14	14	14

Table 5.2. Maximum Orbital Efficiencies and Corresponding Optimal Parameters of Free-running Harmonic Gyrotron Oscillators

s_1	1	2	3	4	5
$\eta_\perp$	0.72	0.71	0.55	0.44	0.37
I_2	0.034	0.034	0.032	0.026	0.029
μ_2	26	28	30	34	40
θ_2	14	14	14	14	14

allows one to significantly increase the efficiency. Most important, however, is the above-mentioned fact that the operating mode, which starts to grow from the signal level, has a preference over parasitic modes, which start to grow from the noise level. Note that this important conclusion is valid for the case of an instant turn-on. For the case of a gradual voltage increase, the situation can be completely different. Indeed, it is not obvious that, when the voltage passes through the self-excitation region of the high-frequency parasite (see Figs. 5.7 and 5.8), while being outside the self-excitation region of the operating mode, the presence of a weak field at the signal frequency will change the situation significantly.

In conclusion, let me mention that in this chapter I outlined only some of the most important issues in mode interaction. In particular, I did not discuss here the self-consistent treatment of mode interaction that was studied in a number of papers (see review paper by Nusinovich 1999). Presently the code MAGY is modified in such a way that makes it possible to use this code for analyzing nonstationary multimode processes in gyrotrons self-consistently (Vlasov et al. 2001).

5.5 Problems and Solutions

Problems

1. It is known that about 10 years ago many gyrotrons operated at such modes as $TE_{22,2}$and $TE_{10,4}$. Presently, high-power gyrotrons operate at such modes as, for instance, $TE_{28,7}$ or $TE_{31,8}$. By using a solver for the roots of the equation $J'_m(\nu_{m,p}) = 0$, which determines the cutoff frequencies of competing modes, evaluate the spectrum density of competing modes in the vicinity of modes used 10 years ago and today. Take into account only the modes with the same radial index p and $p \pm 1$.
2. Consider competition between two modes when one of them is located in the region of soft self-excitation, while another is in the hard self-excitation regime. Assume that the mode intensities can be described by equations

that are similar to (5.3), with specific saturation coefficients:

$$\frac{dM_1}{d\tau} = M_1[\sigma_1 - M_1 - 2M_2],$$

$$\frac{dM_2}{d\tau} = M_2\Big[-\sigma_2 + M_2 - M_2^2 - 2M_1\Big].$$

Determine all equilibrium states and their stability and analyze their dependence on the increment of the first mode σ_1 and the decrement of the second mode σ_2.

Solution

2. There are four types of states of equilibrium. The first one is the state with zero intensity of both modes ($M_1 = M_2 = 0$). In the second one the first mode has a nonzero intensity, while the second mode intensity is equal to zero ($M_1 = \sigma_1 \neq 0$, $M_2 = 0$). Equilibria of the third type have zero intensity of the first mode and two possible equilibrium states with nonzero intensity of the second mode ($M_1 = 0$, $M_2 = 1/2 \pm \sqrt{1/4 - \sigma_2}$). They exist when the decrement is not too large, $\sigma_2 < 1/4$. Finally, the states of equilibrium of the fourth type have nonzero intensities of both modes ($M_1 = \sigma_1 - 2M_2$, $M_2 = 5/2 \pm \sqrt{25/4 - 2\sigma_1 - \sigma_2}$).

Analysis of their stability shows that the equilibrium state with the zero intensity of both modes is always a saddle point (when $\sigma_1 > 0$ and $\sigma_2 > 0$), which is stable with respect to perturbations in the intensity of the second mode but unstable with respect to the first mode perturbations. The second state of equilibrium with the nonzero intensity of the first mode only is always a stable node. The equilibrium states of the third kind have more complicated conditions of stability. The equilibrium with a smaller intensity is always unstable with respect to perturbations of the second mode and it is stable with respect to perturbations of the first mode when the increment of the first mode is small enough: $\sigma_1 < 1$, $\sigma_2 < (2 - \sigma_1)\sigma_1/4$. Under these conditions, this equilibrium is a saddle point. When the latter conditions are violated, this is an unstable node. The equilibrium with a larger intensity of the second mode is always stable with respect to perturbations of its own intensity. It is also stable with respect to the perturbations of the first mode when $\sigma_1 < 1$. In the region $\sigma_1 > 1$ this equilibrium remains stable when $\sigma_2 > (2 - \sigma_1)\sigma_1/4$. Finally, a two-mode equilibrium exists only in a limited region, where $2\sigma_1 + \sigma_2 < 25/4$ and $M_2 < 5/2$. This is an equilibrium state that corresponds to the minus sign in the expression for the stationary intensity of the second mode given above. This equilibrium is a saddle point.

Now one can easily distinguish in the plane of parameters "decrement of the second mode versus increment of the first mode" the following regions with different topology of phase portraits. At $\sigma_2 > 1/4$, the second mode is damped completely and there is only one stable state of equilibrium, which represents single-mode oscillations of the first mode. At $\sigma_2 < 1/4$, but when the first mode has a large increment, $\sigma_1 > 1$, and $\sigma_2 > (2 - \sigma_1)\sigma_1/4$, there are equilibrium states on the M_2-axis, but they are unstable with respect to the first-mode perturbations. Although there are no two-mode states of equilibrium, from any point of the phase portrait (state space) a system arrives at the single-mode oscillations of the first mode. In the region of comparable decrement and increment, where $\sigma_2 < (2 - \sigma_1)\sigma_1/4$, there is a two-mode state of equilibrium, but it is not stable. So, there are two stable states of equilibrium: single-mode oscillations of the first mode and single-mode oscillations of the second mode (the latter corresponds to the plus sign in the definition of the second mode intensity: $M_1 = 0$, $M_2 = 1/2 + \sqrt{1/4 - \sigma_2}$.) Depending on the initial conditions, a system arrives at one of them. Finally, when the first mode increment becomes too small $\sigma_1 < 1$, while the second decrement is large enough $\sigma_2 > (2 - \sigma_1)\sigma_1/4$, a two-mode equilibrium disappears, but there are still the same stable states of equilibrium on both axes. The separatrix, which starts from the equilibrium of the second mode with a smaller intensity ($M_1 = 0$, $M_2 = 1/2 - \sqrt{1/4 - \sigma_2}$), separates the regions of attraction to each of these two stable states.

▪ CHAPTER 6 ▪

Linear Theory of the Gyro-TWT

6.1 Introduction: Instability in Magnetoactive Plasma

The development of gyrotrons, at least in its earlier stage, greatly benefited from the fact that people involved in this development actively interacted with the experts in other fields of physics. One such expert was V. V. Zheleznyakov, who, being active in space plasma research, also discussed with A. V. Gaponov some issues related to the coherent radiation of electromagnetic waves by electron beams in labs.

In a series of papers published in the early 1960s, V. V. Zheleznyakov (1960, 1961) analyzed the instability of magnetoactive plasma with respect to electromagnetic perturbations. For the sake of simplicity, he considered propagation of EM waves along the external constant magnetic field. It was also assumed that, in addition to the magnetoactive plasma in which the presence of ions does not contribute to high-frequency EM perturbations, there is a certain medium with a refractive index n_j. Thus, the waves can propagate along the external magnetic field with the phase velocity different from the speed of light. Zheleznyakov used the method of the linearized kinetic equation for describing the perturbations in the distribution function due to EM perturbations proportional to $\exp\{i(kz - \omega t)\}$. The starting point in his consideration was the dispersion equation for plane EM waves in a nonisotropic medium given elsewhere (Landau and Lifshitz 1960),

$$\det\left| n^2\delta_{i,k} - n_i n_k - \varepsilon_{i,k}(\omega, \vec{k}) \right| = 0. \qquad (6.1)$$

Here $\delta_{i,k}$ is the Kronecker symbol, $\vec{n} = \vec{k}c/\omega$, and $\varepsilon_{i,k}$ is the dielectric tensor, which for relativistic magnetoactive plasma was derived by V. D. Shafranov (1958). In general, all nine components of this tensor are nonzero. However, in the case of a stationary unperturbed distribution function f_0 and propagation of EM perturbations along z ($\vec{H}_0 = H_0\vec{z}_0$), four components of it equal zero, $\varepsilon_{xz} = \varepsilon_{zx} = \varepsilon_{yz} = \varepsilon_{zy} = 0$, and also $\varepsilon_{xx} = \varepsilon_{yy}$ and $\varepsilon_{yx} = -\varepsilon_{xy}$. This allows one to separate in Eq. (6.1) plasma waves described by $\varepsilon_{zz} = 0$ and EM waves given by

$$n^2 - \varepsilon_{xx} \pm i\varepsilon_{xy} = 0. \tag{6.2}$$

Without going into all the details of this derivation, let us mention only that substituting the expressions for the components ε_{xx} and ε_{xy} into Eq. (6.2) yields two fourth-order dispersion equations, which in the case of the δ-distribution of electrons in the orbital and axial momentum, $f_0(\vec{p}) = (1/2\pi p_{\perp 0})\delta(p_\perp - p_{\perp 0})\delta(p_z - p_{z0})$, can be written as

$$c^2k^2 - \omega^2 n^2 + \omega_b^2 \frac{\omega - kv_{z0}}{\omega - kv_{z0} \mp \Omega_0} + \frac{\beta_{\perp 0}^2}{2}\omega_b^2 \frac{c^2k^2 - \omega^2}{(\omega - kv_{z0} \mp \Omega_0)^2} = 0. \tag{6.3}$$

Here $\omega_b^2 = 4\pi e^2 n_b/m$ is the beam plasma frequency. (Here n_b is the electron beam density.) Also here and below we use $n = c/v_{ph}$ instead of n_j. This fundamental equation derived by Zheleznyakov for an unbounded magnetoactive plasma was later rederived for the cylindrical and Cartesian configurations of the interaction space by Friedman et al. (1973) and Ott and Manheimer (1975), respectively. Their treatment has led to the inclusion of some geometrical factors in the two last terms in Eq. (6.3). For those readers who are interested in the derivation of this equation, it is expedient to recommend the paper by Chu and Lin (1988), where the derivation of the dispersion equation for a system with cylindrical geometry is described in detail, or the paper by Chu and Hirshfield (1978).

Most important in (6.3) is the last term on its LHS. This is a new term, which was taken into account by Zheleznyakov (1960) and Gaponov (1959b). This term contains contributions from two mechanisms of electron bunching. Both of them are inversely proportional to the second power of the cyclotron resonance mismatch, $\omega - kv_{z0} \mp \Omega_0$, i.e., corresponding perturbations in electron motion caused by the EM wave evolve in time proportionally to the second power of time. That is why sometimes, as we mentioned above, corresponding bunching mechanisms are called "quadratic" bunching (see, e.g., Gaponov, Petelin and Yulpatov 1967). The first mechanism is the orbital bunching caused by the relativistic dependence of the electron cyclotron frequency on its energy. This bunching is presented by the $-\omega^2$ term in the last term in (6.3). The second is the axial bunching, which can be attributed to the effect of the transverse magnetic field of the wave on the electron axial motion. This bunching is presented by the c^2k^2 term. Sometimes the instability associated with the axial bunching is called Weibel instability (Weibel 1959), although Gaponov analyzed it independently from Weibel in 1959. As was noted by Gaponov (Gaponov 1960), when the wave propagates along the external magnetic field with the phase velocity equal to the speed of light, these two terms cancel each other. Certainly, this cancellation has a common nature with the autoresonance effect considered in Chapter 1. However, it must be noticed that the autoresonance effect is a single-particle effect, while now we are talking about cancellation of two bunching mechanisms in an ensemble of electrons. Clearly, in the case of slow waves that propagate with phase velocities smaller than the speed of light, the axial bunching dominates as it follows from the comparison of these two terms in (6.3). On the contrary, in the case of fast waves $v_{ph} = \omega/kn > c$, the orbital bunching is dominant. Relationships between these two bunching mechanisms have been analyzed elsewhere (Gaponov and Yulpatov 1967; also Chu and Hirshfield 1978).

Note that, in general, the dispersion equation for linear-beam TWTs is also of the fourth order (Pierce 1950). However, as is already known (see, e.g., Pierce 1950), all four waves determined by this equation must be taken into account only in the case of operation at the ends of the passband of a slow-wave structure where the refractive index n, which is proportional to the wave group velocity, is small. In our model, this is the case of operation near cutoff, when $v_{ph} \gg c$. Such an operation is realized in gyrotron oscillators (CRM-monotrons). Therefore, the fourth-order dispersion (6.3) with corresponding boundary conditions was used elsewhere (Bratman and Moiseev 1975, H. Sato et al. 1986) for analyzing the starting conditions in these devices.

In the case of an infinitesimal beam plasma density ($\omega_b \to 0$), the four waves determined by Eq. (6.3) for each sign of the cyclotron frequency in the denominator can be distinguished as two co- and counter-propagating waves of an isotropic medium that obey the equation

$$c^2k^2 - \omega^2n^2 = 0,$$

and two cyclotron waves,

$$\omega - kv_{z0} \mp \Omega_0 = 0, \tag{6.4}$$

with either co- ($\omega - kv_{z0} - \Omega_0 = 0$) or counter- ($\omega - kv_{z0} + \Omega_0 = 0$) rotation. In the limit $\omega_b \to 0$, these cyclotron waves are degenerate in the same way as the space charge waves are degenerate in linear-beam TWTs when $\omega_b \to 0$ (Pierce 1950).

Far enough from cutoff, Eq. (6.3) can be simplified in a way similar to one used by Pierce for linear-beam TWTs. Consider, for instance, the interaction of gyrating electrons with a forward wave. Assuming that

$$k = k_0(1 + \nu), \tag{6.5}$$

where $k_0 = \omega n/c$ is the "cold" wavenumber of a wave in the absence of the beam and the small term $\nu << 1$ accounts for the beam effect, the two first terms in (6.3) can be represented as

$$c^2k^2 - \omega^2 n^2 \simeq 2\omega^2 n^2 \nu.$$

(In other words, we take here into account the small term $\nu << 1$ only in a small difference $ck - \omega n$, but not in a large sum $ck + \omega n \approx 2\omega n$.) Also assume that the interaction takes place under conditions close to the cyclotron resonance, i.e., in denominators of (6.3),

$$\omega - kv_{z0} \mp \Omega_0 = \omega - k_0 v_{z0} \mp \Omega_0 - k_0 v_0 \nu = k_0 v_{z0}(\varepsilon - \nu), \tag{6.6}$$

where $\varepsilon = (\omega - k_0 v_{z0} \mp \Omega_0)/k_0 v_{z0} << 1$ is a small cyclotron resonance mismatch, in which the Doppler term is determined by the "cold" axial wavenumber. Then, assuming that small terms ε and ν are on the order of $(\omega_b^2/\omega^2)^{1/3}$, one can readily reduce (6.3) to the cubic equation

$$(\varepsilon - \nu)^2 \nu + \frac{\beta_{\perp 0}^2}{2\beta_{z0}^2} \frac{\omega_b^2}{\omega^2} \frac{n^2 - 1}{n^4} = 0. \tag{6.7}$$

This equation is similar to the known cubic dispersion equation for TWTs (Pierce 1950) and FELs with negligibly small ohmic losses and space charge effects (Kroll 1978). The last term in the LHS of (6.7) is equivalent to C^3, where C is the Pierce gain parameter (Pierce 1950). Clearly, after an obvious normalization of ε and ν to this last term in the one-third power, this parameter can be eliminated from (6.7), thus reducing it to the known dispersion equation for the linear-beam TWT, which will be discussed in the next section. Since in real devices ω_b^2 is proportional to the beam current, I_b, Eq. (6.7) shows that the spatial growth rate of the *convective instability* in gyro-TWTs, is proportional to $I_b^{1/3}$ in the same fashion as in linear-beam TWTs (Pierce 1950) and FELs in the Compton regime (Kroll 1978). Recall that the convective instability is the instability of the wave whose amplitude increases along the axial distance. This implies that the axial wavenumber is complex. On the contrary, the case when the wave amplitude increases in time at any cross-section of the interaction space is known as the *absolute instability*. In this case the wavenumber is constant, but the wave frequency is complex. (For more details see, e.g., Briggs 1964.)

In addition to the convective instability, the dispersion equation (6.3) allows us also to analyze the absolute instability. As an example, let us consider a simple case of quasi-static waves with a zero axial wavenumber $k = 0$. In such plasma oscillations of beam electrons, the wave electric field rotates in the plane perpendicular to the external magnetic field while the wave magnetic field is absent. So, the absolute instability in this case can be caused only by the relativistic dependence of Ω_0 on the electron energy.

Assuming that $\omega = \Omega_0(1+\delta)$ where $\delta << 1$, and introducing $\bar{\omega}_b^2 = \omega_b^2/n^2$ (the bar will be omitted below), one can readily reduce (6.3) to the quadratic equation

$$\delta^2 - \frac{\omega_b^2}{\Omega_0^2}\delta + \frac{\beta_{\perp 0}^2}{2}\frac{\omega_b^2}{\Omega_0^2} = 0. \tag{6.8}$$

The solutions of this equation are

$$\delta = \frac{\omega_b^2}{2\Omega_0^2} \pm \left[\frac{\omega_b^2}{2\Omega_0^2}\left(\frac{\omega_b^2}{2\Omega_0^2} - \beta_{\perp 0}^2\right)\right]^{1/2}. \tag{6.9}$$

So, the absolute instability occurs at $\beta_{\perp 0}^2 > \omega_b^2/2\Omega_0^2$ and its maximum takes place when $\omega_b^2 = \beta_{\perp 0}^2\Omega_0^2$ and is equal to $(\mathrm{Im}\,\delta)_{\max} = \omega_b^2/2\Omega_0^2 = \beta_{\perp 0}^2/2$. As was noted by Gaponov, Petelin, and Yulpatov (1967), this instability has a common nature with the negative mass instability discussed in Sec. 4.4.

Before closing this introductory section, let us note that the cyclotron waves determined by Eq. (6.4) have phase velocities

$$v_{ph} = \frac{v_{z0}}{1 \mp \Omega_0/\omega}. \tag{6.10}$$

Therefore, the wave with the minus sign is often called the *fast cyclotron wave* and the wave with the plus sign is known as the *slow cyclotron wave*. For each of them, Eq. (6.3) is valid, and hence, when the last term in the LHS of (6.3) does not equal zero, each of them splits into the negative and positive energy cyclotron waves (Sprangle and Drobot 1977). Then, the coupling of the negative energy cyclotron wave to the positive energy waveguide mode is responsible for the instability. Clearly, this instability can be realized for both fast and slow cyclotron waves. However, one should bear in mind that the slow cyclotron wave can be excited in slow-wave structures only, because it requires that the Doppler term in the cyclotron resonance condition is larger than the frequency ω or, in other words, to have $v_{ph} < v_{z0}$. Recall that this is the condition for the anomalous Doppler effect, which was discussed in Sec. 1.4. This situation is illustrated in Fig. 6.1, which shows the dispersion diagram for

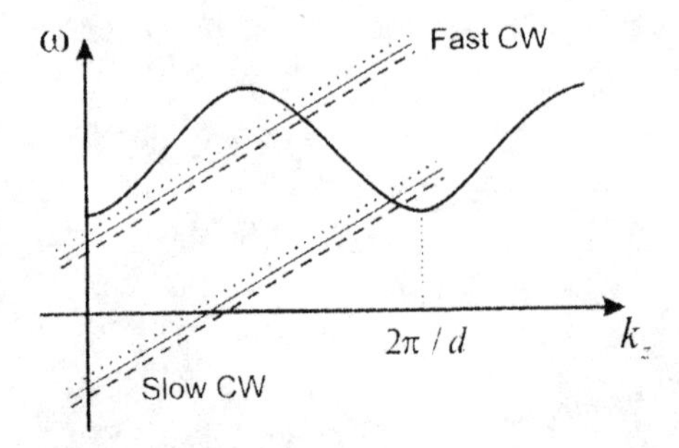

Fig. 6.1. Dispersion diagram for a periodic slow-wave structure (with the period d) and fast ($s > 0$) and slow ($s < 0$) cyclotron waves (CWs). Each of the cyclotron waves is split into the positive energy (dashed) and negative energy (dotted) waves due to the finite beam plasma frequency when electrons have initial orbital velocity.

the waveguide mode in a slow-wave structure and the slow and fast cyclotron waves that become non-degenerate due to the finite beam plasma frequency.

6.2 Derivation of the Dispersion Equation for the Gyro-TWT

The linear theory of the gyro-TWT (Gaponov 1959a,b and 1961), as well as any other microwave tube driven by an electron beam, is based on a self-consistent treatment of the device operation in the framework of the small-signal approximation. The self-consistency in this case means that one should analyze the perturbations in electron motion caused by a weak EM wave, and also consider the excitation of this wave by a synchronous component of an electron beam current that occurs due to electron perturbations.

The wave excitation in a waveguide, as shown in Appendix 2, can be described by the equation

$$\frac{dC_S}{dz} = -\frac{1}{N_S}\int\limits_{S_\perp} \vec{j}_\omega \cdot \vec{E}_S^* ds_\perp. \tag{6.11}$$

Here C_S is the wave amplitude, N_S is the norm of the wave and the function $\vec{E}_S$ describes the spatial structure of the s-th wave (see Appendix 2). This equation is valid for the case when electron coordinates, $\vec{r}$, velocity, $\vec{v}$, and charge density, ρ, are determined in Eulerian coordinates: $\vec{r} = \vec{r}(z,t)$, $\vec{v} = \vec{v}(z,t)$ and $\rho = \rho(z,t)$. Very often, it is more convenient to use Lagrangian coordinates: t_0 or t and τ; here t_0 is the entrance time of a given electron and τ is its transit time in the interaction space, which are related to the time t as $t = t_0 + \tau$. The transition from Eulerian to Lagrangian coordinates can be made with the use of the dependence $z = z(\tau, t)$. This dependence can be found from the equation for electron motion. So, in Lagrangian coordinates the integrand in the RHS of (6.11) should be treated as $\vec{j}_\omega(\tau, t)\vec{E}_s^*\{\vec{r}_\perp(\tau, t), z(\tau, t)\}$.

In the small-signal approximation, one can assume that electron displacement from the stationary trajectory is small, as well as are perturbations in the electron velocity, i.e.

$$\vec{r} = \vec{r}^{(0)}(\tau) + \vec{r}^{(1)}(\tau, t), \qquad \vec{v} = \vec{v}^{(0)}(\tau) + \vec{v}^{(1)}(\tau, t).$$

Here the electron displacement is much smaller than the wavelength and gyro-radius, and the perturbation in electron velocity is much smaller than the unperturbed velocity $\vec{v}^{(0)}$. Below we will consider these perturbations as monochromatic: $\vec{r}^{(1)} = \vec{r}_\omega^{(1)}(\tau)e^{i\omega t}$, $\vec{v}^{(1)} = \vec{v}_\omega^{(1)}(\tau)e^{i\omega t}$. Also the electric field $\vec{E}_s^*$ in the integrand of (6.11) can be expanded in the vicinity of the unperturbed trajectory with the coordinates $\vec{r} = \vec{r}^{(0)}$. We will assume below that the transverse size of an electron beam is small enough, and therefore all electron beamlets (recall that by beamlets we imply electrons gyrating around the same guiding center) experience the action of the identical RF Lorentz force. Note that in the case of a cylindrical system this can be an annular electron beam with a small spread in guiding center radii. This assumption allows us in (6.11) to replace $\int_{S_\perp} \rho ds_\perp$ by $-I_b/v_{z0}$, where I_b is the absolute value of the beam current. (Other variables in the integrand of (6.11) do not depend on the averaging over the cross section under the assumptions made.) Also note that in the reference frame moving with the electrons the electric field acting upon electrons, $\vec{E}_{s,a}$ is equal to

$$\vec{E}_{s,a} = \vec{E}_s + \frac{1}{c}[\vec{v}^{(0)} \times \vec{H}_s]. \tag{6.12}$$

This field, as one can see, has the same structure as the RF Lorentz force acting on electrons. Below we will denote this force as $\vec{G}_{s,\omega}$. With this in mind, one can rewrite the wave excitation equation (6.11) in the following form:

$$\frac{dC_s}{dz} = \frac{i\omega I_b}{v_{z0}N_s}\left(\vec{r}_\omega^{(1)}\vec{G}_{s,\omega}^*\right). \tag{6.13}$$

Note that in the framework of the small-signal theory we assume that, in the presence of an electron beam, the wave propagates proportionally to $\exp\{i(\omega t - k_z z)\}$. For the field represented as $\vec{E} \propto C_s\vec{e}(\vec{r}_\perp)\exp\{i(\omega t - k_{s,z}z)\}$, where $\vec{e}(\vec{r}_\perp)$ describes the transverse structure of the wave field, it means that the wave amplitude C_s varies along z as $\exp\{-i(k_z - k_{z,s})z\}$. Thus, Eq. (6.13) can be rewritten as

$$(k_z - k_{z,s})C_s = -\frac{\omega I_b}{v_{z0}N_s}\left(\vec{r}_\omega^{(1)}\vec{G}_{s,\omega}^*\right). \tag{6.14}$$

In the LHS of (6.14) $(k_z - k_{z,s})$ characterizes the difference between the wave propagation constants in "hot"(beam-loaded) and "cold"(empty) waveguides.

The next step is to determine the perturbations in electron motion caused by this wave. The momentum of a relativistic electron, $\vec{p} = m\gamma\frac{d\vec{r}}{dt}$, obeys the standard equation of motion (3.2). In cylindrical coordinates this equation can be rewritten as

$$\frac{d}{dt}\left(\gamma\frac{dr}{dt}\right) - \gamma r\left(\frac{d\theta}{dt}\right)^2 = -\frac{e}{m}\left\{E_{\sim r} + \frac{1}{c}\left[\frac{d\vec{r}}{dt}\times(\vec{H}_0 + \vec{H}_{\sim})\right]_r\right\}, \quad (6.15)$$

$$\frac{1}{r}\frac{d}{dt}\left(\gamma r^2\frac{d\theta}{dt}\right) = -\frac{e}{m}\left\{E_{\sim\theta} + \frac{1}{c}\left[\frac{d\vec{r}}{dt}\times(\vec{H}_0 + \vec{H}_{\sim})\right]_\theta\right\},$$

$$\frac{d}{dt}\left(\gamma\frac{dz}{dt}\right) = -\frac{e}{m}\left\{E_{\sim z} + \frac{1}{c}\left[\frac{d\vec{r}}{dt}\times(\vec{H}_0 + \vec{H}_{\sim})\right]_z\right\}.$$

In the absence of an EM wave, an electron moves along the helical trajectory with the constant axial velocity, v_{z0}, and orbital velocity, $v_{\perp 0}$. Correspondingly, the gyroradius is equal to $a = v_{\perp 0}/\Omega_0$ and the gyrophase is equal to $\theta^{(0)} = \theta_0 + \Omega_0\tau$, where Ω_0 is the unperturbed electron cyclotron frequency and θ_0 is the gyrophase at the entrance. Assuming that the perturbations with respect to this motion are small, and representing these perturbations as

$$\vec{r}_1 = r_1\vec{r}_0 + \theta_1 a\vec{\theta}_0 + z_1\vec{z}_0,$$

where $\vec{r}_0$, $\vec{\theta}_0$ and $\vec{z}_0$ are unit vectors in the cylindrical reference frame, one can linearize (6.15), which yields the following set of equations (Gaponov 1961):

$$\frac{d^2r_1}{dt^2} - v_{\perp 0}\frac{d\theta_1}{dt} - \beta_{\perp 0}^2\gamma_0^2\left(\Omega_0^2 r_1 + v_{\perp 0}\frac{d\theta_1}{dt} + \frac{v_{z0}}{v_{\perp 0}}\Omega_0\frac{dz_1}{dt}\right) = -\frac{e}{m\gamma_0}C_s G_{\omega,r},$$

$$a\frac{d^2\theta_1}{dt^2} + \Omega_0\frac{dr_1}{dt} + \beta_{\perp 0}^2\gamma_0^2\left(\Omega_0\frac{dr_1}{dt} + a\frac{d^2\theta_1}{dt^2} + \frac{v_{z0}}{v_{\perp 0}}\frac{d^2z_1}{dt^2}\right) = -\frac{e}{m\gamma_0}C_s G_{\omega,r},$$

$$\frac{d^2z_1}{dt^2} + \beta_{\perp 0}\beta_{z0}\gamma_0^2\left(\Omega_0\frac{dr_1}{dt} + a\frac{d^2\theta_1}{dt^2} + \frac{v_{z0}}{v_{\perp 0}}\frac{d^2z_1}{dt^2}\right) = -\frac{e}{m\gamma_0}C_s G_{\omega,z}. \quad (6.16)$$

Here we took into account the fact that $\frac{d^2\vec{r}^{(0)}}{dt^2} = -\Omega_0^2 a\vec{r}^{(0)}$, $(\frac{d\vec{r}_1}{dt})_\theta = \Omega_0 r_1 + a\frac{d\theta_1}{dt}$, $(\frac{d^2\vec{r}_1}{dt^2})_\theta = 2\Omega_0\frac{dr_1}{dt} + a\frac{d^2\theta_1}{dt^2}$, and $(\frac{d^2\vec{r}_1}{dt^2})_r = \frac{d^2r_1}{dt^2} - \Omega_0^2 r_1 - 2\Omega_0 a\frac{d\theta_1}{dt}$. Correspondingly, perturbations in the electron energy $(\gamma = \gamma_0 + \gamma_{(1)})$ are equal to

$$\gamma_{(1)} = \left(\gamma_0^3/c\right)\left[\beta_{\perp 0}\left(\Omega_0 r_1 + a\frac{d\theta_1}{dt}\right) + \beta_{z0}\frac{dz_1}{dt}\right]. \quad (6.17)$$

Assuming that the electric and magnetic fields of the wave are given, respectively, as

$$\vec{E} = C_s \vec{e}(r,\theta) e^{i(\omega t - k_z z)}, \qquad \vec{H} = C_s \vec{h}(r,\theta) e^{i(\omega t - k_z z)},$$

one can also represent the Lorentz force as $\vec{G} = \vec{g}(r^{(0)} = a, \theta^{(0)}) e^{i(\omega t - k_z z)}$, where $\vec{g} = \vec{e} + \frac{1}{c}[\vec{v}^{(0)} \times \vec{h}]$. The solution of the system of linear inhomogeneous equations (6.16) can be represented as $\vec{r}_1 = \vec{r}_1(\tau)e^{i\omega t} + \vec{\rho}_1(\tau)e^{i\omega t}$, where $\vec{\rho}_1(\tau)e^{i\omega t}$ is a general solution of the homogeneous set of equations, which in the absence of an EM wave describes the normal waves of an electron beam. The partial solution of the inhomogeneous set of equations, $\vec{r}_1(\tau)e^{i\omega t}$, as well as the function $\vec{g}(\tau)$, is a periodic function of τ and hence can be expanded in the Fourier series:

$$\vec{r}_1 = \sum_n \vec{r}_n e^{-i(n\Omega_0 + k_z v_z)\tau} e^{i\omega t}, \qquad \vec{G} = \sum_n \vec{G}_n e^{-i(n\Omega_0 + k_z v_z)\tau} e^{i\omega t}, \qquad \textbf{(6.18)}$$

where $\vec{r}_n = r_n \vec{r}_0 + \theta_n a \vec{\theta}_0 + z_n \vec{z}_0$ and $\vec{G}_n = G_{r,n}\vec{r}_0 + G_{\theta,n}\vec{\theta}_0 + G_{z,n}\vec{z}_0$. Substituting these representations into (6.16) yields the following set of linear algebraic equations for r_n, θ_n and z_n:

$$\begin{aligned}
&\left[(\Delta\omega_n)^2 + \gamma_0^2\beta_{\perp 0}^2\Omega_0^2\right] r_n + i\Delta\omega_n\Omega_0\gamma_0^2\left(1-\beta_{z0}^2\right) a\theta_n \\
&\quad + i\Delta\omega_n\Omega_0\gamma_0^2\beta_{\perp 0}\beta_{z0} z_n = \frac{e}{m\gamma_0} C_s G_{r,n}, \\
&-i\Delta\omega_n\Omega_0\gamma_0^2\left(1-\beta_{z0}^2\right) r_n + (\Delta\omega_n)^2\gamma_0^2\left(1-\beta_{z0}^2\right) a\theta_n \\
&\quad + (\Delta\omega_n)^2\gamma_0^2\beta_{\perp 0}\beta_{z0} z_n = \frac{e}{m\gamma_0} C_s G_{\theta,n}, \\
&-i\Delta\omega_n\Omega_0\gamma_0^2\beta_{\perp 0}\beta_{z0} r_n + (\Delta\omega_n)^2\gamma_0^2\beta_{\perp 0}\beta_{z0} a\theta_n \\
&\quad + (\Delta\omega_n)^2\gamma_0^2\left(1-\beta_{\perp 0}^2\right) z_n = \frac{e}{m\gamma_0} C_s G_{z,n}.
\end{aligned} \qquad \textbf{(6.19)}$$

Here $\Delta\omega_n = \omega - k_z v_{z0} - n\Omega_0$ and the components of the Lorentz force are equal to

$$G_{r,n} = \frac{1}{2\pi}\int_{-\pi}^{\pi} \{e_r + \beta_{\perp 0} h_z - \beta_{z0} h_\theta\} e^{in\xi} d\xi, \qquad \textbf{(6.20)}$$

$$G_{\theta,n} = \frac{1}{2\pi}\int_{-\pi}^{\pi} \{e_\theta + \beta_{z0} h_r\} e^{in\xi} d\xi,$$

$$G_{z,n} = \frac{1}{2\pi}\int_{-\pi}^{\pi} \{e_z - \beta_{\perp 0} h_r\} e^{in\xi} d\xi.$$

Here $\xi = \Omega_0 \tau$.

Solving the set of Eqs. (6.20) results in the following expressions for the electron perturbations:

$$r_n = \frac{e}{m\gamma_0} C_s \frac{G_{r,n}\Delta\omega_n - iG_{\theta,n}\Omega_0}{\Delta\omega_{n-1}\Delta\omega_n\Delta\omega_{n+1}}, \tag{6.21}$$

$$a\theta_n = \frac{e}{m\gamma_0} C_s \left\{ \frac{iG_{r,n}\Omega_0 + G_{\theta,n}\Delta\omega_n}{\Delta\omega_{n-1}\Delta\omega_n\Delta\omega_{n+1}} - \frac{G_{z,n}\beta_{z0}\beta_{\perp 0} + G_{\theta,n}\beta_{\perp 0}^2}{(\Delta\omega_n)^2} \right\},$$

$$z_n = \frac{e}{m\gamma_0} C_s \frac{G_{z,n}\left(1-\beta_{z0}^2\right) - G_{\theta,n}\beta_{z0}\beta_{\perp 0}}{(\Delta\omega_n)^2}.$$

Let us now assume that the cyclotron resonance condition is fulfilled for the sth cyclotron harmonic, i.e., for $n = s$ the cyclotron resonance mismatch $\Delta\omega_{n=s}$ is small, while for other n's its absolute value is close to an integer number of Ω_0. Then, substituting the electron perturbations determined by Eqs. (6.21) into the wave excitation equation (6.14) and neglecting all non-resonant terms [note that Eqs. (6.21) contain resonant terms not only at a given harmonics, but also at harmonics $s \pm 1$] yield

$$(k_z - k_{z,s}) = -\frac{\omega I_b}{v_{z0} N_s} \frac{e}{m\gamma_0} \tag{6.22}$$

$$\times \left\{ \frac{(1-\beta_{z0}^2)\left|G_{z,s}\right|^2 - \beta_{\perp 0}^2\left|G_{\theta,s}\right|^2 - \beta_{z0}\beta_{\perp 0}2\mathrm{Re}(G_{\theta,s}G_{z,s}^*)}{(\Delta\omega_s)^2} + \frac{M_s}{2\Omega_0(\Delta\omega_s)} \right\}$$

We denoted by M_s the term, which is responsible for the so-called M-type electron bunching, while the first term in figure brackets is responsible for the so-called O-type electron bunching. This M-type term is equal to

$$M_s = -S_{s-1} + S_{s+1} + T_{s-1} - 2T_s + T_{s+1}. \tag{6.23}$$

Here we used the notations adopted by Petelin and Yulpatov (1975): $S_s = \left|G_{\theta,s}\right|^2 + \left|G_{r,s}\right|^2$ and $T_s = i(G_{r,s}G_{\theta,s}^* - G_{\theta,s}G_{r,s}^*)$. Below we shall assume that the O-type terms play the dominant role, because in the case of operation close to the cyclotron resonance these terms proportional to the inverse cyclotron resonance mismatch squared, $(\Delta\omega_s)^{-2}$, grow faster than M-type terms proportional to $(\Delta\omega_s)^{-1}$. So, the M-type terms will be neglected below. (We will discuss them later in Sec. 13.7.) Correspondingly, Eq. (6.22) can be rewritten, introducing the detuning of propagation constants $\delta k = (k_z - k_{z,s})/(\omega/v_{z0})$ and detuning of the cyclotron resonance $\delta\omega = (\omega - k_{z,s}v_0 - s\Omega_0)/\omega$, where

$|\delta k|, |\delta\omega| << 1$, as follows:

$$\delta k(\delta k - \delta\omega)^2 = \tag{6.24}$$
$$-\frac{eI_b}{mc^3}\frac{c^3}{\gamma_0\omega^2 N_s}\left\{\left(1-\beta_{z0}^2\right)|G_{z,s}|^2 - \beta_{\perp 0}^2|G_{\theta,s}|^2 - \beta_{z0}\beta_{\perp 0}2\mathrm{Re}(G_{\theta,s}G_{z,s}^*)\right\}.$$

Let us call the reader's attention to the fact that the two last terms on the right-hand side of this equation are proportional to the electron velocity components normalized to the speed of light. At the same time, the first term, at first glance, does not vanish in the limit of very small velocities. This was the reason for neglecting the two last terms in the first treatments of this problem, where possibilities of weakly relativistic electrons to produce coherent radiation were analyzed. Later, however, it was realized that, for instance, in the case of transverse-electric waves the first term, in accordance with (6.20), is also proportional to $\beta_{\perp 0}^2$, since it represents only the Lorentz force caused by the transverse magnetic field of the wave. Hence, all terms in figure brackets are of the same order. This conclusion is also valid for transverse-magnetic waves, because the axial electric field of the wave yields zero contribution to the axial component of the Lorentz force synchronous with gyrating electrons.

The right-hand side of (6.24) now plays a role of the cubic Pierce gain parameter, C^3. So, after normalizing our detunings δk and $\delta\omega$ to C, one can easily reduce (6.24) to the known dispersion equation for the linear-beam TWT (Pierce 1950). More exactly, in Pierce notations the dispersion equation has the form

$$\delta_P^2(i\delta_P - b) = 1. \tag{6.25}$$

(Here we denote by δ_P the detuning used by Pierce; we also use his velocity parameter b.) In order to transform our dispersion equation to (6.25), we should introduce $\bar{\gamma} = i(\delta k - \delta\omega) = -i(\omega - k_z v_{z0} - s\Omega_0)/\omega$. Substituting this new variable into (6.24) and denoting the RHS of it by C^3 yields

$$\bar{\gamma}^2(i\bar{\gamma} - \delta\omega) = C^3. \tag{6.26}$$

This equation clearly has the same form as (6.25), thus, after dividing $\bar{\gamma}$ and $\delta\omega$ by C, one can directly apply the well-known results (Pierce 1950) to our study. The roots of Eq. (6.25) are shown in Fig. 6.2, which reproduces Pierce's results. Note that similarity between the dispersion equations for the gyro-TWT and linear-beam TWT is also discussed in detail elsewhere (Yulpatov 1967, Sangster 1980, and Basu 1995). Also note that in the theory of the

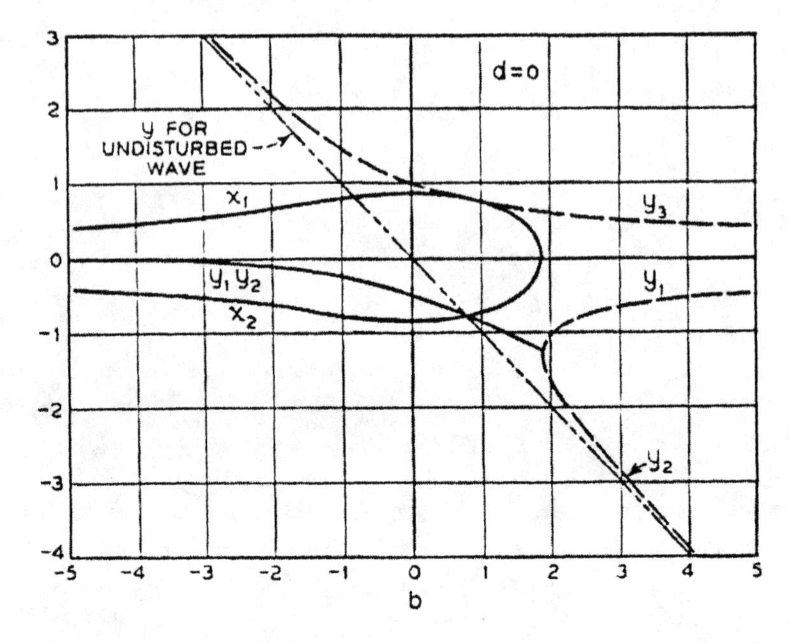

Fig. 6.2. Solutions of the TWT dispersion equation. Real ($y_{1,2,3}$) and imaginary ($x_{1,2}$) parts of wave propagation constants are shown as functions of the detuning in the device without attenuation losses and space charge effects. (Reproduced from Pierce 1950).

gyro-TWT the dispersion equation is often written in the form

$$\gamma^2(\gamma - \delta) + 1 = 0. \tag{6.27}$$

To obtain this equation from (6.26) one should divide $\bar{\gamma}$ and $\delta\omega$ used in (6.26) by C, and then replace this normalized $\bar{\gamma}$ by γ/i and normalized $\delta\omega$ by δ. This general theory yields the results that are independent on C, when, simultaneously with dividing the propagation constants and detunings by C, we introduce the normalized distance as the real axial distance multiplied by $k_z C$. (This issue will be discussed later in more detail.)

So, now we are dealing with a cubic dispersion equation, which has three roots. Therefore all perturbations, including the EM wave, can be considered as superpositions of three partial waves whose propagation constants are determined by the roots of the dispersion equation shown in Fig. 6.2. As follows from Fig. 6.2, the convective instability (imaginary part of two complex conjugate roots of the dispersion equation is not equal to zero) exists at $\delta \leq (3/2)2^{1/3} \approx 1.89$. The maximum increment occurs at the exact resonance ($\delta = 0$) and is equal to $\sqrt{3}/2$. This is the value of the growth rate normalized to C. The maximum nonnormalized wave increment is equal to $(\sqrt{3}/2)(\omega/v_{z0})C$, where C, as follows from (6.24), is proportional to $I_b^{1/3}$.

When the normalized cyclotron resonance detuning δ is larger than $(3/2)(2)^{1/3}$, there are no growing waves: all three roots of the dispersion equation are real. Nevertheless, there is also a possibility of the wave amplification due to the interference of three constant amplitude waves propagating with slightly different axial wavenumbers (a so-called crestatron regime; see Rowe 1959).

As an example, it makes sense to determine the Pierce gain parameter for the case of TE-waves in more detail. Let us start from the expressions for the resonant components of the RF Lorentz force given by (6.20). [We no longer need the radial component of this force because it is absent in (6.24).] The azimuthal component of the electric field and the radial component of the magnetic field, which are now of interest, are determined in Appendix 2 by Eq. (A2.19). Expanding the membrane function in the Fourier series of the periodic argument $\xi = \theta$, keeping the only resonant terms in this expansion, and using for the resonant component of the membrane function the formula (A1.9) derived in Appendix 1 yield

$$G_{\theta,s} = -\frac{1}{\kappa}(1 - n\beta_{z0})J_s'(k_\perp a)L_s, \tag{6.28}$$

$$G_{z,s} = -\beta_{\perp 0}\frac{n}{\kappa}J_s'(k_\perp a)L_s.$$

Here n and κ are, respectively, the axial and transverse wavenumbers normalized to ω/c. Substituting (6.28) into the RHS of (6.24) yields the following expression for the Pierce gain parameter:

$$C^3 = \frac{eI_b}{mc^3}\frac{\beta_{\perp 0}^2}{\gamma_0}\frac{c^3|L_s|^2}{\omega^2 N_s\kappa^2}(1 - n^2)\left[J_s'(k_\perp a)\right]^2. \tag{6.29}$$

Here $(1 - n^2)$ term describes the resulting effect of the orbital and axial bunching mechanisms (or, in other words, the cyclotron maser and Weibel instabilities), which we discussed in Sec. 6.1. The expression for the operator L_s is derived in Appendix 1 for an arbitrary geometry of the interaction space. The wave norm in a cylindrical waveguide is determined by (A2.22) in Appendix 2. When there is a thin annular electron beam in such a waveguide, the squared operator L_s is equal to $|L_s|^2 = J_{m\mp s}^2(k_\perp R_0)$, and the resulting expression for the Pierce gain parameter in this specific case is

$$C^3 = \frac{eI_b}{mc^3}\frac{2\beta_{\perp 0}^2}{\gamma_0}\frac{\kappa^2}{n}\frac{J_{m\mp s}^2(k_\perp R_0)}{(\nu^2 - m^2)J_m^2(\nu)}(1 - n^2)\left[J_s'(k_\perp a)\right]^2. \tag{6.30}$$

When the electron Larmor radius is small in comparison with the transverse wavelength, the Bessel function in the squared brackets in (6.30) can be expanded as a polynomial, which with the use of the cyclotron resonance condition yields

$$[J_s'(k_\perp a)]^2 \simeq \left[\frac{1}{2^s(s-1)!}\right]^2 \left[\frac{\kappa s \beta_{\perp 0}}{1-n\beta_{z0}}\right]^{2(s-1)}. \tag{6.31}$$

6.3 Small-Signal Gain in Single-Stage and Multistage Devices

Single-Stage Gyro-TWT

One of the most important characteristics of any amplifier is its gain, which, being expressed in decibels, is equal to

$$G(dB) = 10\log(P_{out}/P_{in}). \tag{6.32}$$

In the framework of the small-signal theory, the output power, P_{out}, is a linear function of the input power, P_{in}. Therefore, the small-signal gain does not depend on the input power level.

To calculate the small-signal gain, we should start from consideration of an amplified EM wave as a superposition of three partial waves whose propagation constants are given by the roots of a corresponding dispersion equation. So, denoting the complex amplitude of this wave by F, one can write

$$F(\varsigma) = \sum_{l=1}^{3} C_l e^{i\gamma_l \varsigma}. \tag{6.33}$$

Here the normalized axial coordinate ς is equal to $C(\omega z/v_{z0})$, γ_e is determined by (6.27), and therefore, $\gamma_l \varsigma = (\omega - k_{z,l}v_{z0} - s\Omega_0)\tau$.

At the entrance, the initial amplitude F_0 is equally distributed among three waves, thus $C_l = F_0/3$ and

$$F(\varsigma) = \frac{F_0}{3}\sum_{l=1}^{3} e^{i\gamma_l \varsigma}. \tag{6.34}$$

Then, only one of these waves is growing, while another decays and the third one has constant amplitude. When the interaction length is long enough, only the amplified wave is important; thus, the input power deposited into two other waves is lost. This effect is known as the *insertion losses*.

Since the wave power is proportional to the wave intensity, $|F|^2$, we can rewrite (6.32), taking into account (6.34), as

$$G = 10\log\left\{\frac{1}{9}\left|\sum_{l=1}^{3} e^{i\gamma_l \varsigma_{out}}\right|^2\right\} = 10\log\left\{\left|\sum_{l=1}^{3} e^{i\gamma_l \varsigma_{out}}\right|^2\right\} - 9.54\,\text{dB}. \tag{6.35}$$

Here ς_{out} is the normalized interaction length and 9.54 dB is the known value of the insertion losses (Pierce 1950). In the case of a long interaction length, when only the growing wave can be considered, this equation for the gain can be simplified further:

$$G = 8.686(\mathrm{Im}\gamma)\varsigma_{out} - 9.54\,\mathrm{dB}. \tag{6.36}$$

This expression agrees with one known in the small-signal theory of the linear-beam TWTs (Eq. (2.39) in Pierce 1959), when the wave growth rate is maximum ($\mathrm{Im}\,\gamma = \sqrt{3}/2$ in the case of exact synchronism), and the interaction length is expressed in terms of the axial wavelength, N:$\varsigma_{out} = 2\pi C N$. (In our notations, $N = L/\beta_{z0}\lambda$.) Then, $8.686(\sqrt{3}/2)2\pi = 47.263$, which is the number given by Pierce, and $G = 47.263\,CN - 9.54\,\mathrm{dB}$.

Multistage Gyro-TWT

As follows from Eq. (6.36), the gain of the device can be increased proportionally to the interaction length. In reality, however, this cannot be done in such a straightforward way because long waveguides filled with electron beams are prone to parasitic self-excitation. The issue of the stability of the gyro-TWT operation will be discussed in detail in the last section of this chapter. Here, I would like to note only that there are many ways in which this operation can be stabilized. One of them is based on the use of lossy materials in a part of the waveguide. Clearly, the presence of losses reduces the gain per unit length in a lossy part of the waveguide; however, what is much more important, is that it prevents the excitation of parasitic oscillations. This method, which is well known in linear-beam TWTs, was successfully used recently in the gyro-TWT experiments carried out at the National Tsing Hua University, Taiwan (Chu et al. 1999) that resulted in achieving more than 70 dB gain. Another method is the use of multistage devices, in which each of the interaction stages is short enough for providing a stable operation. A schematic of such a two-stage device is shown in Fig. 6.3. As one can see, the input and output waveguides are separated by a drift section, which is similar to that used in gyroklystrons. The purpose of the drift section is to isolate the interaction stages from one

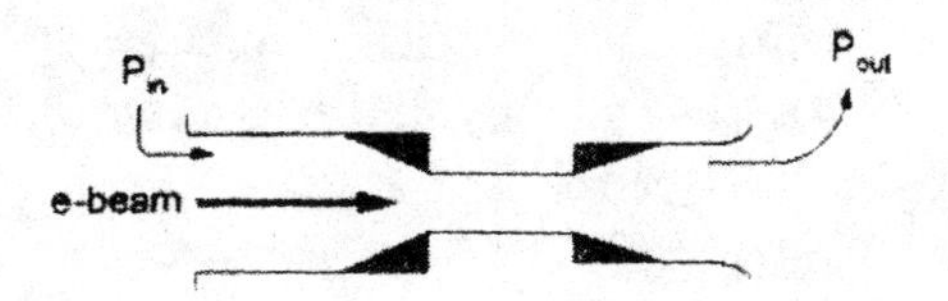

Fig. 6.3. Schematic of a two-stage gyro-TWT with attenuators at the end of the input stage and at the entrance to the output stage.

another. An additional benefit from the use of multistage devices is their ability to operate in frequency-multiplying regimes, which will be discussed later.

A small-signal theory of multistage gyro-TWTs has been developed by Nusinovich and Walter (1999). Therefore, my description of the gain in such devices will be based on those results. Omitting the details of derivation, let us note that this treatment is based on the use of the dispersion equation (6.27), in which the detuning δ and boundary conditions can vary from stage to stage. At the entrance to the input waveguide, the wave amplitude has an initial nonzero value, while the electron energy and phase are nonmodulated. At the exit from this stage, an amplified wave is absorbed by an attenuator and the electron energy and phase are modulated. Then, in the drift section, the electron energy modulation remains unchanged, but the electron phase bunching caused by this modulation evolves. As a result, a bunched electron beam enters the output stage and excites an EM wave. So, the boundary conditions at the entrance to the output waveguide consist of equations defining the energy modulation, the phase bunching and zero amplitude of the wave. (We assume that either this waveguide is well matched at the exit and, hence, there is no reflected wave, or that this reflected wave is absorbed by the attenuator located near the entrance.) The gain is determined by a simple formula for the output waveguide,

$$G = 10 \log \left\{ \left| \sum_{l=1}^{3} C_l^{(2)} e^{i\gamma_l^{(2)} \left(\varsigma_{\text{out}}^{(2)} - \varsigma_{\text{in}}^{(2)} \right)} \right|^2 \right\}, \tag{6.37}$$

which is similar to (6.35). In (6.37) the index (2) denotes the second waveguide. Parameters of the input waveguide and the drift section are present here in the constants $C_l^{(2)}$, which are determined by the boundary conditions (for details, see Nusinovich and Walter 1999).

Some results illustrating the performance of such a device are shown in Fig. 6.4. The figure shows the dependence of the gain on the frequency detuning for the case when the normalized lengths of the first and second waveguides are equal to 3 ($\varsigma_{\text{out}}^{(2)} - \varsigma_{\text{in}}^{(2)} = \varsigma_{\text{out}}^{(1)} - \varsigma_{\text{in}}^{(1)} = 3$) and the normalized length of the drift section is equal to 2.5. The detuning δ shown along the horizontal axis in Fig. 6.4 is the detuning in the first waveguide. The detuning in the second waveguide is represented as $\delta_2 = \delta + \Delta_d$, where d stands for detuning, so this Δ_d describes the difference in parameters of two stages, which can be due to the difference in the waveguide radii or in the external magnetic field values for two stages. As is seen in Fig. 6.4, as Δ_d increases, the gain becomes smaller. However, at the same time, the bandwidth can be increased, first, when the gain curve becomes flat, as shown for $\Delta_d = 2.75$,

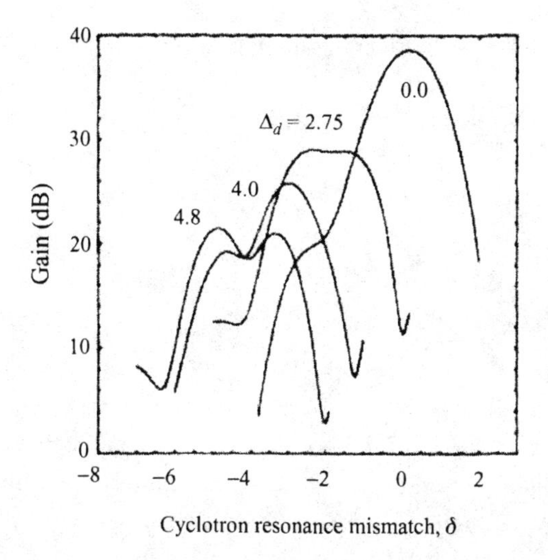

Fig. 6.4. Gain as the function of the cyclotron resonance mismatch in the input waveguide for several values of the detuning parameter Δ_d, characterizing the difference between such mismatches in the input and output waveguides.

and second, when the gain curve exhibits two peaks of approximately the same amplitude with a relatively small valley between them, as shown for $\Delta_d = 4.8$. Note that the change in sign of Δ_d does not affect the results, since in the frame of the linear theory both stages play a similar role in the amplification process. Therefore, the change in sign of Δ_d from plus to minus causes only a shift of the curves to the region of larger detunings δ. (Also note that A. Lin et al. 1992 have analyzed not only two-stage, but also three-stage designs of Ka-band gyro-TWTs.)

So, by making parameters of the input and output stages different, one can increase the bandwidth at the expense of the gain. This effect is quite similar to the effect of the cavity stagger-tuning, which we will analyze later in chapter 8 on gyroklystrons. As is known, such a simultaneous reduction of the gain and the bandwidth enlargement often allows developers to substantially increase the gain-bandwidth product, which is considered a figure-of-merit in various applications.

6.4 Bandwidth

The bandwidth is usually determined as a range of frequencies in which the gain remains relatively constant. Very often, by the bandwidth the developers mean the range of frequencies in which the gain variations are within 3 dB. Sometimes, instead of 3 dB, 1 dB or even 0.5 dB definitions are used.

The general theory, which was outlined in the previous section, allows one to readily determine the bandwidth in terms of the normalized detuning δ. For instance, to evaluate the bandwidth in a single-stage gyro-TWT, one can use Eq. (6.36) and Fig. 6.2. From (6.36) it follows that the gain deviation, ΔG, corresponds to the deviation of the wave increment

$$\Delta(\mathrm{Im}\,\gamma) = \frac{\Delta G}{8.686\varsigma_{\mathrm{out}}}. \tag{6.38}$$

Then, by using Fig. 6.2 one can readily find the range of detunings δ, in which $\Delta(\mathrm{Im}\,\gamma)$ varies within these limits. In a similar fashion, the bandwidth of a two-stage gyro-TWT can be evaluated by using Fig. 6.4 or similar results obtained for different values of the normalized lengths of device sections.

The second step in calculating the bandwidth is the transition from these normalized parameters to real frequencies. Our normalized detuning δ is equal to $(\omega - k_{z,s}v_{z0} - s\Omega_0)/\omega C$, where not only the axial wavenumber $k_{z,s} = (\omega^2 - \omega_{\mathrm{cut}}^2)^{1/2}/c$, but also Pierce gain parameter [see (6.30)] is frequency dependent. The most sensitive to the frequency deviation, $\delta\omega$, of course, is a small cyclotron resonance mismatch $\Delta\omega_s = \omega - k_{z,s}v_{z0} - s\Omega_0$. In the first-order approximation, the deviation of this mismatch is equal to

$$\delta(\Delta\omega_s) = \delta\omega - v_{z0}\frac{dk_{z,s}}{d\omega}\delta\omega = \delta\omega(1 - v_{z0}/v_{gr}). \tag{6.39}$$

Here $v_{gr} = c^2/v_{ph} = c^2k_{z,s}/\omega$ is the group velocity introduced above in chapter 2.

Eq. (6.39) clearly shows that the largest bandwidth can be realized when the group velocity of the wave is equal to the axial velocity of electrons, $v_{gr} = v_{z0}$. This condition corresponds to the grazing of the cyclotron mode to the wave dispersion curve, which is shown in Fig. 6.5. So, if for the nominal operating frequency ω_0 the grazing condition is fulfilled, in the first-order approximation the frequency deviation does not affect the growth rate and, correspondingly, the gain at all. The gain deviation occurs here only when higher-order terms [$(\delta\omega)^2$, etc.] are taken into account, which means that the bandwidth is determined by the dispersion of the microwave circuit or, in other words, by dv_{gr}/dk_z.

The cyclotron resonance mismatch in the case of $v_{gr} = v_{z0}$ is equal to $\Delta\omega_s = \omega(1 - \beta_{z0}^2) - s\Omega_0$. Strictly speaking, an exact grazing occurs when not only $v_{gr} = v_{z0}$, but also the cyclotron resonance mismatch is equal to zero, i.e., in the case of the exact cyclotron resonance. This is the case when a large bandwidth is combined with a large linear increment of the wave, i.e., with the high small-signal gain. Of course, the cyclotron resonance mismatch depends

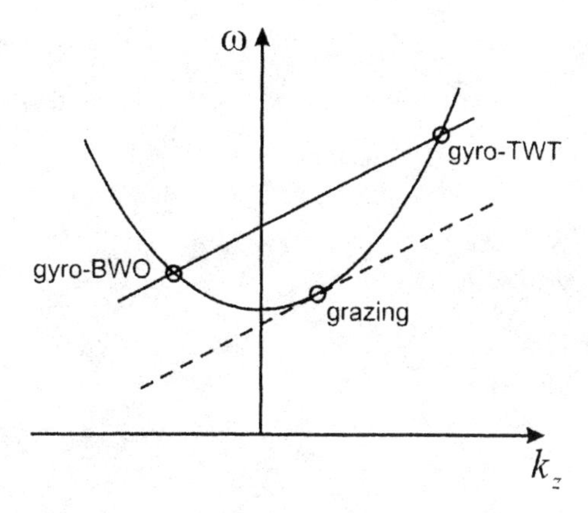

Fig. 6.5. Dispersion diagram showing the case of cyclotron wave grazing (dashed line) and a general case (solid line) when both the gyro-TWT forward-wave interaction and gyro-BWO backward-wave interaction may take place.

on the frequency, so it can be equal to zero only at one specific point, which can be, for instance, the nominal operating frequency.

So far, we have discussed the operation of gyro-TWTs in which an external magnetic field and a waveguide radius are constant in the interaction region or in each stage of it. Of course, it was well understood from the very beginning (Bratman et al. 1973) that a consistent tapering of the magnetic field and waveguide radius (or cross section, in a general case) may lead to the substantial bandwidth enlargement. For instance, Moiseev (1977) analyzed a two-stage tapered gyro-TWT; he also mentioned that the numerical optimization of the waveguide tapering done by Kuraev et al. (1975) showed a possibility to enlarge the bandwidth up to 14%. The concept of such a two-stage tapered gyro-TWT was later successfully realized at NRL by G-S. Park et al. (1995).

In spite of this past work, it is necessary to say that Y. Y. Lau and K. R. Chu (1981) were the first who published a paper in which the concept of the tapered gyro-TWT was described clearly and extensively. To illustrate this concept, let us come back to Fig. 6.5 and assume that the waveguide radius and the external magnetic field vary along the axis in such a way that at any cross section the grazing condition is fulfilled. Clearly, if we consider an up-tapered waveguide, in which the cutoff frequency decreases and, correspondingly, the axial wavenumber increases, then the magnetic field should be down-tapered in order to match the waveguide tapering and provide the cyclotron resonance condition. So, in such a system, the waves of different frequencies will resonantly interact with electrons in different parts of the waveguide and for any

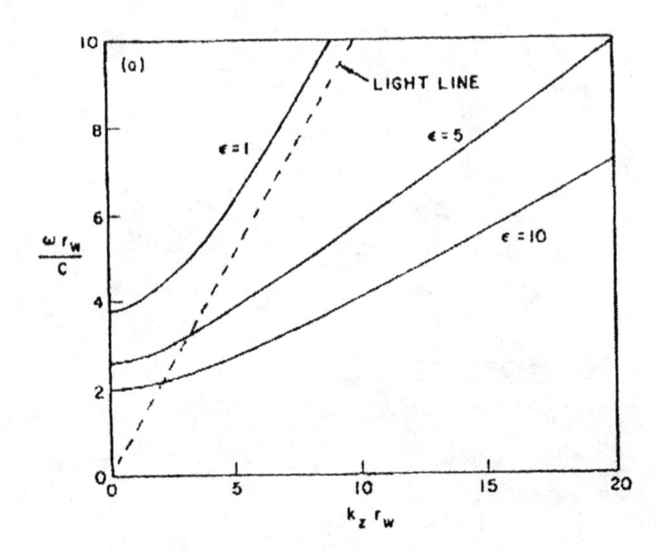

Fig. 6.6. Dispersion diagram for a dielectrically loaded waveguide. Dispersion curves are shown for several values of the dielectric constant ε. (Reproduced from Chu et al. 1981)

frequency the operation may take place close to the grazing condition. Later, Barnett et al. (1981) successfully realized such a gyro-TWT in the experiment, where an operation in a 13% bandwidth has been demonstrated.

There is also another possibility to enlarge the bandwidth of gyro-TWTs. This possibility is based on the use of dielectrically loaded waveguides, since the dielectric loading lowers the group velocity of the propagating EM waves and hence allows one to realize the grazing conditions in a larger bandwidth. Figure 6.6 shows an example illustrating the effect of a dielectric on the dispersion characteristics of the $TE_{0,1}$-wave in the cylindrical, dielectric loaded waveguide. This figure is reproduced from the paper by Chu et al. (1981b), where the theory of such slow-wave gyro-TWTs is developed. Note that, in principle, such a device can operate in the fast wave regime, when the cyclotron wave line is close to the waveguide mode in that part of the dispersion curve, which is located above the light line. A larger bandwidth, however, can be realized in the slow wave regime, as is obvious from Fig. 6.6 (see also Fig. 3 in Chu et al. 1981b).

6.5 Stability

Stability is a key issue for the design and operation of any amplifier. There are several possible reasons for making the gyro-TWT operation unstable. Below we shall consider most of them in a successive manner.

Backward-Wave Excitation

Since a beam of gyrating electrons can interact with EM waves not only at the fundamental cyclotron resonance, but also at cyclotron harmonics, it is impossible to completely eliminate the beam interaction with backward waves. This statement is illustrated by Fig. 6.7. Here the dispersion diagram is shown for a tube designed for operation at the TE_{21}-wave at the second harmonic, which can also operate at the lowest-order TE_{11}-wave at the fundamental resonance. As one can see, there are numerous intersections of the harmonic cyclotron wave lines (cases of the second and third harmonic interaction are shown in the figure) with dispersion curves in the region of negative axial wavenumbers. Of course, when a device is designed for operation at one of higher-order waves, there can be intersections of the fundamental cyclotron beam line with dispersion curves of lower-order waves as well. These intersections show that the resonant interaction of electrons moving forward with EM perturbations of various waves propagating backward is possible.

So, while the *interaction* with backward waves is unavoidable, it is necessary to figure out whether we can avoid the *excitation* of these waves. Self-excitation conditions for backward-wave oscillations in gyro-TWTs can easily be evaluated based on two arguments. First, we have already shown in Sec. 6.2 that the dispersion equation for the gyro-TWT can be reduced to that for the conventional linear-beam TWT. Second, it is well known (see, e.g., Johnson 1955), how to use this dispersion equation for analyzing the self-excitation conditions in backward-wave oscillators (BWOs). To do this, we should first

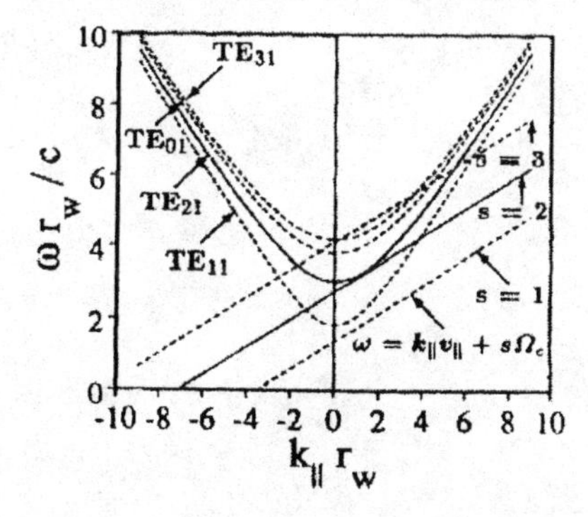

Fig. 6.7. Dispersion diagram for several modes showing possible excitation of backward waves at cyclotron harmonics in a gyro-TWT designed for operation at the TE_{21}-wave at the second cyclotron harmonic. Although no waves are resonant at the fundamental cyclotron harmonic, there are possible backward-wave oscillating modes at the second and third cyclotron harmonics; the most dangerous among them is the second harmonic resonance with the TE_{11}-wave. (Reproduced from Wang et al. 1995)

change the sign of the last term in the LHS of (6.27) from plus to minus, because the backward wave propagates in the opposite direction, and hence the wave norm [see RHS of (6.24)] derived in Appendix 2 has an opposite sign. Then we should add proper boundary conditions to the dispersion equation (6.27). For the case of a nonmodulated electron beam entering the waveguide, these are the absence of modulation in electron energies and uniform distribution in gyrophases at the entrance. Also, for the case of a waveguide with a well-matched output, the boundary condition for the backward wave is its zero amplitude at the exit. The latter yields the transcendental equation, which, in accordance with (6.33), can be given as

$$\sum_{l=1}^{3} C_l e^{i\gamma_l \varsigma_{\mathrm{out}}} = 0.$$

Simultaneous solution of these equations yields the eigenvalues of ς_{out} and δ. This value of the normalized length should be treated as the starting length, and this value of the detuning δ determines the oscillation frequency. In our simple case (no space charge and no attenuation in waveguide walls), as follows from the results by Johnson (1955), these values are equal to 1.97355 and 1.522, respectively. This normalized length, $\varsigma_{\mathrm{out}} = C(\omega/v_{z0})L$, is proportional to the Pierce gain parameter, which, in turn, is proportional to $I_b^{1/3}$. So, in other words, the starting current for backward-wave oscillations is inversely proportional to the third power of the interaction length: $I_{st} \propto 1/L^3$.

When the beam current is determined by parameters of an electron gun, one should choose the interaction length taking into account that for any parasitic backward wave a real critical length is inversely proportional to the Pierce gain parameter of this specific wave. Then, the waveguide length should be chosen as one that is smaller than critical lengths of all parasitic backward waves.

Reflections

Above, we considered excitation of parasitic backward waves in a waveguide with no reflections at the output. Real waveguides that are in use in gyro-TWTs typically have a cutoff narrowing at the input, which protects the gun region from microwaves, and the output up-taper, through which an amplified output signal propagates. It is very difficult, especially in large-bandwidth gyro-TWTs, to avoid reflections of the outgoing radiation from this taper back to the interaction region. These reflections can form a feedback loop, and hence lead to the self-excitation of the operating wave.

This case is illustrated by Fig. 6.8, which shows the operating wave propagating forward and the reflected backward wave in the dispersion diagram.

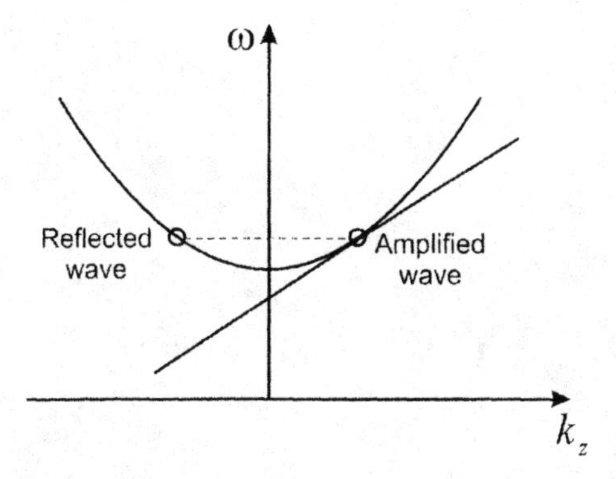

Fig. 6.8. Dispersion diagram showing the operating forward wave and the reflected backward wave that may cause a feedback loop leading to instability.

When the frequency of this wave is far enough from the cutoff, the reflected wave does not interact synchronously with electrons. Correspondingly, the amplitude of this wave, A_r, which is determined by the reflection coefficient at the output, R, $A_r = RA_f(L)$ (here $A_f(L)$ is the amplitude of the forward wave at the exit) remains constant. Let us now assume that at the exit the growing forward wave dominates over two other partial forward waves, thus the wave amplitude at the exit can be determined as $A_f(L) = (A_0/3)\exp\{|\mathrm{Im}\,k_z|\,L\}$. Also we can assume that at the entrance the backward wave is completely reflected, i.e. $A_0 = A_r$. (Here, for the sake of simplicity, we ignore phase relations between the forward and backward waves at both ends.) Correspondingly, the condition for self-excitation, which implies that a wave, after passing via a feedback loop, reproduces itself, can be written as

$$A_0 = A_r = \frac{RA_0}{3}e^{|\mathrm{Im}k_z|L},$$

or

$$\mathrm{Im}\,|k_z|\,L = \ln(3/R). \tag{6.40}$$

Certainly, this condition is frequency dependent, since both the reflection coefficient and the wave growth rate depend on frequency. So, in all of the operating bandwidth, the reflection coefficient R should be smaller than its critical value given by (6.40).

Gyrotron-Type Interaction

We have already mentioned in Sec. 6.1 that each cyclotron wave, once a finite beam current is taken into account, splits into two waves known as the positive and negative energy waves. A negative energy wave has a slightly

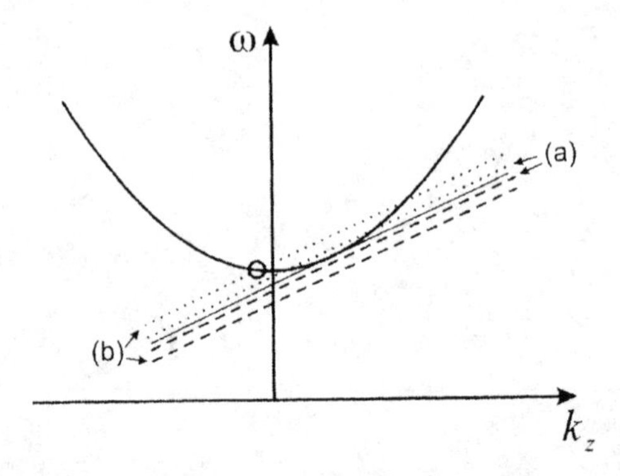

Fig. 6.9. Dispersion diagram showing that a large splitting of the degenerate cyclotron wave due to the finite beam plasma frequency may cause self-excitation of backward waves.

higher frequency than a positive one. Thus, when the operating frequency does not greatly exceed the cutoff, which can be, for instance, the case of grazing at low voltages, the negative energy cyclotron wave can intersect the waveguide dispersion curve at $k_z \leq 0$, as shown in Fig. 6.9, where the cases (a) and (b) correspond to small and large currents, respectively. Certainly, this intersection can give rise to the absolute instability of a device.

Since the excitation of waves at frequencies near cutoff requires the lowest current, special care must be taken to avoid this sort of excitation. Note that the wave growth rate and self-excitation conditions for the case of operation near cutoff were studied elsewhere (Bratman et al. 1973, Sato et al. 1986, Nusinovich and Walter 1999).

Cross-Talks in Multistage Gyro-TWTs

In multistage gyro-TWTs, of course, each stage must be stable with respect to self-excitation of parasitic oscillations in it. However, when these stages are not isolated well enough, EM waves can penetrate from one stage into another and can be a reason for a global parasitic excitation of a whole microwave circuit or two or more stages in it. In a certain sense, gyro-TWTs are more vulnerable in this regard than linear-beam TWTs, because an electron beam diameter in gyro-TWTs is significantly larger than in their linear-beam counterparts operating in the same frequency range. (This is the price to be paid for the superior power handling capabilities.) Therefore, the drift sections must be carefully designed for avoiding the propagation of any wave and loaded with lossy materials in order to isolate the device stages well enough.

▪ CHAPTER 7 ▪

Nonlinear Theory of the Gyro-TWT

7.1 Self-Consistent Set of Equations. Energy Conservation Law. Low-Current Limit

Self-Consistent Set of Equations

Operation of the gyro-TWT can be described by a self-consistent set of equations. This set consists of equations for electron motion and an equation describing the wave excitation by an electron beam. The equation for the wave excitation is derived in Appendix 2 and reproduced in Sec. 6.2 as Eq. (6.11). Gyro-averaged equations for electron motion in resonators are derived in Appendix 1. Now these equations should be modified for the case of electron motion in waveguides. Also, the equation for the wave excitation should be rewritten in new variables describing an electron beam current density.

Let us start from recalling that in our consideration of the electron motion in resonators we ignored the effect of the resonator field on the axial momentum of electrons. This was done because we considered the interaction with TE-waves excited near cutoff frequencies, in which case the axial momentum of electrons remains practically constant. (As was shown in Problem 2 of Chapter 3, this statement is also valid for TM-modes excited near cutoff.) However, in the case of interaction with traveling waves having an arbitrary phase velocity, the changes in the axial momentum must be taken into account.

Below we shall again restrict our consideration by TE-waves only. In this case, the axial component of the Lorentz force in the equation for electron axial momentum contains only contributions from the transverse magnetic field or, more exactly, from its radial component, because the gyrating electrons have azimuthal velocity in their transverse motion:

$$\frac{dp_z}{dt} = e\beta_\perp H_r. \tag{7.1}$$

The equation for electron energy [(1.2) in Chapter 1] can also be rewritten as

$$\frac{d\mathcal{E}}{dt} = -ev_{\perp}E_{\varphi}. \tag{7.2}$$

As follows from Eqs. (A2.19), once we neglect the difference between the amplitudes of electric and magnetic fields, the radial magnetic field and the azimuthal electric field of the wave are related as $H_r = -(k_z/k)E_{\varphi}$. Then, Eqs. (7.1) and (7.2) readily yield the autoresonance integral

$$\frac{d}{dt}\mathcal{E} - v_{\mathrm{ph}}p_z = \mathrm{const.} \tag{7.3}$$

Eq. (7.3) coincides with Eq. (1.17) derived in Chapter 1 from simple quantum-mechanical consideration. This makes the use of an equation for the axial momentum of electrons unnecessary, once the changes in electron energy are known. These changes, in turn, can be expressed in terms of changes in the electron orbital momentum, because the validity of the autoresonance integral (7.3) allows us to use the relation between the electron energy and orbital momentum given by Eq. (1.20). The latter equation can be rewritten as

$$p_{\perp}^2 = p_{\perp 0}^2 - 2\gamma_0(\gamma_0 - \gamma)(1 - n\beta_{z0})\left[1 - \frac{1-n^2}{2(1-n\beta_{z0})}\frac{\gamma_0 - \gamma}{\gamma_0}\right]. \tag{7.4}$$

Here we use notations adopted in Chapter 1; now the refractive index is $n = 1/\beta_{\mathrm{ph}} = k_z/(\omega/c)$. As was shown in Sec. 1.6, the efficient interaction between electrons and the wave during many cyclotron orbits is possible only when the refractive index obeys limitations given by Eq. (1.25). (We do not discuss here some possibilities associated with tapering of the parameters of the interaction space, which were discussed in Sec. 1.6.) Eq. (1.25) indicates, however, that the last term in the RHS of (7.4) can be neglected (Ginzburg, Zarnitsyna, and Nusinovich 1981). Correspondingly, the orbital momentum can be described by the variable

$$w = p_{\perp}^2/p_{\perp 0}^2 = 1 - \left[2(1 - n\beta_{z0})/\beta_{\perp 0}^2\right][(\gamma_0 - \gamma)/\gamma_0]. \tag{7.5}$$

As in Chapter 3, this variable determines the ratio of the squared orbital momentum to its initial value. However, being expressed via the changes in electron energy, it differs from (3.33) by the factor $(1 - n\beta_{z0})$, which now takes into account possible changes in the electron axial velocity. Using this variable in (7.3), one can represent the axial momentum of an electron as

$$p_z = p_{z0}[1 - b(1 - w)]. \tag{7.6}$$

Here parameter b, which is often called a recoil parameter, since it describes the changes in electron axial momentum in the process of radiation, is equal to

$n\beta_{\perp 0}^2/2\beta_{z0}(1 - n\beta_{z0})$. Eq. (7.6) is essentially the same as Eq. (3.10), rewritten in new variables. Note that above we assumed that the amplitudes of the electric and magnetic fields of the wave are the same. Strictly speaking, for the waves, whose amplitude varies along the axis, this is not correct. When the difference between the amplitudes of $\vec{E}$ and $\vec{H}$ is taken into account, the axial momentum (as shown in Problem 2) can be represented as the term given by (7.6) and an additional term proportional to dC/dz (here $C(z)$ is the wave amplitude). This additional term may cause some changes in the phase relations between the wave and electrons. Corresponding equations were derived and analyzed by Ginzburg, Zarnitsyna, and Nusinovich (1981) and Fliflet (1986).

As follows from the definition of the variable w given by (7.5), the electronic efficiency can be expressed in terms of the averaged changes in w as

$$\eta = \left\langle \frac{\gamma_0 - \gamma(L)}{\gamma_0 - 1} \right\rangle = \frac{\beta_{\perp 0}^2}{2\left(1 - \gamma_0^{-1}\right)(1 - n\beta_{z0})} \eta_{\perp}, \tag{7.7}$$

where

$$\eta_{\perp} = \langle 1 - w(L) \rangle. \tag{7.8}$$

Here angular brackets denote the averaging over electron entrance phases. The value $\eta_{\perp}$ is known as the orbital efficiency not only in the theory of gyrotron oscillators presented above, but also in the theory of gyro-TWTs, because, as it follows from (7.5) it describes the average changes in the squared orbital momentum of electrons. Note that the electronic efficiency given by (7.7) differs from its definition for gyrotron oscillators given by (3.37) only by the term $(1 - n\beta_{z0})$ in the denominator. This term, which characterizes the amount of the beam energy withdrawn from electron axial motion, indicates that the efficiency of the gyro-TWT can be higher than the efficiency of the gyrotron oscillator, when both devices utilize the beams with the same parameters (beam voltage and the orbital-to-axial velocity ratio).

As is shown in Appendix 3, with the use of (7.5) and (7.6), a self-consistent set of equations for the gyro-TWT can be reduced to the following equations:

$$\frac{dw}{d\varsigma} = 2\frac{w^{s/2}}{1 - b(1 - w)} \operatorname{Re}(F e^{-i\vartheta}), \tag{7.9}$$

$$\frac{d\vartheta}{d\varsigma} = \frac{1}{1 - b(1 - w)} \{1 - w + s w^{(s/2)-1} \operatorname{Im}(F e^{-i\vartheta})\}, \tag{7.10}$$

$$\frac{dF}{d\varsigma} - i\Delta F = -I_0 < \frac{w^{s/2}}{1 - b(1 - w)} e^{i\vartheta} >, \tag{7.11}$$

with the boundary conditions at the entrance: $w(0) = 1$, $F(0) = F_0$, and $\vartheta(0) = \vartheta_0$, where the entrance phase ϑ_0 is uniformly distributed from 0 to 2π. Here we introduced the normalized axial coordinate

$$\varsigma = \frac{\beta_{\perp 0}^2}{2\beta_{z0}} \frac{1-n^2}{1-n\beta_{z0}} kz, \tag{7.12}$$

the normalized cyclotron resonance detuning

$$\Delta = \frac{2}{\beta_{\perp 0}^2} \frac{1-n\beta_{z0}}{1-n^2} \left(1 - n\beta_{z0} - s\frac{\Omega_0}{\omega}\right), \tag{7.13}$$

the normalized wave amplitude

$$F = \frac{eC}{mc\omega} J_{m\mp s}(k_\perp R_0) e^{i\Delta\varsigma} \frac{(1-n\beta_{z0})^{3-s}}{\gamma_0 \beta_{\perp 0}^{4-s} \kappa^{4-s}} \left(\frac{s}{2}\right)^{s-1} \frac{1}{(s-1)!}, \tag{7.14}$$

and the normalized beam current parameter

$$I_0 = 16 \frac{eI_b}{mc^3} G \frac{\kappa^{2(s-2)}}{n} \frac{(1-n\beta_{z0})^{5-2s}}{\gamma_0 \beta_{\perp 0}^{2(3-s)}} \left[\frac{s^{s-1}}{(s-1)!2^s}\right]^2. \tag{7.15}$$

Eqs. (7.9) and (7.10) can also be represented in the quasi-Hamiltonian form

$$\frac{dw}{d\varsigma} = \frac{1}{1-b(1-w)} \frac{\partial H}{\partial \vartheta}, \qquad \frac{d\vartheta}{d\varsigma} = -\frac{1}{1-b(1-w)} \frac{\partial H}{\partial w},$$

with the Hamiltonian $H = \frac{(1-w)^2}{2} - 2w^{s/2}\mathrm{Im}(Fe^{-i\vartheta})$.

The set of equations (7.9)–(7.15) is given for the simplest case of the gyro-TWT with a cylindrical waveguide and a thin annular ideal electron beam with a negligibly small velocity spread. (General equations, which are valid for an arbitrary geometry of the waveguide and beam configuration, are given in Appendix 3.) Therefore, in the definition of the normalized wave amplitude (7.14) we used the Bessel function defining a coupling of a thin beam to a wave of a cylindrical waveguide. Correspondingly, in the definition of the normalized beam current parameter (7.15) we used the coupling impedance G determined by (3.59). This set of equations can also be used for analyzing the interaction of gyrating electrons with the oppositely propagating wave, which is the case of a gyro-backward-wave oscillator. To treat these devices it is necessary to change the sign of the normalized beam current in (7.11), because in the case of the wave propagation in the opposite direction the normalized axial wavenumber n in (7.15) is negative.

Linearization of Eqs. (7.9)–(7.11) yields the dispersion equation

$$\Gamma^2(\Gamma - \Delta) - I_0(s+b)\Gamma + I_0 = 0. \tag{7.16}$$

To derive this equation we first linearized (7.9)–(7.11) with respect to perturbations caused by the wave field (i.e., we assumed that $w = 1 + \tilde{w}$ and $\vartheta = \vartheta_0 + \tilde{\vartheta}$ where $\tilde{w}$ and $\tilde{\vartheta}$ are small perturbations). Then we introduced the perturbations in energy and phase averaged over the entrance phase, $\bar{w} = \langle \tilde{w} e^{i\vartheta_0} \rangle$ and $\bar{\vartheta} = \langle \tilde{\vartheta} e^{i\vartheta_0} \rangle$, and assumed that the wave amplitude and perturbations $\bar{w}$ and $\bar{\vartheta}$ propagate along the axis as $\exp(i\Gamma\varsigma)$. Clearly, (7.16) contains in its left-hand side not only the last term, which is responsible for the "O"-type electron bunching, but also the preceding term, which is linearly proportional to the propagation constant. The latter term describes the "M"-type bunching effects (cf. (6.22) and corresponding discussion in Chapter 6).

The wave attenuation in the waveguide wall can be taken into account in (7.11) by considering a complex detuning instead of the real one. The imaginary part of this detuning is associated with the imaginary part of the axial wave number in the cyclotron resonance mismatch, $\omega - k_z v_{z0} - s\Omega_0$. In the normalized cyclotron resonance mismatch determined by (7.13), it corresponds to representation of n in parenthesis as $n' + in''$, where $n'' = \mathrm{Im}(k_z)/(\omega/c)$. (Other n's in (7.13) originate from the real part of the axial wavenumber, thus only the real part of k_z should be taken into account there as well as in other places.)

Energy Conservation Law

The set of equations (7.9)–(7.11) has a number of integrals (Yulpatov 1967, Ginzburg, Zarnitsyna, and Nusinovich 1981). First, averaging Eq. (7.9) yields

$$\frac{d\langle w \rangle}{d\varsigma} = \left\{ F \left\langle \frac{w^{s/2}}{1 - b(1-w)} e^{-i\vartheta} \right\rangle + F^* \left\langle \frac{w^{s/2}}{1 - b(1-w)} e^{i\vartheta} \right\rangle \right\}. \qquad (7.17)$$

At the same time, Eq. (7.11) being multiplied by the complex conjugate amplitude F^* and combined with its complex conjugate equation yields

$$\frac{d|F|^2}{d\varsigma} + 2\Delta'' |F|^2 = -I_0 \left\{ F^* \left\langle \frac{w^{s/2}}{1 - b(1-w)} e^{i\vartheta} \right\rangle + F \left\langle \frac{w^{s/2}}{1 - b(1-w)} e^{-i\vartheta} \right\rangle \right\}. \qquad (7.18)$$

Here in the left-hand side we included into consideration the imaginary part of the cyclotron resonance detuning, which takes into account the wave attenuation in a waveguide with certain losses of the microwave power in the wall. So, as one can see, these two equations yield

$$\frac{d|F|^2}{d\varsigma} + 2\Delta'' |F|^2 = I_0 \frac{d\langle 1 - w \rangle}{d\varsigma}. \qquad (7.19)$$

This equation is equivalent to the energy conservation law for the beam-wave system. In the absence of attenuation in the wall, this equation can be readily integrated, yielding

$$|F|^2 - |F_0|^2 = I_0 \eta_\perp. \tag{7.20}$$

Here we took into account the definition of the orbital efficiency given by (7.8). Eqs. (7.9)–(7.11) have two more integrals (Yulpatov 1967, Zhurakhovskiy 1972, Ginzburg, Zarnitsyna, and Nusinovich 1981). These integrals, however, are not as useful as the energy conservation law. Therefore they will not be considered here.

Low Current Limit

As was discussed in Sec. 3.2, in gyrodevices with a small beam current the field amplitude can be rather small. Therefore, in order to improve the electron phase bunching in a wave of small amplitude, it is reasonable to increase the interaction length. In such a case, in the RHS of Eq. (7.10) the last term proportional to the wave amplitude can be ignored because the first term describing the energy modulation is proportional to the wave amplitude multiplied by a large interaction distance. The neglect of this term allows one to reduce (7.9) and (7.10) to one second-order equation

$$\frac{d^2\vartheta}{d\varsigma^2} = -2\frac{w^{s/2}}{[1 - b(1-w)]^3}\mathrm{Re}(Fe^{-i\vartheta}). \tag{7.21}$$

Since we consider now the case of a weak energy modulation, which at long distances leads to substantial phase bunching, we can assume that $|1 - w| \ll 1$, which reduces (7.21) to

$$\frac{d^2\vartheta}{d\varsigma^2} = -2\mathrm{Re}(Fe^{-i\vartheta}). \tag{7.22}$$

Correspondingly, the wave excitation equation (7.11) can be reduced to

$$\frac{dF}{d\varsigma} - i\Delta F = -I_0\langle e^{i\vartheta}\rangle. \tag{7.23}$$

Repeating the same linearization procedure that was used for deriving Eq. (7.16), one can readily derive from (7.22) and (7.23) the following dispersion equation:

$$\gamma^2(\gamma - \Delta) + I_0 = 0. \tag{7.24}$$

As one can see from comparing (7.24) with (7.16), Eq. (7.24) does not contain terms responsible for the "M"-type electron bunching. So this equation is essentially the same as (6.24) derived in Chapter 6. After normalizing the propagation constant γ and cyclotron resonance detuning Δ to $I_0^{1/3}$, this equation coincides with (6.27). So, $I_0^{1/3}$ plays here a role of Pierce gain parameter C, and Eq. (7.24) is essentially the linear-beam TWT dispersion equation. In the same fashion as in the theory of the linear-beam TWT, this parameter I_0 can be eliminated from Eqs. (7.22) and (7.23). For doing this, one should introduce a new set of normalized parameters and variables: $\bar{\varsigma} = I_0^{1/3}\varsigma$, $\bar{\Delta} = \Delta / I_0^{1/3}$ and $\bar{F} = F / I_0^{2/3}$. In these variables, Eqs. (7.22) and (7.23) have the same form as before, but now they do not contain the beam current parameter. Correspondingly, the orbital efficiency, in accordance with its definition and Eq. (7.9), in new variables is equal to

$$\eta_{\perp} = -2I_0^{1/3}\mathrm{Re}\left\{ \int_0^L \bar{F}\langle e^{-i\vartheta}\rangle d\bar{\varsigma} \right\},$$

i.e., it is proportional to the equivalent of the Pierce gain parameter in the same way as the efficiency of the linear-beam TWT. Also $\bar{\Delta}$ is equivalent to the Pierce velocity parameter b, as we discussed in Sec. 6.3, and the normalized distance $\bar{\varsigma}$ is equivalent to the number of axial wavelengths N multiplied by $2\pi I_0^{1/3}$.

7.2 Beam-Wave Interaction

As we discussed above several times, in spite of the difference in the physical mechanisms of coherent EM radiation from electrons in linear-beam TWTs and gyro-TWTs, the equations describing these two classes of microwave sources are quite similar. Therefore, the results of the analysis of these equations also have a certain similarity.

Typical processes of the electron energy modulation and phase bunching and the wave amplification are shown in Fig. 7.1. These results are obtained by integrating Eqs. (7.9)–(7.11) with the following set of parameters: $s = 1$, $b = 0.1$, $\Delta = 0.5$, $I_0 = 0.1$, and $F_0 = 0.01$. These plots show how the energy (Fig. 7.1a) and phase (7.1b) of electrons with different entrance phases evolve along the z-axis. Also, the process of the wave amplification is shown in Fig. 7.1c. Initially, the changes in the electron energy and phase are linear and the wave amplitude grows exponentially, which is the linear stage of interaction. Then, the saturation effects become pronounced, when the changes in the electron energy and phase are significant and the wave amplitude reaches its maximum. As follows from the comparison of Fig. 7.1b with Fig.7.1c, the maximum of the wave amplitude corresponds to the shift of an electron bunch in phase on the order of π. Since for realizing the wave amplification the

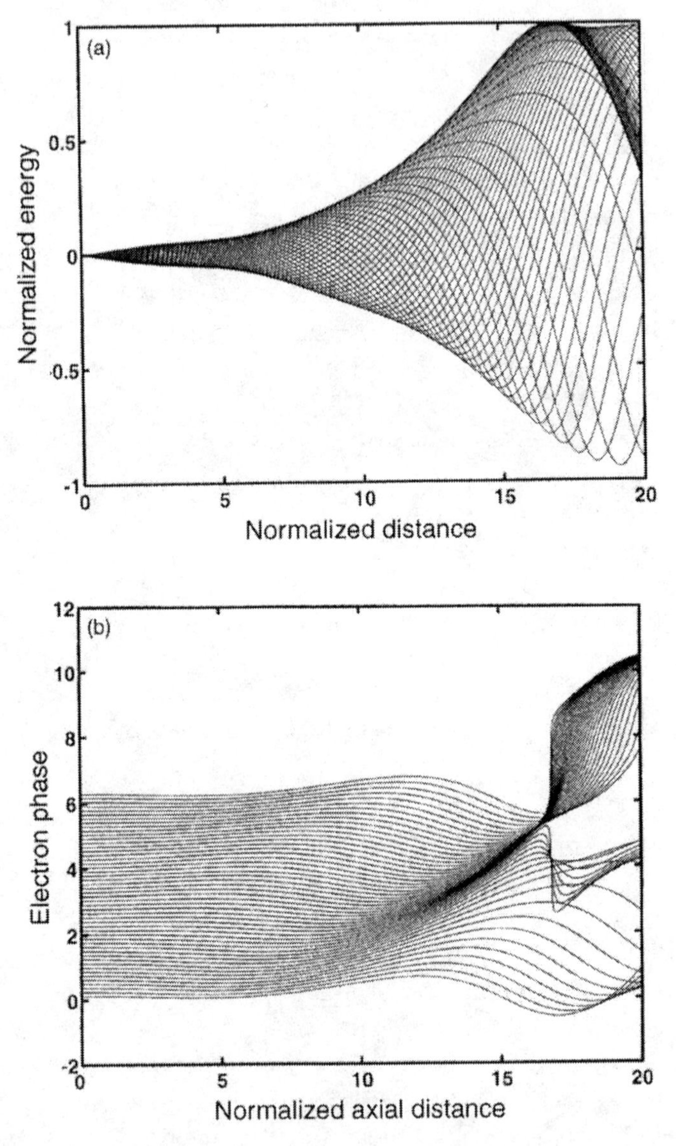

Fig. 7.1. Electron energies (a) and phases (b) and the wave amplitude and orbital efficiency—shown in (c) on facing page by the solid and dashed lines, respectively—as functions of the normalized axial coordinate.

bunch should be formed in the decelerating phase, this shift indicates that electrons move from the decelerating phase into the accelerating one. In the latter phase, the electrons start to withdraw the EM wave energy back, thus decreasing the wave amplitude. Such a behavior is quite typical for linear-beam TWTs (Nordsieck 1953, Weinstein 1957).

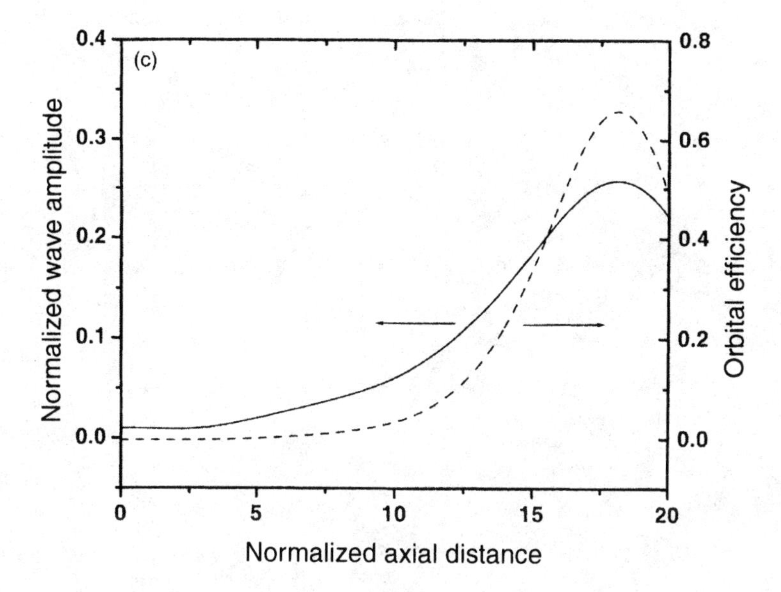

When the interaction space is long enough, this energy exchange between the wave and the beam continues more or less periodically, as is shown in Fig. 7.2. This process is associated with oscillations of a trapped bunch in the potential well formed by the wave field. Certainly, the bunch quality degrades in this process, therefore the amplitude and efficiency oscillations shown in Fig. 7.2 are not completely regular. The axial dependence of the wave amplitude shown in this figure is very similar to axial dependencies of the wave amplitude in the linear-beam TWT studied elsewhere (Weinstein 1957). Such processes are also known in various active systems, for instance, in the case of the beam interaction with plasma waves (see, e.g., Shapiro and Shevchenko 1969).

The interaction efficiency of the gyro-TWT is a very sensitive function of the cyclotron resonance mismatch. Several typical axial dependencies of the wave amplitude calculated for different values of this mismatch are shown in Fig. 7.3. Here Figs. 7.3a and b correspond to different values of the normalized current parameter and show that, as the beam current increases, the saturation occurs at shorter distances since the wave increment increases with the current. Recall that the cyclotron resonance mismatch depends on both the signal frequency and the electron cyclotron frequency. As follows from (7.15), the normalized beam current parameter I_0 also depends on the signal frequency; the most important for this dependence is the term

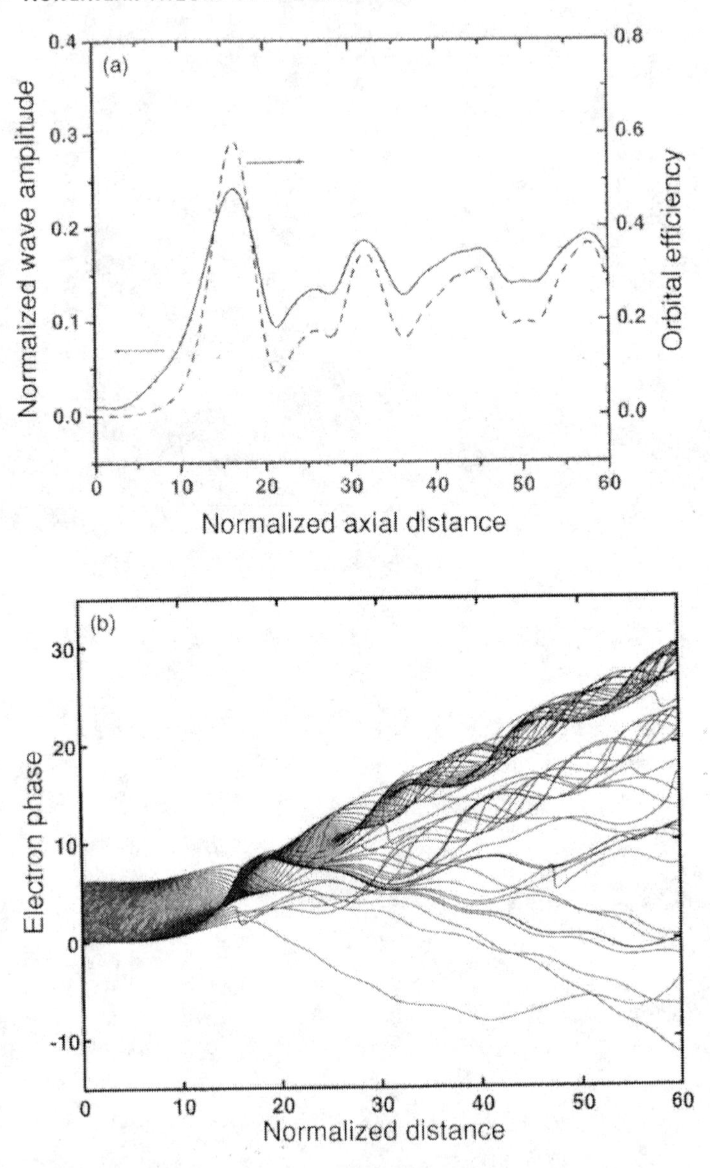

Fig. 7.2. Energy exchange in a device with a long interaction region: (a) wave amplitude and orbital efficiency; (b) electron phase evolution.

$n = (1 - \omega_{\text{cut}}^2/\omega^2)^{1/2}$ in the denominator of (7.15). Therefore, the dependencies shown in Fig. 7.3 for given values of I_0 but different Δ's can be directly used for interpreting the device operation at different values of the magnetic field, but not the signal frequency. In order to analyze the device operation at a fixed magnetic field but different signal frequencies, it is necessary to

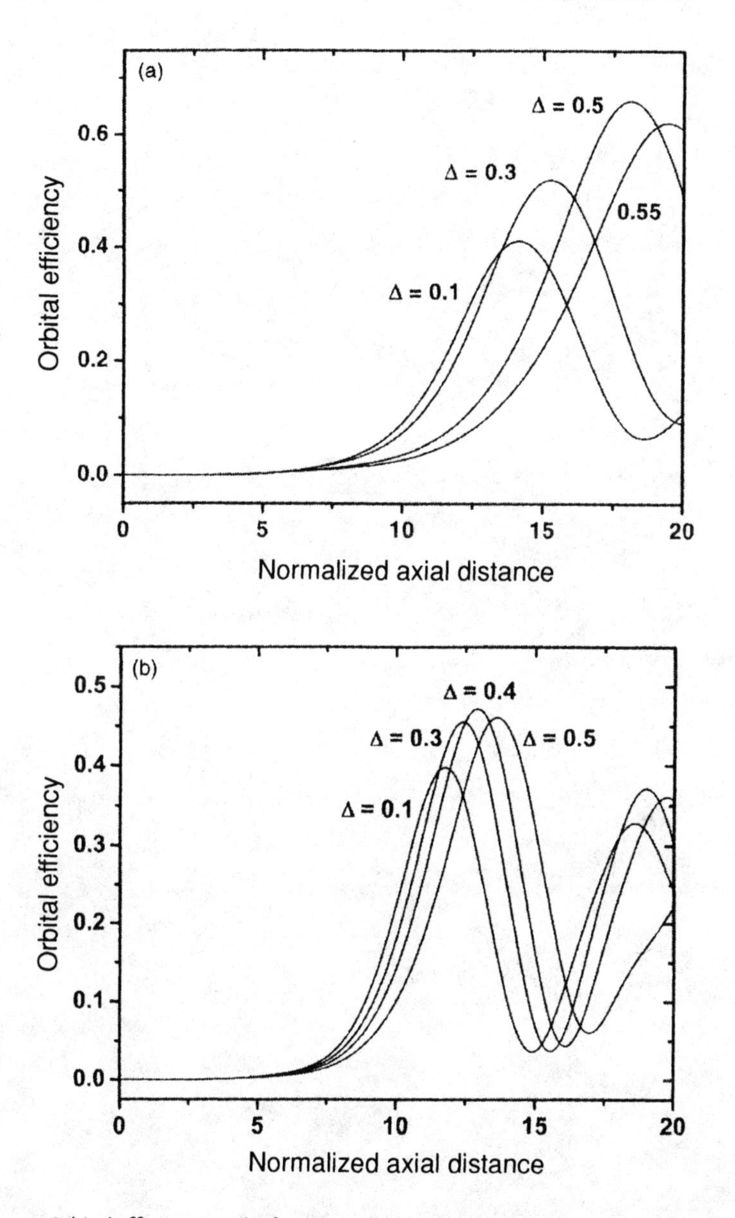

Fig. 7.3. Orbital efficiency as the function of the normalized axial coordinate for several values of the cyclotron resonance mismatch and two values of the normalized beam current: (a) $I_0 = 0.1$; (b) $I_0 = 0.3$. (In both cases the input wave amplitude F_0 is equal to 0.01.)

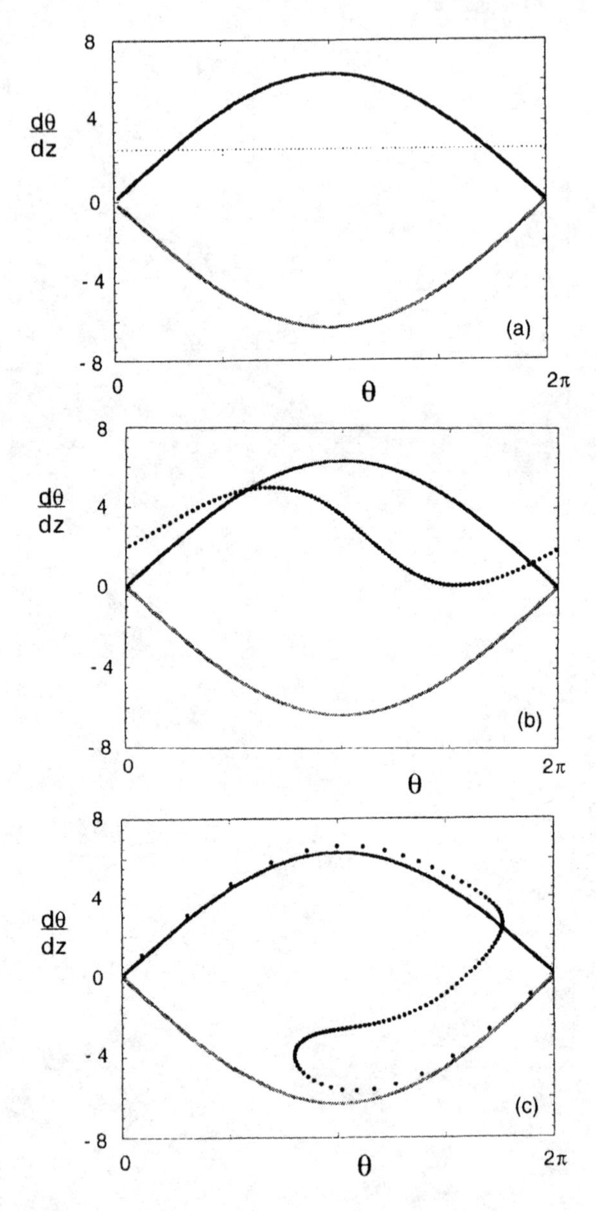

Fig. 7.4. Phase space at the entrance (a), in the process of energy modulation (b), and after bunching and energy extraction (c).

simultaneously vary the detuning Δ and the current parameter I_0. Such a consistent variation of Δ and I_0 should be done for a device with specified parameters of the beam and waveguide.

The process of electron bunching and trapping can also be illustrated with the use of the phase space characterizing the electron phase and energy, which is widely used for studying various physical problems (see, e.g., Lichtenberg). In such a space, an example of which is shown in Fig. 7.4, the particle trapping is determined by the vertical width of the separatrix, which increases with the wave amplitude.

Since our set of equations (7.9)–(7.11) contains a relatively small number of parameters, there were a number of attempts to more or less fully investigate these equations in order to find the range of parameters providing the high-efficiency operation. Resulting plots showing the contours of equal orbital efficiency in the plane of normalized parameters can be found elsewhere (see, e.g., Ginzburg, Zarnitsyna, and Nusinovich 1981, Nusinovich and Li 1992). In these references, however, a set of parameters was used slightly different from Eqs. (7.9)–(7.11). Therefore, it is more reasonable to represent here the results by Bratman et al. (1979), where the studies were done for the same set of parameters. These results are shown in Fig. 7.5; here, (a), (b), and (c) correspond to different values of the recoil parameter b: 0, 0.5, and 1.0, respectively. (Fig. 7.5b, with additional dashed lines showing the optimal values of the interaction length for the case of the initial wave amplitude equal to 0.025, was also shown by Bratman et al. 1981.) As follows from the definition of this parameter, at relatively low voltages (below 100 kV, let's say) and operation not too far from the cutoff, this parameter is rather small, so just Fig.7.5a is applicable to most of the gyro-TWTs. Other two figures can be used in the case of relativistic gyro-TWTs driven by high-voltage electron beams and operating far from the cutoff, which is the case of CARMs discussed in Chapter 1 (these devices will also be discussed in Chapter 13). As is seen in Fig. 7.5, the orbital efficiency decreases with the increase in the recoil parameter b. However, this increase in b causes an extraction of some energy from the electron axial motion. Therefore, the total electronic efficiency, in accordance with (7.7), can be increased in spite of the degradation of the orbital efficiency. For instance, it was shown that the total electronic efficiency of the ultra-relativistic gyro-TWT operating in the CARM regime is maximal when the recoil parameter is equal to 0.5 (Ginzburg, Zarnitsyna, and Nusinovich 1981). In this case, the total efficiency is equal to half of the orbital efficiency, which means that its maximum value, in accordance with Fig. 7.5b, exceeds 32%.

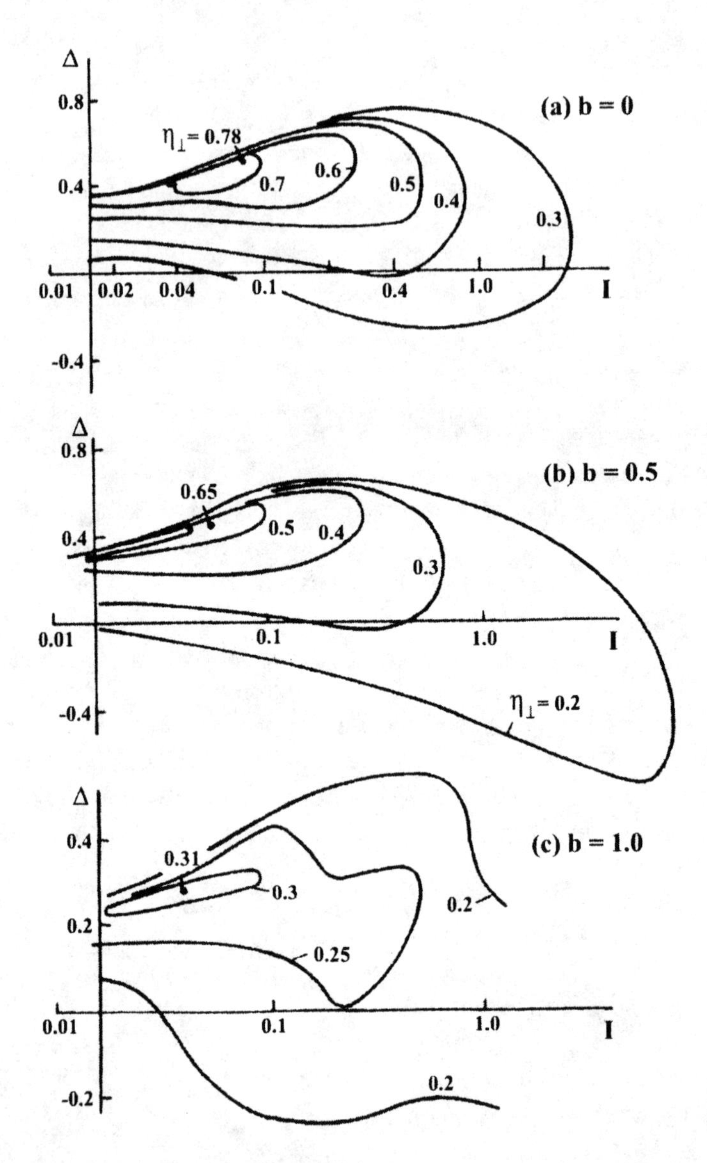

Fig. 7.5. Lines of equal orbital efficiencies in the plane of parameters—cyclotron resonance mismatch versus normalized beam current—for several values of the recoil parameter *b:* (a) 0; (b) 0.5; (c) 1.0.

7.3 Gain and Bandwidth

In Sec. 7.2 we presented results of the studies focused on the efficiency optimization of the gyro-TWT. Very often, equally or even more important are such performance characteristics as the gain and bandwidth of the gyro-TWT. The gain and bandwidth of the gyro-TWT operating in the small-signal regime were analyzed in Secs. 6.3 and 6.4, respectively. Therefore, in this section we will discuss only nonlinear effects, which are important for these characteristics.

Gain

In Sec. 6.2 we already discussed the small-signal gain in single-stage and multistage gyro-TWT. So far, however, we considered only the multistage gyro-TWT, in which all stages operate at the same cyclotron harmonic. In such a tube, as well as in a single-stage gyro-TWT, at low input power levels the output power is linearly proportional to the input power. This case is of interest to most of the communication systems, which typically require a high degree of linearity. When a multistage gyro-TWT operates in the regime of frequency multiplication that implies that an electron beam prebunched by an input wave at the signal frequency in the first stage then excites the wave at the harmonic of this frequency in the output waveguide, the operation is purely nonlinear. Therefore, this kind of operation will be considered separately below.

For gyro-TWTs operating without frequency multiplication, the most important nonlinear effect is the saturation of the amplification at high power levels. This saturation leads to the decrease of the gain in comparison with its initial value, which is constant in the small-signal regime. Some gain curves illustrating this saturation are shown in Fig. 7.6. In accordance with results

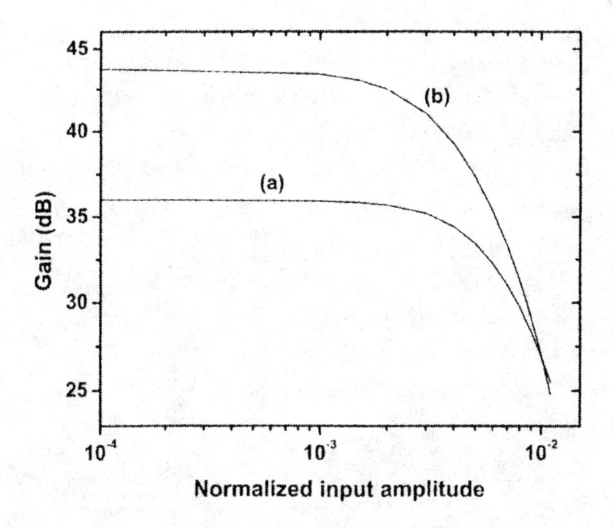

Fig. 7.6. Gain as the function of the input amplitude. Curves labeled by (a) and (b) correspond to the following sets of normalized parameters: (a) $I_0 = 0.1$, $\Delta = 0.5$, $\zeta_{out} = 20$; (b) $I_0 = 0.3$, $\Delta = 0.4$, $\zeta_{out} = 15$. In both cases $b = 0$.

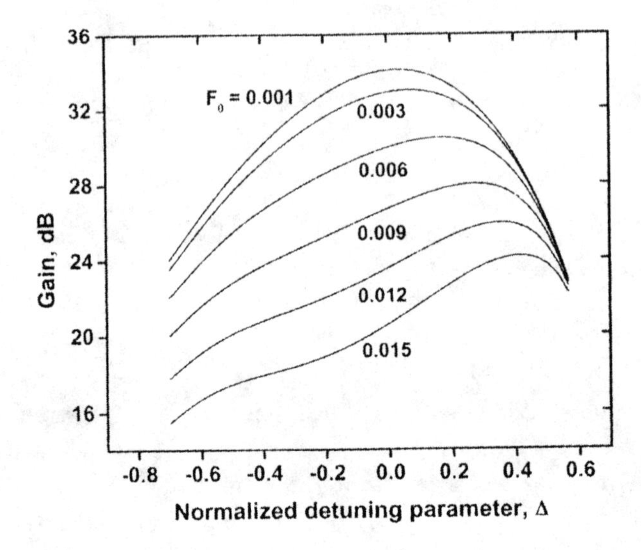

Fig. 7.7. Gain as the function of the frequency detuning for several input amplitudes; ($I_0 = 0.1$, $\zeta_{out} = 20$, $b = 0$).

shown in Fig. 7.3, here calculations for a larger current were performed for a shorter distance. Nevertheless, even at a shorter distance the gain marked by (b) is higher than the gain marked by (a), which was calculated for a longer distance.

Bandwidth

Saturation effects also cause a certain deformation of the gain curve, which results in the dependence of the bandwidth on the input power at high enough input power levels. An example of such a deformation is shown in Fig. 7.7, which shows the gain curves for several values of the input wave amplitude in a system of a fixed length. By using this figure, one can easily estimate the corresponding bandwidth as the function of the input wave amplitude. Note that the bandwidth, which can be found from Fig. 7.7, is expressed in terms of the normalized detuning. As we discussed in Sec. 6.4, for different operating conditions the same bandwidth expressed in terms of the normalized detuning corresponds to different values of the real bandwidth. So, in order to use these results for evaluating the bandwidth in real devices, it is necessary to specify the ratio of the external magnetic field to its grazing value and also specify other parameters.

Gain and Bandwidth in Frequency-Multiplying Gyro-TWTs

As we pointed out above, the regime of frequency multiplication is based on the purely nonlinear effect of harmonic generation in the electron current modulated by the wave of a signal frequency. As known (see, e.g., Bates and Ginzton 1957), the small-signal gain in such regimes is proportional to P_{in}^{M-1}, where M is the harmonic number. In other words, this dependence shows that at low levels of the input power, the output power is proportional to P_{in}^{M}. Then, at higher levels of the input power, this polynomial dependence saturates. An example of the dependence of the output power on the input power is shown in Fig. 7.8. This dependence is shown for a frequency-doubling gyro-TWT. Therefore, at low powers, the dependence shown in Fig. 7.8 is parabolic. The calculations are done for a device consisting of two cylindrical waveguides separated by a drift section and driven by a thin annular electron beam with negligibly small velocity spread. The calculations are performed for the device with the input frequency 16.6 GHz; the lengths of the first waveguide, drift section, and output waveguide are equal to 15.2 cm, 9.0 cm, and 14.6 cm, respectively. It is assumed that the tube is driven by a 65 kV, 5 A electron beam with the orbital-to-axial velocity ratio of 1.0 and guiding center radius of 0.58 cm. It is also assumed that the magnetic field is profiled in such a way

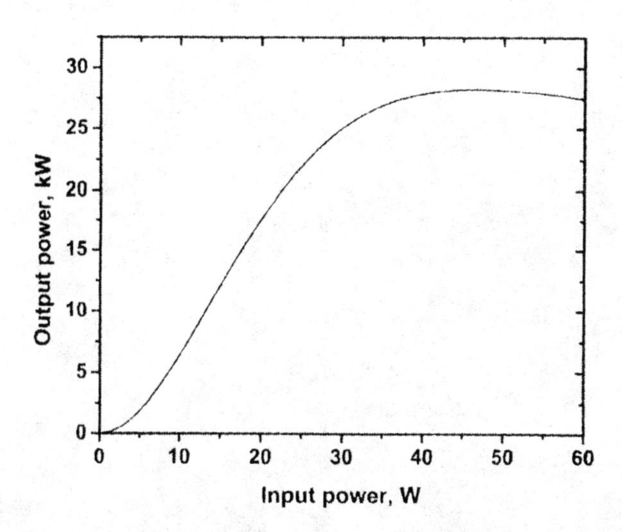

Fig. 7.8. Output power as the function of input power in the frequency-doubling gyro-TWT. Beam parameters are: 65 kV, 1A, $v_{\perp}/v_z = 1.0$, $R_0 = 0.58$ cm. The input waveguide of a 15.2 cm length and the output waveguide of a 14.6 cm length are separated by a 9.0 cm long drift section. The input waveguide operates at the TE_01-wave (16.6 GHz frequency, $s_1 = 1$), the output waveguide operates at the TE_02-wave at the doubled frequency ($s_2 = 2$). It is assumed that the normalized cyclotron resonance mismatch in the input waveguide equals zero ($\Delta_1 = 0$), while in the output waveguide, due to the magnetic field tapering in the drift region, $\Delta_2 = 0.05$.

that the cyclotron resonance mismatch is equal to 0.00 and 0.05 in the input and output waveguides, respectively.

Let us note that the paper by Gates and Ginzton mentioned above is devoted to the frequency-multiplying linear-beam TWT. However, as we pointed out several times above, linear-beam TWTs and gyro-TWTs have so many features in common that it is not surprising that frequency-multiplying gyro-TWTs, as was shown by Chu, Guo, and Granatstein (1997), exhibit the same nonlinear dependence of the output power on P_{in}.

7.4 Concluding Remarks

We have outlined above only some of the most important features in the gyro-TWT operation in the large-signal regime. Of course, to design real devices it is necessary to use much more precise methods. By those we mean numerical codes (see, e.g., Ganguly and Ahn 1982), which take into account the velocity spread, the issue of stability (Lin et al. 1992), and other important factors.

At the same time, some of the performance characteristics can be calculated by using our simple formalism accurately enough. An example of a reasonable agreement between the predictions of the self-consistent code MAGY and our formalism, in which the velocity spread was taken into account, is shown in Fig. 7.9 reproduced from Sinitsyn et al. (2002). Fig. 7.9 shows results of calculations done for the Ka-band gyro-TWT designed at NRL (Nguyen et al. 2001). These results demonstrate that the interaction processes in a waveguide of a constant wall radius can be described by our simple theory reasonably well.

7.5 Problems and Solutions

Problems

1. Derive nonaveraged and gyro-averaged equations for electron energy and axial momentum in the case of interaction with a TM-wave. Show that the autoresonance integral is not valid for nonaveraged equations, but is valid after averaging.
2. Derive the dispersion equation for the transversely homogeneous model of the gyro-TWT considered in Sec. 3.1 for two cases: (a) the amplitudes of the wave electric and magnetic fields are equal, (b) these amplitudes are different, in accordance with Maxwell's is equations. Find the difference in dispersion equations and explain it.

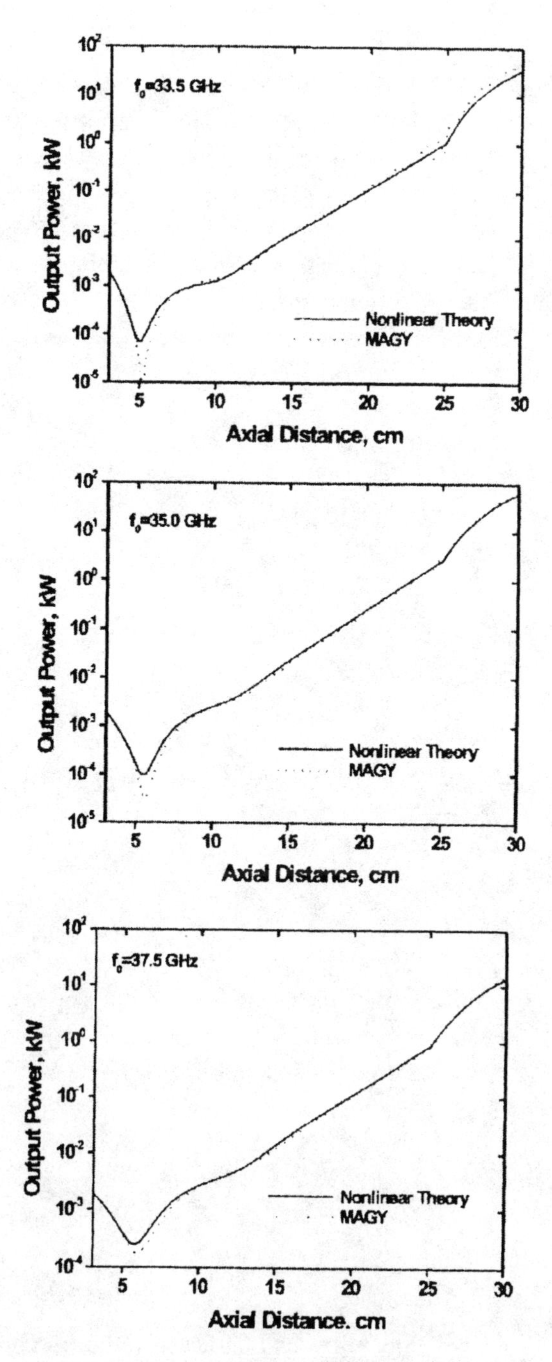

Fig. 7.9. Comparison of simulation results obtained by using a simple semianalytical formalism considered above (solid lines) and the code MAGY (dotted lines) for a uniform waveguide serving as a main part of the interaction region in the Ka-band gyro-TWT designed at NRL. Figures correspond to different operating frequencies.

3. Estimate the normalized parameters adopted in (7.9)–(7.11) for a gyro-TWT, which operates at the fundamental cyclotron resonance and is driven by a 70 kV, 6 A electron beam with the orbital-to-axial velocity ratio 0.7. Assume that the device operates in the TE_{21}-wave and the beam radius corresponds to the maximum coupling to the co-rotating wave. Also assume that the operating frequency exceeds the cutoff frequency by 10%, the cyclotron frequency to operating frequency ratio is 0.85, the interaction length is equal to 8 wavelengths and the input power is equal to 100 W.

Solutions

1. Let us follow in our derivations the method used by Zhurakhovskiy (1972). In accordance with (1.2), the equation for the electron energy can be written as

$$\frac{d\mathcal{E}}{dt} = -e(v_\perp E_\varphi + v_z E_z).$$

Equation for the electron axial velocity follows from the equation for electron axial momentum:

$$\frac{dp_z}{dt} = \frac{1}{c^2}\left(\frac{d\mathcal{E}}{dt}v_z + \mathcal{E}\frac{dv_z}{dt}\right) = -e\{E_z - \beta_\perp H_r\}.$$

From these two equations it follows that

$$\frac{dv_z}{dt} = -\frac{ec^2}{\mathcal{E}}[E_z - \beta_\perp H_r - \beta_z(\beta_\perp E_\varphi + \beta_z E_z)].$$

Electric and magnetic fields of the wave can be represented in the same fashion as in Appendix 3. Then, taking into account the symmetry between the fields of TE- and TM-waves and the expressions for field components of TE-waves given by (A2.19), we can represent two combinations of fields in the equation for axial velocity as

$$E_z - \beta_\perp H_r = \mathrm{Re}\left\{Ce^{i(\omega t - k_z z)}\left(i\Psi + \beta_\perp \frac{k}{k_\perp^2 r}\frac{\partial \Psi}{\partial \varphi}\right)\right\},$$

$$\beta_z E_z + \beta_\perp E_\varphi = \mathrm{Re}\left\{Ce^{i(\omega t - k_z z)}\left(i\beta_z\Psi + \beta_\perp \frac{k_z}{k_\perp^2 r}\frac{\partial \Psi}{\partial \varphi}\right)\right\}.$$

These expressions are different, which indicates that in such a case the autoresonance integral does not hold.

Now, let us represent the membrane function as superposition of angular harmonics, as it was done in Appendix 3, and use the cyclotron resonance condition. After averaging over fast electron gyrations, our first field combination yields

$$\tilde{E}_z - \beta_\perp \tilde{H}_r = \operatorname{Re}\left\{ iCe^{i(\omega t - k_z z - s\theta)} \left(1 - s\beta_\perp \frac{k}{k_\perp^2 a}\right) \Psi_s \right\},$$

where $s\beta_\perp \frac{k}{k_\perp^2 a} = s\Omega \frac{k}{k_\perp^2 c} \approx \frac{k^2}{k_\perp^2}(1 - n\beta_z)$, and hence, $1 - s\beta_\perp \frac{k}{k_\perp^2 a} = -\frac{n(n-\beta_z)}{1-n^2}$ (recall that $\frac{k_\perp^2}{k^2} = 1 - n^2$). Similarly, the second combination reduces to

$$\beta_z \tilde{E}_z + \beta_\perp \tilde{E}_\varphi = \operatorname{Re}\left\{ iCe^{i(\omega t - k_z z - s\theta)} \left(\beta_z - s\beta_\perp \frac{k_z}{k_\perp^2 a}\right) \Psi_s \right\},$$

where $\beta_z - s\beta_\perp \frac{k_z}{k_\perp^2 a} \approx -\frac{n-\beta_z}{1-n^2}$. Correspondingly, the gyro-averaged equations for the electron energy and axial velocity can be written as

$$\frac{d\mathcal{E}}{dt} = ec\frac{n - \beta_z}{1 - n^2}\operatorname{Re}\left\{iC\Psi_s e^{i\vartheta}\right\},$$

$$\frac{dv_z}{dt} = \frac{ec^2}{\mathcal{E}}\frac{(n - \beta_z)^2}{1 - n^2}\operatorname{Re}\left\{iC\Psi_s e^{i\vartheta}\right\}.$$

From these two equations one can readily derive the autoresonance integral (7.3). From these equations it also follows that, when the electron axial velocity is equal to the wave group velocity ($\beta_z = n$), there is no energy exchange between electrons and the wave on average. This effect can be easily understood by considering a reference frame moving with the wave group velocity. In such a reference frame, the wave propagates perpendicular to the z-axis ($k_z' = 0$) and only the axial component of its electric field is not equal to zero ($\vec{E}' = E\vec{z}_0$). At the same time, for electrons in this frame $v_z' = 0$, and therefore $(\vec{v}' \cdot \vec{E}') = 0$.

2. Assume that the vector potential of the wave is given by (3.1), where the amplitude A slowly varies along z due to the wave interaction with electrons. Then, electric and magnetic fields of a circularly polarized plane wave can be given as

$$E_+ = E_x + iE_y = -i\frac{\omega}{c}Ae^{i(\omega t - k_z z)},$$

$$H_+ = H_x + iH_y = \left(k_z A + i\frac{dA}{dz}\right)e^{i(\omega t - k_z z)}.$$

Correspondingly, the transverse component of the Lorentz force can be written as

$$F_+ = F_x + iF_y = e\left\{i\frac{\omega}{c}(1 - n\beta_z)A + \beta_z\frac{dA}{dz}\right\}e^{i(\omega t - k_z z)}.$$

Using the cyclotron resonance condition, one can readily show that the last term in figure brackets is negligibly small when the axial scale of the wave envelope variation, L_A, is much larger than the pitch length of electron orbit: $L_A >> v_z/\Omega$. Below we will assume this condition to be fulfilled and take the derivative dA/dz into account in the small cyclotron resonance mismatch only. This means that the axial velocity in the Doppler term present in the cyclotron resonance mismatch should be calculated accounting for this effect.

Equation for the electron axial momentum being written with the account for this axial dependence is

$$\frac{dp_z}{dz} = -\frac{e\,|p_+|}{\mathcal{E}\beta_z}\mathrm{Im}\left\{\left(k_z A + i\frac{dA}{dz}\right)e^{i\vartheta}\right\},$$

while the equation for electron energy is

$$\frac{d\mathcal{E}}{dz} = -\frac{e\,|p_+|}{\mathcal{E}\beta_z}\omega\mathrm{Im}\{Ae^{i\vartheta}\}.$$

(Here $\vartheta = \omega t - k_z z - \theta$ is a slowly variable gyrophase with respect to the phase of the wave.) Thus, we can represent the axial momentum as a sum of two terms, $p_z = \hat{p}_z + q$, where the first term, $\hat{p}_z$, obeys the equation for axial momentum in the field of a constant amplitude and the second one, q, is determined by the last term in the figure brackets of the equation for axial momentum. Correspondingly, for the axial momentum, $\hat{p}_z$, and electron energy the autoresonance integral holds again.

The wave envelope equation can be obtained from the wave equation

$$\left(\frac{\partial^2}{\partial z^2} - \frac{n^2}{c^2}\frac{\partial^2}{\partial t^2}\right)A_+ = -\frac{4\pi}{c}j_+,$$

where the electron current density is presented in the same fashion as the electron velocity, $j_+ = -|\rho|\,v_+$. Taking into account that $k_z^2 = (n^2/c^2)\omega^2$ and our assumption about slow axial variation of the wave envelope, one can readily reduce this wave equation to the first-order equation for the wave amplitude:

$$\frac{dA}{dz} = i\frac{|j_0|}{k_z c}\int_0^{2\pi}\frac{|p_+|}{p_z}e^{-i\vartheta}d\vartheta_0.$$

By using the formalism described in Appendix 3, we can transform these equations into a self-consistent set of equations similar to (A3.29), (A3.30), and (A3.31):

$$\frac{dw}{dZ} = \frac{\sqrt{w}}{1-b(1-w)}\mathrm{Re}(Fe^{i\vartheta}),$$

$$\frac{d\vartheta}{dZ} = \frac{1}{1-b(1-w)}\left\{\delta - \mu(1-w) - bg - \frac{1}{2\sqrt{w}}\mathrm{Im}(Fe^{i\vartheta})\right\},$$

$$\frac{dg}{dZ} = \frac{w}{1 - b(1 - w)} \text{Re}\left\{ i\frac{dF}{dZ} e^{i\vartheta} \right\},$$

$$\frac{dF}{dZ} = -I_0 \frac{1}{\pi} \int_0^{2\pi} \frac{\sqrt{w}}{1 - b(1 - w)} e^{-i\vartheta} d\vartheta_0.$$

Here a set of normalized variables and parameters is used, which is quite similar to that in Appendix 3: $Z = \omega z / v_{z0}$, $F = i\frac{2(1 - n\beta_{z0})}{\beta_{\perp 0}} \frac{eA}{mc^2 \gamma_0}$, $g = (n/b)(q/mc\gamma_0)$, $I_0 = (1 - n\beta_{z0}) \frac{\omega_p^2}{\omega^2} \frac{\beta_{z0}}{2n}$. (In deriving the last expression for the normalized beam current we used relations $|j_0| = en_e v_{z0}$ and $\omega_p^2 = 4\pi e^2 n_e / m\gamma_0$.)

These equations can be linearized in the same fashion as Eqs. (7.9)–(7.11), whose linearization procedure is described after (7.16). The resulting dispersion equation can be written as

$$\Gamma^2(\Gamma - \delta) - I_0(1 - 2b)\Gamma + I_0(\mu - 2b\delta) = 0.$$

Here Γ is the wave propagation constant in the presence of electrons.

If we ignore in our equations the term g caused by the wave envelope inhomogeneity, the resulting dispersion equation has the form

$$\Gamma^2(\Gamma - \delta) - I_0(1 - b)\Gamma + I_0(\mu - b\delta) = 0.$$

So, the account for this extra term doubles the recoil effect described by the parameter b. This is understandable because just the transverse magnetic field, where this additional term caused by the wave envelope inhomogeneity is present, determines the axial component of the RF Lorentz force acting on electron axial momentum.

Note that the dispersion equation written above for the case with nonzero g can be rewritten as

$$(\Gamma - \delta)(\Gamma^2 + I_0 b) - I_0(1 - b)\Gamma + I_0(\mu - b\delta) = 0.$$

Here in combination $(\Gamma^2 + I_0 b)$ the term $I_0 b$ resembles the space charge parameter $4QC$ in the dispersion equation for the linear-beam TWT (see (7.14) in Pierce 1950). However, the nature of this similarity is not well understood so far.

3. In normalized parameters of Eqs. (7.9)–(7.11) the beam parameters are present via $\beta_{\perp 0}$, β_{z0}, γ_0, and R_0. For a 70 kV beam the normalized energy is $\gamma_0 \approx 1.137$. Then, for a given value of α the normalized axial and orbital velocities are equal to 0.39 and 0.273, respectively. The optimal beam radius, as follows from the definition of the coupling impedance given by (3.59), is equal to $0.6R_w$, where the waveguide wall radius R_w should

be determined by the operating wavelength. The coupling impedance for our choice of beam radius and the wave is equal to 0.27. Our operating frequency to cutoff frequency ratio yields for the normalized axial (n) and transverse (κ) wavenumbers the values 0.3787 and 0.909, respectively. Hence, with the account for a given waveguide length to the wavelength ratio, the normalized interaction length equals to 4.827. A given ratio of the cyclotron frequency to the operating frequency, in combination with other parameters determined above, yields the cyclotron resonance mismatch, $\Delta \approx 0.0615$. The normalized beam current is close to 0.12 and the initial value of the normalized wave amplitude, as follows from (A3.35), is equal to 0.009.

▪ CHAPTER 8 ▪

Theory of Gyroklystrons

8.1 Introductory Remarks

As was pointed out in Sec. 2.4, for any microwave source driven by a linear electron beam one can find its gyro-counterpart. One of the most important members of the family of gyro-devices is the gyroklystron, which was schematically shown in fig. 2.8. Like a conventional klystron, the gyroklystron consists of a chain of cavities separated by drift sections. The input cavity is excited by an input signal. The field excited in this cavity modulates the energy of electrons that leads to the electron phase orbital bunching, which proceeds when electrons move along the device axis. These modulation and bunching processes cause an appearance of high-frequency components at the signal frequency and its harmonics in the electron current density. These high-frequency components excite oscillations in other cavities that improve the quality of electron bunches and increase the gain and efficiency. The microwave power is extracted from the output cavity.

There are many important issues for the gyroklystron operation. First, as any amplifier, the device should be zero-drive stable, which means that in the absence of the input signal there should be no signal at the output. Another issue, which is closely related to the first one, is the absence of microwaves in the drift sections. So, it is necessary to avoid parasitic self-excitation of these sections. Also, a parasitic effect such as leaking of the microwave energy from a cavity into a drift region should not lead to "cross-talk" between cavities. This means that drift sections should first be made cutoff for the operating frequencies, and second, these sections should contain some lossy materials for absorbing the leaking microwave power in order to prevent penetration of this power from one cavity into another. One more issue, which is very important for gyroklystrons operating at high levels of the average power, is a cooling of the penultimate cavity. Indeed, if we assume that each stage of the device operates with a certain gain, the microwave power of oscillations increases with the cavity number. However, while the microwave power can be extracted from the output cavity due to some diffractive losses, as we discussed in Sec. 2.3, it is practically impossible to extract the power from other cavities in this way because these cavities are axially closed. Therefore, the main source of losses in these cavities is the ohmic losses in the cavity walls, which lead to the wall heating causing a number of technical problems. To avoid this heating, the cavity should be intensively cooled, etc.

These technical difficulties will be discussed later. In the present chapter we will consider the most important physical effects in the gyroklystron operation. To do this we will first develop a simple theory describing the gyroklystron operation, and we will then apply this theory for describing the physical effects and such important performance characteristics as efficiency, gain, and bandwidth. This theory is based on the method originally suggested by V. K. Yulpatov (1968).

8.2 General Formalism. Point-Gap Model

General Formalism

The formalism describing the operation of gyroklystrons can be based on simple equations, which were derived in Chapter 3 for the gyromonotron. In accordance with our discussion in Sec. 8.1, we can assume that Q-factors of gyroklystron cavities are much larger than their minimum diffractive values. This allows us to use a cold-cavity approximation for the axial distribution of cavity fields. We will also assume that the space charge effects and velocity spread are negligibly small. Under these assumptions, the electron motion can be described by Eqs. (3.34) or (3.35) for the electron energy and the slowly variable phase.

Strictly speaking, in the case of short cavities typical for gyroklystrons, it is also necessary to take into account the changes in the electron axial momentum. Recall that we neglected these changes in the theory of the gyromonotron, because in the framework of this theory we considered long cavities operating at near-cutoff frequencies. Correspondingly, the magnetic field of a TE-mode, which was under consideration in Sec. 3.2, as follows from (3.23), had only a z-component, and therefore, the axial component of the RF Lorentz force acting upon electrons was equal to zero. In general, the electric and magnetic fields of a TE-mode with an arbitrary axial wavenumber can be determined via the Hertz potential as

$$\vec{E}_s = \frac{\omega/c}{k_\perp^2}[\nabla_\perp H_s \times \vec{z}_0], \; \vec{H}_s = i\left\{H_s\vec{z}_0 + \frac{1}{k_\perp^2}\nabla \times [\nabla(H_s\vec{z}_0)]\right\}. \tag{8.1}$$

This magnetic field also has transverse components, which are determined by the axial nonuniformity of the resonator field. If, as in Sec. 3.2, we represent the Hertz potential as $H_s = \Psi_s(\vec{r}_\perp)f(z)$, then these components become proportional to df/dz. Here the function $f(z)$ describes the axial structure of the resonator field. When the resonator length is on the order of the wavelength, this nonuniformity can be strong enough, and hence the changes in the electron axial momentum caused by the transverse components of the RF magnetic field can be important (see, e.g., Nusinovich, Latham, and Dumbrajs 1995). Below, however, for the sake of simplicity, these changes will be neglected.

Let us now come back to Eqs. (3.34). Since in gyroklystrons an electron beam can interact with the fields of different cavities at different cyclotron harmonics, instead of (3.34) we will use equivalent Eqs. (3.35). In these variables, the susceptibility that determines the interaction of an electron beam with the cavity mode, as follows from (3.58) and (3.61), is equal to

$$\hat{\chi} = \frac{2}{F}\frac{1}{2\pi}\int_0^{2\pi}\left\{\int_{\varsigma_{in}}^{\varsigma_{out}} w^{s/2}e^{is\vartheta} f^*(\varsigma)d\varsigma\right\} d\vartheta_0. \tag{8.2}$$

In accordance with (3.61) (see also the relation between the orbital efficiency and the real part of the gain function given after (3.62)), the imaginary part of this susceptibility relates to the electron orbital efficiency as

$$\eta_\perp = |F|^2\,\hat{\chi}''. \tag{8.3}$$

Also the balance equation of active powers (3.62) can be rewritten as

$$I_0\hat{\chi}'' = 1, \tag{8.4}$$

and the balance equation for the reactive power, as it follows from (3.56), can be written as

$$I_0\hat{\chi}' = -\delta. \tag{8.5}$$

In (8.5) we introduced the normalized detuning between the operating frequency and the real part of a cold-cavity frequency, $\delta = (\omega - \omega_s')/(\omega/2Q)$. The balance equations (8.4) and (8.5) are valid for all cavities except for the first one, where the input signal excites oscillations. For the input cavity, one can use a similar expression for the susceptibility, which originates from the source term written in general form in (3.50). In the presence of the drive signal, the current density in this source term can be represented as a sum of the electron current density, which we considered so far, and an equivalent current density in an element exciting the input cavity. Such a representation allows one to use for the susceptibility of the input cavity the following expression (Ergakov and Moiseev 1975):

$$\hat{\chi}_{\Sigma} = \hat{\chi} + \frac{A}{I_0 |F|} e^{-i\psi}. \tag{8.6}$$

Here we represented the complex amplitude of the cavity field as $F = |F| e^{i\psi}$. Assuming that the phase of this field describes the phase difference between the cavity field and the drive signal phase, one can consider the normalized amplitude of the drive signal, A, as a real value. The intensity of this signal, A^2, relates to the input power of the driver, P_{dr}, as (Ergakov and Moiseev 1975, Nusinovich, Danly, and Levush 1997)

$$A^2 = 4I_{01} \frac{P_{\mathrm{dr}} Q_{1,T}}{P_{b\perp} Q_{\mathrm{cpl}}}. \tag{8.7}$$

Here I_{01} is the normalized beam current parameter for the first input cavity, $Q_{1,T}$ is its total quality factor ($1/Q_{1,T} = 1/Q_{\mathrm{ohm}} + 1/Q_{\mathrm{cpl}}$), here Q_{cpl} is the coupling or external Q-factor of this cavity, Q_{ohm} is ohmic Q-factor (typically there is no diffraction loss in input cavities), and $P_{b\perp} = [\beta_{\perp 0}^2/2(1 - \gamma_0^{-1})] V_b I_b$ is the beam power associated with the electron gyration.

So, after replacing the susceptibility in the balance equations (8.4) and (8.5) by its expression (8.6), the balance equations for the input cavity can be written as

$$|F_1|^2 = \frac{A^2}{(1 - I_{01}\hat{\chi}_1'')^2 + (\delta_1 + I_{01}\hat{\chi}_1')^2}, \tag{8.8}$$

and

$$\tan\psi = \frac{\delta_1 + I_{01}\chi_1'}{1 - I_{01}\chi_1''}. \tag{8.9}$$

Note that the input cavity operates in the small-signal regime, and at the entrance to this cavity an electron beam is not modulated. Therefore, as follows from (8.2), one can determine the susceptibility of the beam to this cavity field as

$$\hat{\chi}_1 = \frac{2}{F_1} \frac{1}{2\pi} \int_0^{2\pi} \left\{ \int_0^{\mu_1} f^*(\varsigma) e^{is(\vartheta_0 - \Delta\varsigma)} \left(\frac{s}{2} w_{(1)} + is\vartheta_{(1)} \right) d\varsigma \right\} d\vartheta_0. \tag{8.10}$$

Here the perturbations in electron energy, $w_{(1)}$, and phase, $\vartheta_{(1)}$, are given by (3.67) and (3.68), respectively. (Note that in (8.10) we dropped the zero-order term in parentheses because it yields zero after averaging over entrance phases.) Then, using the same procedure as in Sec. 3.4, one can readily derive from (8.10), (3.67), and (3.68) the following expression for $\hat{\chi}_1$:

$$\hat{\chi}_1 = -2i\left(s + \frac{\partial}{\partial \Delta}\right)\int_0^{\mu_1} f'^*(\varsigma)\left[\int_0^{\varsigma} f'(\varsigma')d\varsigma'\right]d\varsigma. \tag{8.11}$$

As one can see, the imaginary part of $\hat{\chi}_1$ given by (8.11) coincides with the linearized orbital efficiency (3.71) divided by $|F|^2$, in accordance with the general relation (8.3).

Point-Gap Model

As in conventional klystrons, many important physical effects in gyroklystrons can be described analytically in the framework of the point-gap model, which was originally used in gyroklystrons by V. K. Yulpatov (1968). (Detailed description of this formalism can be found in Nusinovich, Danly, and Levush 1997.) Such a model implies that the drift regions are much longer than the cavities. Correspondingly, inside short cavities one can take into account only the electron energy modulation, neglecting the phase perturbations. At the same time, in long drift regions even a weak energy modulation that took place in the preceding cavities can lead to significant ballistic phase bunching. This means that the electron phase at the entrance to the ith cavity, as follows from the equation for the phase in (3.35), relates to the electron phase at the exit from *(i-1)* th cavity as

$$\vartheta(\varsigma_{i,\text{in}}) = \vartheta(\varsigma_{i-1,\text{out}}) - \Delta\mu_{i-1,\text{dr}} - [w(\varsigma_{i-1,\text{out}}) - 1]\mu_{i-1,\text{dr}}. \tag{8.12}$$

Here $\mu_{i-1,\text{dr}} = \varsigma_{i,\text{in}} - \varsigma_{i-1,\text{out}}$ is the normalized length of the corresponding drift section and $\Delta\mu_{i-1,\text{dr}}$ is the electron transit angle in this section. The last term in the RHS of (8.12) describes the ballistic bunching in this drift section. The bunching processes in the preceding stages of the device are contained in the phase at the exit from the preceding cavity, $\vartheta(\varsigma_{i-1,\text{out}})$. As a result, all ballistic processes can be described by a superposition of bunching parameters, which characterize the effect of each cavity on the electron ballistic bunching in all successive drift sections. For the case of operation in different stages at different cyclotron harmonics, this formalism was developed elsewhere (Nusinovich, Saraph, and Granatstein 1997). Simplifications that follow from adopting the point-gap model will be obvious from the considerations done below.

8.3 Gain, Bandwidth, and Efficiency

Typically, gyroklystrons, as well as conventional klystrons, are designed in such a way that each stage operates with a certain gain, and therefore the field intensity in each successive cavity is much larger than in the preceding one. This means that it is reasonable to assume that, even in the case of device operation in the large-signal regime only the last output cavity operates in this regime while the small-signal approach can be applied to all preceding cavities.

Let us start to develop this theory by considering Eqs. (3.35). When the field amplitude is small, the electron energy and phase in any cavity can be represented as

$$w = w(0) + \tilde{w}, \vartheta = \vartheta(0) - \Delta\varsigma + \tilde{\vartheta}. \tag{8.13}$$

Here $w(0)$ and $\vartheta(0)$ are, respectively, the values of the electron energy and phase at the cavity entrance, and $\tilde{w}$ and $\tilde{\vartheta}$ are the perturbations caused by the cavity field. As follows from (3.35), these perturbations obey the following equations:

$$\frac{d\tilde{w}}{d\varsigma} = 2\mathrm{Im}\{Ff(\varsigma)e^{-is\vartheta_{(0)}}\}, \tag{8.14}$$

$$\frac{d\tilde{\vartheta}}{d\varsigma} + \tilde{w} = -\mathrm{Re}\{Ff(\varsigma)e^{-is\vartheta_{(0)}}\}. \tag{8.15}$$

Here we introduced a slowly variable phase $\vartheta_{(0)} = \vartheta(0) - \Delta\varsigma$, which describes the phase unperturbed by the field of a given resonator. We also assume, first, that perturbations in electron energy are small, and second, that energy perturbations caused by the fields of preceding cavities are even smaller. These assumptions allow us to rewrite the susceptibility (8.2) in a simplified form:

$$\hat{\chi} = \frac{2}{F}\frac{1}{2\pi}\int_0^{2\pi}\left\{\int_{\varsigma_{in}}^{\varsigma_{out}} f^*(\varsigma)e^{i\vartheta_{(0)}}\left(1 + is\tilde{\vartheta} + \frac{s}{2}\tilde{w}\right)d\varsigma\right\}d\vartheta_0. \tag{8.16}$$

Eq. (8.16) is similar to (8.10). However, we now consider a prebunched beam and the electron phase at the entrance to a given cavity, $\vartheta_{(0)}$, is determined by (8.12).

Since (8.14) and (8.15) are linearized equations, they can readily be integrated. Substituting their solutions into (8.16) allows one to rewrite (8.16) as

$$\hat{\chi} = \hat{\chi}_{(0)} + \hat{\chi}_{(1)}\langle e^{is\vartheta'(0)}\rangle + \hat{\chi}_{(2)}\langle e^{i2s\vartheta'(0)}\rangle. \quad \textbf{(8.17)}$$

Here the first term in the RHS, $\hat{\chi}_{(0)}$, is the susceptibility without accounting for prebunching effects, which is the same as the linearized susceptibility (8.11) determined above for the input cavity; angular brackets in (8.17) denote averaging over ϑ_0.

The second term in the RHS of (8.17), which originates from averaging $e^{i\vartheta_{(0)}}$ over the entrance phase, is inversely proportional to the cavity field amplitude:

$$\hat{\chi}_{(1)} = -2i\frac{|u|}{|F|}. \quad \textbf{(8.18)}$$

Finally, the last term there is

$$\hat{\chi}_{(2)} = -2se^{i2\alpha}\int_{\varsigma_{in}}^{\varsigma_{out}} f'^{*}(\varsigma)\left[\int_{\varsigma_{in}}^{\varsigma} u^{*}d\varsigma'\right]d\varsigma. \quad \textbf{(8.19)}$$

In (8.18) and (8.19) we introduced $u = \int_{\varsigma_{in}}^{\varsigma} f'(\varsigma')d\varsigma'$ and denoted $u(\varsigma_{out}) = |u|\,e^{i\alpha}$. Also the phase $\vartheta'(0)$ in (8.17) is equal to $\vartheta(0) + (\alpha + \psi)/s$.

In the first cavity, there are no prebunching effects. Therefore, the last two terms in the RHS of (8.17) are equal to zero, and correspondingly, the balance equations (8.8) and (8.9) for this cavity can be used implying $\hat{\chi}_1 = \hat{\chi}_{1(0)}$ given by (8.11). Note that in the limiting case of very short cavities, the second term in the RHS of (8.11), which is proportional to the cavity length, is negligibly small and this susceptibility is simply $\lim_{\mu_i \to 0}\hat{\chi}_{(0)} = -is$.

In all other cavities these two last terms in (8.17) are nonzero. However, the second one, $\hat{\chi}_{(2)}$, as follows from (8.19), is proportional to the normalized cavity length squared, and hence, is negligibly small in the case of short cavities. As was pointed out by Zasypkin et al. (1995), the electron velocity spread makes this term even smaller, therefore anything further on this term will be neglected. This simplification allows us to rewrite the balance equations (8.4) and (8.5) for all cavities starting from the second one as

$$|F_i|^2 = \frac{4I_{0i}^2\,|u_i|^2\,|\mathcal{F}_i|^2}{\left(I_{0i}\,\hat{\chi}''_{i(0)} - 1\right)^2 + \left(I_{0i}\,\hat{\chi}'_{i(0)} + \delta_i\right)^2}, \quad \textbf{(8.20)}$$

$$\tan\phi_i = \frac{\delta_i + I_{0i}\,\hat{\chi}'_{i(0)}}{1 - I_{0i}\,\hat{\chi}''_{i(0)}}. \quad \textbf{(8.21)}$$

Here we introduced the function

$$\mathcal{F}_i = |\mathcal{F}_i|\, e^{i\phi_i} = \langle e^{is_i\vartheta_i(0)}\rangle. \tag{8.22}$$

This function takes into account all bunching processes in the preceding cavities and drift sections and characterizes a corresponding high-frequency harmonic in the electron current density. The phase ϕ_i is the phase of the ith cavity field taken with respect to the phase of the high-frequency harmonic in the electron current density. For a particular case, this phase will be determined below in Sec. 8.4.

Being determined for the output cavity, this function $\mathcal{F}$ can also be used for characterizing the orbital electron efficiency. Indeed, taking into account (8.3) and a simplified version of (8.17), in which the last term is neglected and two other terms are determined by (8.11) and (8.18), respectively, one can rewrite the orbital efficiency as

$$\eta_\perp = |F_N|^2\, \hat{\chi}''_{N(0)} - 2\,|u_N|\,|F_N|\,|\mathcal{F}_N| \cos\phi_N. \tag{8.23}$$

As we already derived, for the point-gap model $\hat{\chi}''_{(0)} = -s$, and hence the efficiency can be positive only due to the prebunching effects described by the last term in (8.23). This fact indicates that the point-gap model should predict a lower efficiency than a real gyroklystron can demonstrate. Indeed, it was shown in Chapters 1 (see Fig. 1.4 and corresponding text) and 3 (Fig. 3.1 and Eq. (3.72)) that, for realizing the coherent radiation from a nonprebunched electron beam in gyrotron cavities, it is necessary to have the normalized cavity length, μ, large enough. However, the point-gap model corresponds to the limiting case of very short cavities, so all possibilities to enhance the induced radiation due to the finite length of cavities are out of the scope of this model. Note that even first studies of the large-signal operation of gyroklystrons revealed that the account for the finite length of the output cavity substantially increases the efficiency (Koshevaya 1970). Optimal parameters and corresponding maximum efficiencies of two-cavity gyroklystrons with cavities of a finite length were calculated by Belousov, Ergakov, and Moiseev (1978; see also Nusinovich [1992b]). In particular, for the case of fundamental cyclotron resonance in both cavities, the maximum orbital efficiency is equal to 0.72. In the case of tapered magnetic fields, it exceeds 0.8 (Kuraev 1979).

It is also necessary to note that the fact that $\hat{\chi}''_{(0)} = -s$ means the positive beam loading of the cavities. In this case, the term $(I_{0i}\,\hat{\chi}''_{i(0)} - 1)^2$ in the denominator of (8.8) and (8.20) is equal to $(1 + s_i I_{0i})^2$, i.e., it is larger than in the case of the negative beam loading, which is the case of $\hat{\chi}''_{(0)} > 0$. Correspondingly, the field amplitudes in cavities of the point-gap model are smaller than in the cavities of a finite length once the negative beam loading takes place in such finite-length cavities. Clearly, the largest amplitude of the cavity field in gyroklystrons, as well as in conventional klystrons, can be realized in the regime of a so-called regenerative amplification, which is the case of $\hat{\chi}''_{(0)} > 0$ when the device operates near the border of the cavity self-excitation. The self-excitation condition is $I_{0,\mathrm{st}} = 1/\hat{\chi}''_{(0)}$, and typical safety margins require having the beam current not larger than (0.8–0.9) of the start current.

When $\hat{\chi}''_N < 0$, one can easily find from (8.23) that, first, the optimal phase of the output cavity field ϕ_N is equal to π, and second, there is an optimal value of the output cavity field amplitude, $|F_N|_{\mathrm{opt}} = |u_N|\,|\mathcal{F}_N|/|\hat{\chi}''_{N(0)}|$. This yields the maximum orbital efficiency

$$\eta_{\perp,\max} = \frac{|u_N|^2}{\left|\hat{\chi}''_{N(0)}\right|}\,|\mathcal{F}_N|^2. \tag{8.24}$$

As follows from the balance equations (8.20) and (8.21), this optimal choice of the output cavity field amplitude and phase corresponds to the normalized beam current parameter

$$I_{0N,\mathrm{opt}} = 1/\left|\hat{\chi}''_{N(0)}\right|, \tag{8.25}$$

and the detuning between the signal frequency and the cold-cavity frequency of the output cavity

$$\delta_{N,\mathrm{opt}} = -I_{0N,\mathrm{opt}}\,\hat{\chi}'_{N(0)}. \tag{8.26}$$

In the limiting case of very short cavities, Eqs. (8.25) and (8.26) reduce to $I_{0N,\mathrm{opt}} = 1/s_N$ and $\delta_{N,\mathrm{opt}} = 0$, respectively.

The gain of a gyroklystron can be characterized by the following definition:

$$G(\mathrm{dB}) = 10\log\left(\frac{|F_N|^2}{A^2}\right). \tag{8.27}$$

In multicavity devices it is reasonable to represent this ratio of field intensities, $|F_N|^2/A^2$, as the products of ratios characterizing the gain in each stage:

$$\frac{|F_N|^2}{|F_{N-1}|^2}\,\frac{|F_{N-1}|^2}{|F_{N-2}|^2}\cdots\frac{|F_1|^2}{A^2}.$$

Each ratio in this chain can be expressed in terms of the corresponding balance equations that follow from (8.8), (8.9), (8.4), and (8.5). Correspondingly, the gain can be represented as the sum of gains in all stages. Note that in the case of stagger tuning, which is the case when cold-cavity frequencies in different cavities are different, the detunings, δ_i, are also different.

In the framework of the small-signal theory, all processes in an electron beam initiated by the high-frequency field of all cavities are small, including the ballistic bunching effects in all stages of a device. Therefore, the function $\mathcal{F}_i$ in (8.20) can be linearized with respect to the cavity fields.

As known, the bandwidth is the range of frequencies in which the gain deviation does not exceed a certain limit. Below, we will interpret the bandwidth in terms of normalized signal frequency detunings and assume that the gain deviation within the bandwidth should not exceed 3 dB.

Very often, the figure-of-merit of the performance of any microwave amplifier is not the gain or the bandwidth itself, but the gain-bandwidth product. To maximize this product one can use a stagger tuning of gyroklystron cavities, which means the use of cavities with slightly different eigenfrequencies. Since the resonance properties of each cavity are determined by its Q-factor, the frequency stagger tuning should, apparently, be within the limit of the width of resonance curves: $|\omega_i - \omega_j| < \omega/Q$. Below, we will analyze this kind of operation with stagger-tuned cavities in the simplest case of a two-cavity gyroklystron operating at the fundamental cyclotron resonance in both cavities.

8.4 Two-cavity Gyroklystron

Efficiency

Interaction of electrons with the field of the input cavity is quite similar to that in the gyromonotron, which was analyzed in Chapter 3. Let us consider the input cavity with a simple axial structure of the cavity field, $f(\varsigma) = 1/\mu_1 (0 \leq \varsigma \leq \mu_1)$. The energy modulation in this cavity, as follows from (3.67) for given $f(\varsigma)$ and $s = 1$, is equal to

$$\tilde{w}(\mu_1) = -2\frac{|F_1|}{\Delta\mu_1}\{\cos(\Delta\mu_1 + \psi_1 - \vartheta_0) - \cos(\psi_1 - \vartheta_0)\}.$$

In the short-cavity limit ($\Delta\mu_1 << \pi$) this expression reduces to

$$\tilde{w}(\mu_1) = -2|F_1|\sin(\vartheta_0 - \psi_1). \tag{8.28}$$

(Note that the coefficient $u_i = \int_{\varsigma_{i,in}}^{\varsigma_{i,out}} f'(\varsigma)d\varsigma$, used in section 8.3, in the case of short cavities with $f(\varsigma) = 1/\mu$, is equal to one.) The field amplitude and phase in the input cavity of a point-gap model, as follows from (8.8)–(8.9) and $\hat{\chi}_{(0)} = -i$, are equal, respectively, to

$$|F_1|^2 = \frac{A^2}{(1 + I_{01})^2 + \delta_1^2} \quad \text{and} \quad \tan\psi_1 = \frac{\delta_1}{1 + I_{01}}.$$

Introducing the normalized input amplitude $A' = A/\sqrt{1 + I_{01}}$ and the normalized frequency detuning $\delta_1' = \delta_1/(1 + I_{01})$ reduces these equations to

$$|F_1|^2 = \frac{(A')^2}{1 + (\delta_1')^2}, \tag{8.29}$$

$$\tan\psi_1 = \delta_1'. \tag{8.30}$$

The electron phase shift in the drift region caused by the energy modulation in the first cavity and the cyclotron resonance mismatch, as follows from (3.35) and (8.28), is equal to

$$\vartheta(\mu_{dr}) = \vartheta_0 - \Delta\mu_{dr} + 2\,|F_1|\,\mu_{dr}\sin(\vartheta_0 - \psi_1). \tag{8.31}$$

As one can see, this equation is quite similar to the known expression for the electron phase in the drift region of a linear-beam klystron. Therefore, similar to the klystron theory, it makes sense to introduce the bunching parameter, $q = 2\,|F_1|\,\mu_{dr}$, which will describe the ballistic bunching effects.

Substituting (8.31) into (8.22) determines the function $\mathcal{F}_2$, which characterizes the high-frequency harmonic in the electron current density at the entrance to the output cavity, as

$$\mathcal{F}_2 = e^{-i(\Delta\mu_{dr}+\psi_1)}J_1(q). \tag{8.32}$$

(In the case of operation in the frequency-multiplying regime, the first-order Bessel function, which describes the saturation effects in (8.32), should be replaced by $J_M(s_2 q)$, where $M = s_2/s_1$ is the frequency-multiplication factor (Nusinovich, Danly and Levush 1997).) Correspondingly, the susceptibility of a prebunched beam with respect to the second cavity field, as follows from substituting (8.32) into (8.17), is equal to

$$\hat{\chi}_2 = -i - \frac{2i}{|F_2|}e^{-i\phi}J_1(q). \tag{8.33}$$

Here the phase $\phi = \psi_2 - \psi_1 + \pi + \Theta$ takes into account the difference in phases of the fields in two cavities and the transit angle $\Theta = \Delta\mu_{dr}$ in the drift region. Correspondingly, the balance equations for the output cavity, as follows from (8.20) and (8.21), can be written as

$$|F_2|^2 = \frac{4I_{02}^2 J_1^2(q)}{(1 + I_{02})^2 + \delta_2^2}, \tag{8.34}$$

and

$$\tan\psi_2 = \frac{\delta_2}{1 + I_{02}}. \tag{8.35}$$

Also, the maximum orbital efficiency, as follows from substituting (8.32) into (8.24), is equal to

$$\eta_{\ell,\max} = J_1^2(q). \tag{8.36}$$

Thus, the maximum value of the orbital efficiency is equal to 34%.

It is interesting to point out that (8.36) slightly differs from the maximum efficiency of the point-gap model of a linear-beam klystron (Webster 1939, Savelyev 1940): $\eta_{\max} = J_1(q)$. This result was obtained for linear-beam klystrons under assumptions that, first, the susceptibility of the output gap with respect to a nonmodulated electron beam can be ignored, and second, that the AC voltage in this gap is close to the DC beam voltage. In (8.23) these two assumptions correspond to the neglect of the first term in the RHS and to the assumption that in the second term $2|F_2| = 1$, respectively. At the same time, the optimal value of the bunching parameter responsible for the ballistic bunching in both cases is the same: $q_{\text{opt}} = 1.84$.

This bunching parameter, in accordance with (8.29), can be rewritten as $q = q_0/\sqrt{1+(\delta_1')^2}$, where $q_0 = 2A'\mu_{\text{dr}}$. Therefore, one can distinguish two stages in the dependence of the efficiency on the detuning δ_1'. When the input power (and, correspondingly, the signal amplitude) is small and $q_0 < q_{\text{opt}} = 1.84$, the zero detuning, $\delta_1' = 0$, is optimal for the efficiency. However, when the input power is high enough and, hence, $q_0 > q_{\text{opt}} = 1.84$, a too strong signal causes an overbunching in the case of the zero detuning. In this case, it is necessary to slightly detune the signal and cavity frequencies in order to avoid this effect. The optimal detuning here is equal to

$$(\delta_1')_{\text{opt}} = \pm\sqrt{\left(\frac{q_0}{1.84}\right)^2 - 1}. \tag{8.37}$$

Fig. 8.1 illustrates this dependence of the orbital efficiency on the detuning. As one can see, when the bunching parameter at the zero detuning exceeds the optimal value, the valley in the efficiency curve appears near the axis that increases the range of detunings, which corresponds to reasonably high efficiencies. For example, when $q_0 = 1.84$, the range of δ's where the orbital efficiency is larger than $\eta_{\perp,\max}/2$ is equal to $2\sqrt{3}$. At the same time, at larger q_0's, the minimum efficiency in the valley, $\eta_\perp(\delta_1 = 0)$ is equal to $\eta_{\perp,\max}/2$ when $q_0 \simeq 2.79$. In this case, the range of δ's where $\eta_\perp(\delta) \geq \eta_{\perp,\max}/2$, is equal to 5.726, i.e., it is about 1.65 times larger than for $q_0 = 1.84$. The bunching parameter q_0 can be increased either by lengthening the drift section or by enlarging the signal amplitude. Of course, as the bunching parameter increases

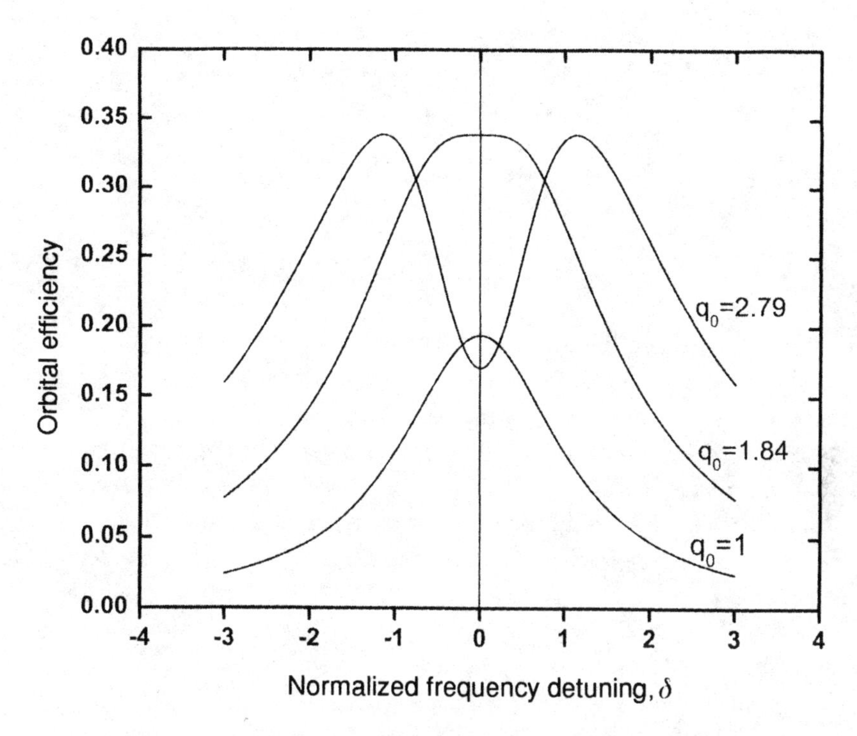

Fig. 8.1. Orbital efficiency as the function of the normalized signal frequency detuning for several values of the bunching parameter in a two-cavity gyroklystron.

due to the increase in the input power, this reduces the gain (8.27). All these effects are known in linear-beam klystrons.

Gain

The gain of such a gyroklystron can easily be determined by substituting (8.29) and (8.34) into (8.27), which results in the following expression:

$$G(\mathrm{dB}) = 10\log\left\{\frac{16I_{02}^2\mu_{dr}^2}{\left[(1+I_{01})^2+\delta_1^2\right]\left[(1+I_{02})^2+\delta_2^2\right]}\frac{J_1^2(q)}{q^2}\right\}. \quad \textbf{(8.38)}$$

To obtain this expression we first represented the ratio of the intensity of the output cavity field to the intensity of the input signal as the product of two ratios, $|F_2|^2/|F_1|^2$ and $|F_1|^2/A^2$, as we discussed above, and then replaced the field amplitude in the first cavity by the bunching parameter.

For studying the effect of frequency detuning on the gain, it is convenient to represent the gain as (Nusinovich, Danly, and Levush 1997)

$$G = G^{(\mathrm{const})} - G^{(\mathrm{var})}. \quad \textbf{(8.39)}$$

Here the first term, $G^{(\text{const})}$,

$$G^{(\text{const})} = 20 \log \left\{ \frac{2 I_{02} \mu_{\text{dr}}}{(1 + I_{01})(1 + I_{02})} \right\}, \tag{8.40}$$

does not depend on the frequency detuning, while the second one

$$G^{(\text{var})} = 10 \log \left\{ \frac{q^2}{4 J_1^2(q)} \frac{\left[(1 + I_{01})^2 + \delta_1^2\right]\left[(1 + I_{02})^2 + \delta_2^2\right]}{(1 + I_{01})^2 (1 + I_{02})^2} \right\} \tag{8.41}$$

is the variable part of the gain that is dependent. These expressions for the gain components are normalized in such a way that in the small-signal regime, when the bunching parameter is small ($q \to 0$), and in the case of the zero detunings in both cavities ($\delta_1 = \delta_2 = 0$), the variable term vanishes. Thus, (8.40) determines the maximum value of the small-signal gain. When the signal frequency differs from the cold-cavity frequencies, the variable term determines the gain degradation, which limits the amplification bandwidth of the device.

Bandwidth

The bandwidth of the gyroklystron can be determined by using (8.41). In the small-signal regime, (8.41) can be rewritten as

$$G_{ss}^{(\text{var})} = 10 \log \left\{ \left(1 + \delta_1'^2\right)\left(1 + \delta_2'^2\right) \right\}. \tag{8.42}$$

Here we used the same notations for primed detunings as above in (8.29) and (8.30). When both cavities are identical, i.e., when they have the same cold-cavity frequencies and Q-factors, $\delta_1' = \delta_2' = \delta'$, so (8.42) reduces to

$$G_{ss}^{(\text{var})} = 20 \log\{(1 + \delta'^2)\}. \tag{8.43}$$

The normalized bandwidth of such an ideally tuned device, being expressed in terms of δ', is equal to $2(\sqrt{2} - 1)^{1/2}$. This corresponds to the frequency band related to the carrier frequency equal to $(\sqrt{2} - 1)^{1/2}(1 + I_0)/Q$ that for small values of the normalized current parameter I_0 is close to $0.6436/Q$. This means that in this case the bandwidth is about 2/3 of the width of the cavity resonance curves.

This bandwidth, however, can be substantially increased when the eigenfrequencies of the cavities are slightly different. Let us assume, for the sake of simplicity, that, again, such cavities have the same Q-factors and the normalized beam current parameter in both cavities is also the same $I_{01} = I_{02} = I_0$. Then, one can introduce the mean frequency for eigenfrequencies of two cavities, $\omega_0 = (\omega_1 + \omega_2)/2$, the stagger tuning parameter,

$\xi = (\omega_2 - \omega_1)/(\omega_2/2Q)(1 + I_0)$, and the detuning of the signal frequency with respect to ω_0, $\delta = (\omega_{sig} - \omega_0)/(\omega_2/2Q_2)(1 + I_{02})$. In these notations, the small-signal gain (8.42) can be rewritten as

$$G_{ss}^{(var)} = 10\log\left\{\left[1 + \left(\delta + \frac{\xi}{2}\right)^2\right]\left[1 + \left(\delta - \frac{\xi}{2}\right)^2\right]\right\}. \quad \textbf{(8.44)}$$

This formula can easily be studied analytically. From the analysis it follows that in the presence of the stagger tuning the variable gain has a local maximum at the zero detuning, where $G_{ss.max}^{(var)}(\delta = 0) = 10\log\{1 + (\xi^2/4)\}$, and two symmetric minima where $G_{ss,min}^{(var)} = 10\log(\xi^2)$. (Recall that, in accordance with (8.39), minimum of the variable gain means the maximum gain.) The difference between these minimum and local maximum values of the gain correspond to the 3 dB gain variation when the stagger tuning parameter ξ is equal to its maximum value: $|\xi|_{max} = 2[3 + \sqrt{8}]^{1/2} \approx 4.83$. This means that the ratio of the frequency separation of cavity eigenfrequencies to the signal frequency should be about $2.4(1 + I_0)/Q$. A corresponding bandwidth expressed in terms of δ' is now equal to $4(\sqrt{2} + 1)$. So, the stagger tuning of two cavities can increase the bandwidth by $2[(\sqrt{2} + 1)/(\sqrt{2} - 1)]^{1/2} \approx 4.83$ times. However, simultaneously the maximum gain calculated for optimum values of the detuning δ' is 13.7 dB smaller. As is shown by Chu, Latham, and Granatstein (1988), in multicavity gyroklystrons the frequency detuning of intermediate cavities can also result in efficiency enhancement; this effect is also known in linear-beam klystrons.

Dependencies of the gain and bandwidth on the stagger tuning parameter are shown in Fig. 8.2. Also shown is the ratio of the gain-bandwidth product in the presence of stagger tuning to that in its absence:

$$\Phi = \frac{BW(\xi)G(\xi, \delta = 0)}{BW_0 G_0(\delta = 0)} = \frac{BW(\xi)}{BW_0}\left(1 - \frac{\Delta G}{G_0}\right). \quad \textbf{(8.45)}$$

Here subindex 0 designates the values of the gain and bandwidth that correspond to the absence of stagger tuning. For the gain it means that $G_0(\delta = 0) = G^{(const)}$ given by (8.40). These ratios are shown in Fig. 8.2 as functions of the stagger-tuning parameter $\bar{\xi} = (\xi/2)^2$ for several values of G_0. As can be seen from Fig. 8.2, the maximum gain-bandwidth product corresponds to $\bar{\xi}$ smaller than $\bar{\xi}_{max}$ and, as the constant gain decreases, the value of the stagger tuning parameter, which yields the maximum gain-bandwidth product, decreases as well.

Note that (8.41) allows one to analyze in a similar fashion also the large-signal gain of the gyroklystron. We will not analyze this large-signal gain here (this was done by Nusinovich, Danly, and Levush 1997), but only mention that

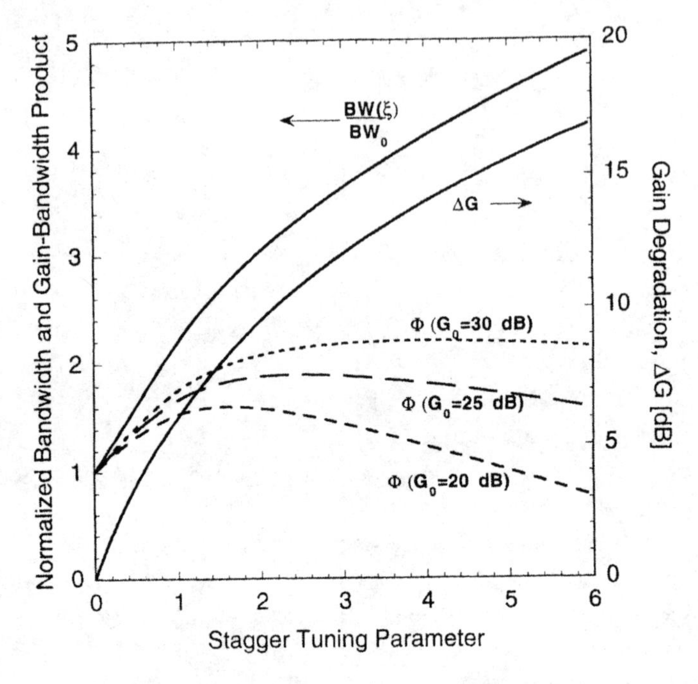

Fig. 8.2. Normalized bandwidth, gain-bandwidth product, and the gain degradation as functions of the stagger-tuning parameter in a two-cavity gyroklystron. (Reproduced from Nusinovich, Danly, and Levush 1997)

in the large-signal regime the variable gain given by (8.41) can be represented as a sum of two terms, first of which,

$$G_{LS}^{(\mathrm{var})} = 10 \log \left\{ \frac{q^2}{4J_1^2(q)} \right\}, \tag{8.46}$$

describes the saturation effects, while the second one is the variable part of the small-signal gain given by (8.42), which was analyzed above. Let us emphasize that the large signal gain (8.46) also depends on the signal frequency because, in accordance with (8.29), $q = q_0/\sqrt{1 + (\delta_1')^2}$.

As mentioned above, our general formalism is also applicable to the case when different cavities operate at different cyclotron harmonics (see Nusinovich, Levush, and Danly 1999 and references therein).

Before closing this chapter, let us emphasize that here we have focused the readers' attention only on the issues associated with the basic physics of gyroklystrons. For carrying out the gyroklystron experiments, it is necessary to do the design of gyroklystrons. This design methodology for gyroklystrons driven by weakly relativistic and relativistic electron beams can be found, respectively, in Ganguly and Chu (1981) and Chu et al. (1985).

8.5 Problems and Solutions

Problems

1. Derive Eq. (8.7).
2. Electron beams used in gyroklystrons have a certain velocity spread. In order for electron energy modulation in the input cavity to dominate over this spread, the input power should be large enough. Find the minimum input power and corresponding limitation on the gain as functions of the orbital velocity spread. Make estimates for the input cavity of a W-band (94 GHz) gyroklystron operating at the fundamental cyclotron resonance and driven by a 70 kV, 5 A electron beam with the orbital-to-axial velocity ratio equal to 1. Assume that the operating mode of a cylindrical cavity is $TE_{0,1,1}$, the axial structure of the mode is sinusoidal, the cavity length L is equal to 2 wavelengths, the cavity Q-factor is 100, and the beam power is 0.9 of the threshold, which corresponds to the self-excitation of the input cavity. Also, for the estimate of the gain, assume that the device operates with 30% efficiency.
3. Estimate the maximum orbital efficiency of a two-cavity gyroklystron with the output cavity operating at the second cyclotron harmonic for two cases: (a) the input cavity operates at the fundamental cyclotron resonance, and (b) the input case operates at the second harmonic. Assume that both cavities are short and the axial structure of the RF field there is constant, i.e., $f(\varsigma) = 1/\mu$, where μ is the normalized length of the cavity. Take into account that the ballistic bunching in short cavities can be neglected that reduces equations for electron motion (3.34) or (3.35) in such a cavity to a set of linear differential equations. (*Note*: This problem was originally formulated and studied by V. K. Yulpatov.)

Solutions

1. Consider an input cavity coupled to the input waveguide assuming that the diffractive losses in the cavity are negligibly small. The power flow into the cavity, P, relates to the power flow in the waveguide fed by the driver, P_{dr}, as (Slater, 1950, Eq. (3.4) in Chapter 5)

$$P = \frac{4Q_{cpl}/Q_{ohm}}{(1 + Q_{cpl}/Q_{ohm})^2 + \delta_{Sl}^2} P_{dr}. \tag{1}$$

Here the detuning $\delta_{Sl} = Q_{cpl}[\omega^2 - (\omega_1')^2]/\omega^2 \approx 2Q_{cpl}(\omega - \omega_1')/\omega$ is the detuning used by Slater.

In the stationary regime, the power flow into the cavity compensates for the microwave losses there, i.e., $P = (\omega/Q_{ohm})W$, where W is the microwave energy stored in the cavity. Expressing in (1) P via W yields

$$\frac{\omega}{Q_{1,T}}W = \frac{4P_{dr}Q_{1,T}}{Q_{cpl}(1+\delta^2)} \tag{2}$$

Here $Q_{1,T}$ is the total Q-factor of the input cavity, which is determined by the obvious relation (cf. [3.41]) $Q_{1,T}^{-1} = Q_{ohm}^{-1} + Q_{cpl}^{-1}$, and $\delta = (Q_{1,T}/Q_{cpl})\delta_{Sl}$ is our normalized detuning used in the text of Chapter 8.

Note that Slater considered the input cavity without a beam. Representing the input cavity as an equivalent circuit, in which the cold-cavity frequency is given by $\omega_1^2 = 1/LC$, one can treat the beam as a dielectric medium with the susceptibility χ in the capacitor, so in the presence of the beam $C = C_0(1 + 4\pi\chi)$. Assuming that the susceptibility χ is small enough and using (3.61), one can express the detuning in the presence of the beam as $\delta + I_0\hat{\chi}$. Clearly, in the case of the complex susceptibility, the denominator in (1) and (2) should be properly modified as was done in Chapter 8.

In the presence of the beam, the microwave losses and the power withdrawn from the beam are related in the stationary regime as

$$\frac{\omega}{Q_{1,T}}W = \eta P_b = \eta_\perp P_{b\perp}. \tag{3}$$

Thus, combining (2) and (3), using Eq. (8.3) and Eq. (8.4) given in the text and taking into account the modification of δ discussed above result in

$$|F_1|^2 = \frac{4P_{dr}Q_{1,T}I_0/Q_{cpl}P_{b\perp}}{|1+\delta+I_0\hat{\chi}|^2}. \tag{4}$$

Comparing (4) with Equation (8.8), one can readily get the definition of A^2 given by (8.7).

2. When electrons interact with TE-modes at the fundamental resonance, the energy modulation, as follows from (1.2), is on the order of

$$\Delta\mathcal{E} \sim ev_\perp AL/v_z. \tag{1}$$

Here A is the amplitude of the RF field $\vec{E} = \mathrm{Re}\{A\vec{E}_s(\vec{r})e^{i\omega t}\}$. Since most of the gyroklystrons operate at relatively low voltages and the interaction changes mainly the electron orbital momentum, one can assume that $\Delta\mathcal{E} \sim (m/2)\Delta(v_\perp^2)$ and derive from (1) the following estimate for the orbital velocity modulation:

$$\left|\frac{\Delta v_\perp}{v_\perp}\right| \sim \frac{eAL}{mc^2}\frac{1}{\beta_\perp\beta_z}. \tag{2}$$

So, for this modulation to dominate over the initial spread, the field amplitude should be larger than

$$A_{\min} \approx \frac{mc^2}{eL}\beta_\perp\beta_z\left|\frac{\Delta\beta_\perp}{\beta_\perp}\right|_{\text{in}}. \tag{3}$$

In accordance with the definition of the normalized field amplitude given in Subsection 3.2, the minimal field amplitude given by Eq. (3) corresponds to the minimal normalized field amplitude:

$$\left|F_{I,\min}\right| = \frac{eA_{\min}}{mc\omega\gamma_0}\beta_\perp^{-3}. \tag{4}$$

Taking into account the balance equation (8.8), in which we will ignore the frequency pulling and detuning effects, and the definition of the normalized intensity of the field excited by a driver given by (8.7), one can readily get the following relation between the minimum power of a driver required for overcoming the effect of velocity spread:

$$P_{\text{dr,min}}(\text{kW}) = 8.77\cdot 10^3\frac{Q_{\text{cpl}}}{Q_{1,T}^2}\left(1 - I_b/I_{b,st}\right)^2\frac{\beta_{z0}^4}{\beta_{\perp 0}^2}\frac{1}{G}\left(\frac{\lambda}{L_1}\right)^3\left|\frac{\Delta\beta_\perp}{\beta_\perp}\right|_{\text{init}}^2. \tag{5}$$

Here we took into account that $m^2c^5/e^2 = 8.7\cdot 10^6\ kW$ and used the assumptions about operation at the fundamental resonance and about the sinusoidal axial structure of the cavity field. The coupling impedance in (5) is determined by (3.59).

The maximum gain, as obvious, is equal to

$$G_{\max} = 10\log P_{\text{out}} - 10\log P_{\text{dr,min}}. \tag{6}$$

Now, we can make the estimates. For the $\text{TE}_{0,1}$-mode, in the case of the optimal positioning of a thin annular electron beam, the coupling impedance is equal to 0.1433. For given values of the beam voltage and orbital-to-axial velocity ratio, $\beta_\perp^2 = \beta_z^2 \simeq 0.113$. Therefore, for given values of the Q-factor (assume critical coupling of the input cavity, i.e., $Q_{\text{cpl}} = 2Q_{1,T}$), cavity length and the beam current to start current ratio, (5) yields

$$P_{\text{dr,min}}(W) \simeq 172.8\left|\frac{\Delta\beta_\perp}{\beta_\perp}\right|_{\text{init}}^2.$$

So, for instance, for the case of a 5% spread, the minimum drive power is about 0.432W. Correspondingly, the maximum gain, as follows from (6) and the data for the beam power and efficiency, is about 54 dB.

3. For our axial structure, a weak energy modulation at the exit from the input cavity, as follows from (3.35), is approximately equal to

$$w(\mu_1) = 1 - 2|F_1|\sin(s_1\vartheta_0 - \psi_1).$$

Due to cyclotron resonance mismatch and the ballistic bunching in a long drift section, the electron phase at the entrance to the output cavity/exit from the drift section is equal to

$$\vartheta(\mu_{\rm dr}) = \vartheta_0 - \Delta\mu_{\rm dr} + q\sin(s_1\vartheta_0 - \psi),$$

where $q = 2 \mid F_1 \mid \mu_{\rm dr}$ is the bunching parameter and ψ_1 is the phase of the input cavity field. This equation is a generalization of (8.31) for the case of an arbitary harmonic s_1.

Let us now rewrite (3.35) for the output cavity assuming $s_2 = 2$ and $\mid w - 1 \mid << 1$. This yields

$$\frac{dw}{d\varsigma} = 2w\,\mathrm{Im}\{Ff(\varsigma)e^{-i2\vartheta}\},$$

$$\frac{d\vartheta}{d\varsigma} + \Delta = -\mathrm{Re}\{Ff(\varsigma)e^{-i2\vartheta}\}.$$

Introducing $a = \sqrt{w}e^{-i\vartheta}$ reduces these equations to linear equations

$$\frac{da}{d\varsigma} - i\Delta a = i(Ffa)^*,\ \frac{da^*}{d\varsigma} + i\Delta a^* = -iFfa.$$

For $f(\varsigma) = 1/\mu_2$ this yields a characteristic equation (assume a and $b = a^*$ proportional to e^{pt})

$$p^2 = \Delta^2 - A^2,$$

where $A = |F_2|/\mu_2$. Then, the solution is

$$a(\varsigma) = e^{-i\vartheta_{\rm dr}}\left\{\cos p\varsigma + i\frac{\sin p\varsigma}{p}[\Delta + Ae^{i(2\vartheta_{\rm dr} - \psi_2)}]\right\}$$

for $p^2 > 0$ and

$$a(\varsigma) = e^{-i\vartheta_{\rm dr}}\left\{\cosh|p|\varsigma + i\frac{\sinh|p|\varsigma}{|p|}[\Delta + Ae^{i(2\vartheta_{\rm dr} - \psi_2)}]\right\}$$

for $p^2 < 0$.

In our notations, the orbital efficiency (3.36) is equal to

$$\eta_\perp = \left\langle 1 - |a(\mu_{\rm out})|^2 \right\rangle.$$

Here for $p^2 > 0$

$$|a(\mu_{\text{out}})|^2 = 1 + 2\left(\frac{\sin p\mu_2}{p}\right)^2 (A^2 + \Delta A \cos \vartheta'_{\text{dr}})$$
$$-2A\frac{\sin p\mu_2}{p} \cos p\mu_2 \sin \vartheta'_{\text{dr}}$$

and for $p^2 < 0$

$$|a(\mu_{\text{out}})|^2 = 1 + 2\left(\frac{\sinh |p|\, \mu_2}{|p|}\right)^2 (A^2 + \Delta A \cos \vartheta'_{\text{dr}})$$
$$-2A\frac{\sinh |p|\, \mu_2}{|p|} \cosh |p|\, \mu_2 \sin \vartheta'_{\text{dr}}.$$

In these equations we introduced $\vartheta'_{\text{dr}} = 2\vartheta_{\text{dr}} - \psi_2$. Averaging over the entrance phase results in
(a) for $s_1 = 1$

$$\langle \cos \vartheta'_{\text{dr}} \rangle = \langle \cos\{2[\vartheta_0 - \Delta\mu_{\text{dr}} + q \sin(\vartheta_0 - \psi_1)] - \psi_2\} \rangle$$
$$= J_2(2q) \cos(2\psi_1 - \psi_2 - 2\Delta\mu_{\text{dr}})$$

and

$$\langle \sin \vartheta'_{\text{dr}} \rangle = J_2(2q) \sin(2\psi_1 - \psi_2 - 2\Delta\mu_{\text{dr}}).$$

Thus, introducing $\Delta\psi = 2\psi_1 - \psi_2 - 2\Delta\mu_{\text{dr}}$ and $x = (A/p) \sin p\mu_2$, we can rewrite the orbital efficiency in the case of $p^2 > 0$ as

$$\eta_\perp = 2x\{J_2(2q)[\cos p\mu_2 \sin \Delta\psi - \frac{\Delta}{A} x \cos \Delta\psi] - x\}$$

and, correspondingly, introducing in the case of $p^2 < 0$ $x = (A/|p|) \sinh |p|\, \mu_2$ yields

$$\eta_\perp = 2x\{J_2(2q)[\cosh |p|\, \mu_2 \sin \Delta\psi - \frac{\Delta}{A} x \cos \Delta\psi] - x\}.$$

In both cases, the orbital efficiency being optimized with respect to the phase difference $\Delta\psi$ is determined by the same expression:

$$\eta_\perp(\Delta\psi_{\text{opt}}) = 2x\{J_2(2q)\sqrt{1 + x^2} - x\}.$$

From this equation one can readily find the optimum value of the parameter x:

$$2x_{\text{opt}}^2 = \frac{1}{\sqrt{1 - J_2^2(2q)}} - 1.$$

Correspondingly, the maximized orbital efficiency is equal to

$$\eta_{\perp}(\Delta\psi_{\mathrm{opt}}, x_{\mathrm{opt}}) = 1 - \sqrt{1 - J_2^2(2q)}.$$

So, the optimal value of the bunching parameter is equal to $q_{opt} \simeq 1.55$ and the corresponding maximum orbital efficiency is approximately equal to 0.126.

In the same fashion we can analyze the case (b) $s_1 = 2$ and find that in this case $\langle \cos \vartheta'_{\mathrm{dr}} \rangle = \langle \cos\{2[\vartheta_0 - \Delta\mu_{\mathrm{dr}} + q\sin(2\vartheta_0 - \psi_1)] - \psi_2\} \rangle = -J_1(2q)\cos(\psi_1 - \psi_2 - 2\Delta\mu_{\mathrm{dr}})$ and $\langle \sin \vartheta'_{\mathrm{dr}} \rangle = -J_1(2q)\sin(\psi_1 - \psi_2 - 2\Delta\mu_{\mathrm{dr}})$.

Correspondingly, for the case of $p^2 > 0$ the orbital efficiency is now equal to

$$\eta_{\perp} = 2x\left\{J_1(2q)\left[\frac{\Delta}{A}x\cos\Delta\psi - \cos p\mu_2 \sin\Delta\psi\right] - x\right\},$$

(as above, here $x = (A/p)\sin p\mu_2$) and a similar expression can be written for the case of $p^2 < 0$. Then, repeating the same procedure of optimization with respect to $\Delta\psi$ and x, one can readily find that the maximized efficiency is equal to

$$\eta_{\perp}(\Delta\psi_{\mathrm{opt}}, x_{\mathrm{opt}}) = 1 - \sqrt{1 - J_1^2(2q)}.$$

So, in the case of operation in both cavities at the second harmonic, the optimal value of the bunching parameter is equal to $q_{\mathrm{opt}} \simeq 0.92$ and the maximum orbital efficiency is equal to 0.186. Note that our treatment is close to the point-gap model of the gyroklystron with cavities operating at different harmonics analyzed elsewhere (Nusinovich, Saraph, and Granatstein 1997). In that reference, the orbital efficiency is determined by a simple equation $\eta_{\perp} = J^2_{s_2/s_1}(s_2 q)/s_2$. Therefore, the optimal values of bunching parameters obtained in our consideration are the same as in the cited reference paper. Also, the maximum values of the orbital efficiency obtained in our consideration (0.126 and 0.186, for the 1-2 and 2-2 schemes, respectively) are close to the 0.118 and 0.17 that were obtained earlier. A slight increase in the maximum efficiency in our treatment can be attributed to the fact that we took into account the transit effects (mismatch Δ) in the last cavity.

■ CHAPTER 9 ■

Fluctuations: Intrinsic and Extrinsic Noise

9.1 Radiation Linewidth, Sources of Noise

When a source of electromagnetic radiation generates or amplifies a harmonic signal with the constant amplitude, A, and phase, φ,

$$E = Ae^{i(\omega t+\varphi)}, \tag{9.1}$$

the spectral line of this signal is the delta-function. However, such ideal sources do not exist. In real devices, the radiation spectral line always has a finite thickness. More exactly, the radiation spectrum of a real device typically consists, as shown in Fig. 9.1, of a very narrow signal spike and a broadband pedestal. Clearly, the finite width of the signal spike as well as the presence of the pedestal degrade the signal and reduce the amount of information that can be conveyed in communication systems. These factors also restrict the capabilities of the radar systems used for detecting the moving targets by receiving the Doppler-shifted reflected signal as well as the capabilities of the active microwave diagnostics based on collective Thomson scattering of electromagnetic waves from plasma.

For the field representation given by (9.1), this spectral broadening can be interpreted as the fact that the amplitude and phase of what can be called a "monochromatic" radiation are slowly variable functions of time, $A = A(t)$, and $\varphi = \varphi(t)$, for which

$$\left|\frac{1}{A}\frac{dA}{dt}\right|, \left|\frac{d\varphi}{dt}\right| << \omega. \tag{9.2}$$

This time dependence of the amplitude and phase, being Fourier transformed, indicates the presence of spectral components of the signal in the vicinity of the signal frequency ω that is equivalent to broadening of the radiation spectrum.

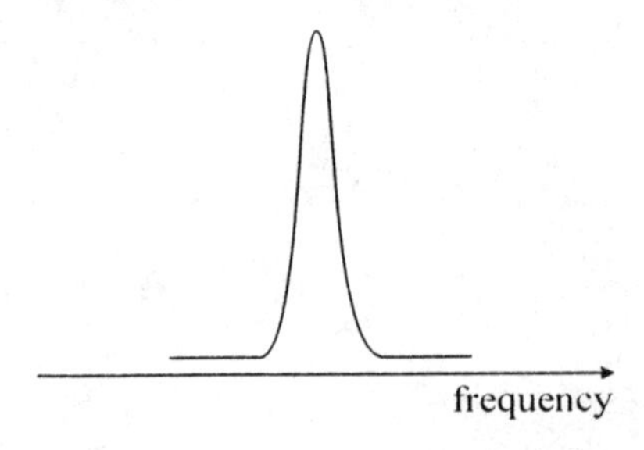

Fig. 9.1. Radiation line in a source of electromagnetic radiation.

In microwave sources driven by electron beams, there are many reasons for broadening the radiation linewidth (see, e.g., Smullin and Haus 1959). All these reasons can be broken into two categories. This first group can be classified as the sources of the *intrinsic noise*. This noise mostly originates from processes directly associated with the formation and transport of an electron beam. There are known several sources of such a noise. First, there is a *shot noise*, which is caused by the discreteness of the electronic charge. Second, there is a *flicker noise*, which is associated with imperfections in the emission process (presence of "foreign" atoms at the emitter surface; see Schottky 1926). Third, there is also *velocity noise*, which originates from the distribution of electrons emitted from a hot cathode in velocities. Finally, the *thermal agitation noise*, which originates from the finite temperature of microwave circuit walls and the Kirchhoff radiation caused by it, can be attributed to this group. The sources of the intrinsic noise, which are known for practically all microwave tubes with thermionic cathodes, determine what is usually called a natural line width of a device. In addition to them, in gyrotrons, oscillations of electrons trapped in the region of the beam magnetic compression, which we discussed in Chapter 2, can also lead to parasitic noise. As shown by Sukhorukov and Sheludchenkov (2002), the oscillations of trapped electrons cause the appearance of additional noise at frequencies of about 100 MHz. These frequencies correspond to oscillations of trapped electrons with periods on the order of 10 nsec. The authors also found that these noiselike oscillations, which were also detected by other experimentalists, may leak through the cathode ceramic and their power level may exceed an admissible level of parasitic radiation in this frequency range, which is used for radio and TV broadcasting.

The second group includes the sources of the *extrinsic* or *technical noise*, which is the noise caused by fluctuations in the operating parameters of a device. These parameters can fluctuate about their nominal values due to the fluctuations in external power supplies attached to the device. For gyrodevices, the most important operating parameters are the mod-anode and beam voltages, the beam current, and external magnetic fields. (Of course, in the case of using the superconducting solenoids the fluctuations in magnetic fields can be ignored.) Let us note that our classification as well as any other classification is, to some extent, artificial because, for instance, fluctuations in the beam current can be attributed to any of the two categories. More exactly, they can be caused by different reasons, part of which can be treated as the sources of the intrinsic noise, while others (such as, for instance, the noise generated by the filament heating a cathode; see, e.g., Gilmour 1994) can be called the sources of the extrinsic noise. Also note that these sources of noise can make contributions to different parts of the spectral line width of a microwave source.

In this chapter, we will consider the sources of noise listed above and their effect on the radiation line width of gyrodevices. Our consideration will be primarily focused on gyrodevices using the cavities (gyrotron oscillators and gyroklystrons). We will also analyze the phase stability in gyroklystrons and gyro-traveling-wave tubes (gyro-TWTs).

9.2 General Formalism

In previous chapters we developed a general formalism describing the operation of gyrotron oscillators, gyroklystrons, and gyro-TWTs in the absence of noise. This theory was developed with the use of a small number of normalized parameters. These normalized parameters, in turn, depend on the real operating parameters, such as the mod-anode and beam voltages, beam current, magnetic fields, etc. Therefore, in order to apply our formalism for describing the noise in gyrodevices, we should know how fluctuations in the technical parameters cause corresponding fluctuations in the normalized parameters. Then, since the amplitude and phase (or frequency, in the case of free-running oscillators) of the radiated microwave field depend on the normalized parameters, we can show how fluctuations in these parameters result in the appearance of what is known as the amplitude and phase noise.

Let us show how this effect can be studied in a free-running gyrotron oscillator (Dumbrajs and Nusinovich 1997). We shall start our consideration from the general formalism, which was developed by Malakhov (1968) and is applicable to any microwave oscillator.

General Form

In any microwave oscillator, where an electron beam excites oscillations in a cavity, the amplitude and phase of these oscillations, in accordance with

Eq. (3.50), can be determined by the equations

$$\frac{dA}{dt} = \frac{\omega_S}{2} A \left(4\pi \chi'' - \frac{1}{Q}\right), \tag{9.3}$$

$$\frac{d\varphi}{dt} = -\frac{\omega_S}{2} \left(4\pi \chi' - 2\frac{\omega - \omega_S}{\omega_S}\right). \tag{9.4}$$

Here ω_S is again the real part of the cold-cavity frequency and the susceptibility χ, in accordance with (3.50), is equal to

$$\chi = \chi' + i\chi'' = -\frac{i}{2\pi \omega_S AN} \frac{1}{T} \int_t^{t+T} \left[\int_V \vec{j} \cdot \vec{E}_S^* e^{-i(\omega t + \varphi)} dv \right] dt. \tag{9.5}$$

Let us assume that in the absence of fluctuations the device exhibits stationary oscillations with the amplitude A_0 and phase φ_0, which obey the balance equations

$$4\pi \chi''(A_0, \omega) = \frac{1}{Q}, \; 4\pi \chi'(A_0, \omega) = 2\frac{\omega_S - \omega}{\omega_S}. \tag{9.6}$$

Note that, since we are considering a free-running oscillator, these equations determine the amplitude and frequency of oscillations, but not the phase.

Then, fluctuations in technical parameters $\xi_i = \xi_{i0} + \delta\xi_i(t)$ (here i is the parameter index and $|\delta\xi_i| << \xi_{i0}$) of the device, should cause some variations of the amplitude and phase about stationary values:

$$A(t) = A_0[1 + \alpha(t)], \quad \varphi(t) = \varphi_0 + \delta\varphi(t). \tag{9.7}$$

Below we will assume that these variations are small, i.e., $|\overline{\alpha^2}| << 1$ and $[d(\delta\varphi)/dt] = \nu$ with $\overline{\nu^2} << \omega^2$. (Here the horizontal bar means time averaging.) Then, the effect of fluctuations on the susceptibility can be taken into account by expanding this susceptibility in the Taylor series about stationary values of the field amplitude and frequency and the operating parameters

$$\chi = \chi(A_0, \omega) + \left.\frac{\partial \chi}{\partial A}\right|_{A_0} A_0 \alpha + \frac{\partial \chi}{\partial \omega} \nu + \sum_i \left[\frac{\partial \chi}{\partial \xi_i}\right]_{\xi_{i0}} \delta\xi_i. \tag{9.8}$$

This expansion allows us to reduce (9.3) and (9.4) to two linear equations for the perturbations in the field amplitude and frequency:

$$\frac{d\alpha}{dt'} + p\alpha = \delta\chi'', \tag{9.9}$$

$$\overline{\nu} + q\alpha = -\delta\chi'. \tag{9.10}$$

Here $t' = 2\pi\omega_S t$ and $\overline{\nu} = \nu/2\pi\omega_S$ are the dimensionless time and frequency, respectively, parameter $q = A_0 \frac{\partial \chi'}{\partial A}|_{A_0} (1 + 2\pi\omega_S \frac{\partial \chi'}{\partial \omega})^{-1}$ characterizes the dependence of the oscillation frequency on the oscillation amplitude (it is often called a "nonisochronism" of an oscillator), and parameter $p = q2\pi\omega_s \frac{\partial\chi''}{\partial\omega} - A_0 \frac{\partial\chi''}{\partial A}|_{A_0}$ describes the restoring of an oscillator to its steady state with the amplitude of oscillations $A = A_0$. The perturbations in the susceptibility, which stand in the right-hand sides of (9.9) and (9.10), determine the direct effect of perturbations in the device parameters:

$$\delta\chi'' = \sum_i A_i \frac{\delta\xi_i}{\xi_{i0}}, \ \delta\chi' = \sum_i B_i \frac{\delta\xi_i}{\xi_{i0}}. \tag{9.11}$$

Here coefficients A_i and B_i are, respectively, equal to

$$A_i = \xi_{i0} \left(\frac{\partial\chi''}{\partial\xi_i} - \frac{2\pi\omega_s (\partial\chi''/\partial\omega)}{1 + 2\pi\omega_s (\partial\chi'/\partial\omega)} \frac{\partial\chi'}{\partial\xi_i} \right)_{\xi_{i0}}, \tag{9.12}$$

and

$$B_i = \xi_{i0} \left. \frac{\partial\chi'}{\partial\xi_i} \right|_{\xi_{i0}} \left(1 + 2\pi\omega_s \frac{\partial\chi'}{\partial\omega} \right)^{-1}. \tag{9.13}$$

Eqs. (9.12) and (9.13) as well as definitions of parameters q and p given above contain some terms that under normal conditions are negligibly small. Indeed, the derivative $\partial\chi'/\partial\omega$ is on the order of $\chi'T$ since the electron transit time through the resonator, T, determines the spectrum width of the resonant interaction band (for gyrotrons this is the cyclotron resonance band). At the same time, real and imaginary parts of the susceptibility are, in general, of the same order, and the imaginary part of the susceptibility, as it follows from (9.6), is on the order of $1/Q$. This means that the term $\omega_s \partial\chi'/\partial\omega$ in (9.12), (9.13), and definitions of q and p is on the order of $\omega T/Q$. Typically, in microwave oscillators utilizing high-Q cavities (this sort of device includes gyrotrons) the electron transit time is much smaller than the cavity decay time. Therefore, this term can be neglected, and, correspondingly, the equations given above can be reduced to

$$q = A_0 \left. \frac{\partial\chi'}{\partial A} \right|_{A_0}, \ p = -A_0 \left. \frac{\partial\chi''}{\partial A} \right|_{A_0}, \ A_i = \xi_{i0} \left. \frac{\partial\chi''}{\partial\xi_i} \right|_{\xi_{i0}}, \ B_i = \xi_{i0} \left. \frac{\partial\chi'}{\partial\xi_i} \right|_{\xi_{i0}}. \tag{9.14}$$

The spectral line of a microwave source, however, is determined not by the perturbations in the amplitude and phase of oscillations, but by their correlation functions. In order to determine these functions and corresponding spectral densities, it is necessary to add to Eqs. (9.9) and (9.10) the same equations written for another instant of time, $t + \tau$ (here and below the prime for dimensionless time t and the bar for dimensionless frequency ν are omitted, τ is the delay time):

$$\frac{d\alpha_\tau}{dt} + p\alpha_\tau = \delta\chi''_\tau, \tag{9.15}$$

$$\nu_\tau + q\alpha_\tau = -\delta\chi'_\tau. \tag{9.16}$$

Here the subscript τ designates the fact that corresponding variables are taken for the instant of time $t + \tau$.

Below we will consider only stationary random processes for which the correlation functions depend on the delay time τ, but not on t (see, e.g., Malakhov 1968, Stratonovich 1963). Multiplying (9.9) by (9.15) and averaging this product over all possible realizations yield

$$\left\langle \frac{d\alpha}{dt}\frac{d\alpha_\tau}{dt} \right\rangle + p\left\langle \alpha\frac{d\alpha_\tau}{dt} + \alpha_\tau\frac{d\alpha}{dt} \right\rangle + p^2\langle \alpha\alpha_\tau \rangle = \Phi_{22}. \tag{9.17}$$

Here $\Phi_{22}(\tau) = \langle \delta\chi''(t)\delta\chi''(t+\tau) \rangle$ and for any variable $x(t)$ such an averaged product $\langle x(t)x(t+\tau) \rangle$ means $\iint x_1 x_2 W(x_1, t_1; x_2, t_2)dx_1 dx_2$, where the function W describes the probability of the variable $x(t)$ equal to x_1 at $t = t_1$ to be located in the interval $[x_2, x_2 + dx_2]$ at $t = t_2$. Using the relations between the functions and their derivatives, which are valid for stationary random processes,

$$\left\langle \alpha\frac{d\alpha_\tau}{dt} \right\rangle = \frac{d\Phi_\alpha}{d\tau}, \left\langle \frac{d\alpha}{dt}\alpha_\tau \right\rangle = -\frac{d\Phi_\alpha}{d\tau}, \left\langle \frac{d\alpha}{dt}\frac{d\alpha_\tau}{dt} \right\rangle = -\frac{d^2\Phi_\alpha}{d\tau^2},$$

one can reduce (9.17) to

$$-\frac{d^2\Phi_\alpha}{d\tau^2} + p^2\Phi_\alpha = \Phi_{22}. \tag{9.18}$$

Here $\Phi_\alpha(\tau) = \langle \alpha(t)\alpha(t+\tau) \rangle$ is the correlation function for amplitude fluctuations.

In a similar fashion, we can multiply (9.10) by (9.16) and after averaging obtain

$$\Phi_\nu + q(\Phi_{\alpha\nu} + \Phi_{\nu\alpha}) + q^2\Phi_\alpha = \Phi_{11}, \tag{9.19}$$

where $\Phi_{11}(\tau) = \langle \delta\chi'(t)\delta\chi'(t+\tau) \rangle$, and $\Phi_\nu(\tau) = \langle \nu(t)\nu(t+\tau) \rangle$ is the correlation function for the frequency fluctuations, while $\Phi_{\alpha\nu}(\tau) = \langle \alpha(t)\nu(t+\tau) \rangle$ and $\Phi_{\nu\alpha}(\tau) = \langle \nu(t)\alpha(t+\tau) \rangle$ are cross-correlation functions. The two latter functions also obey two more equations, which follow from multiplying, respectively, (9.9) by (9.16) and (9.10) by (9.15) and averaging over all possible realizations:

$$-\frac{d\Phi_{\alpha\nu}}{d\tau} + p\Phi_{\alpha\nu} - q\frac{d\Phi_\alpha}{d\tau} + pq\Phi_\alpha = -\Phi_{21}, \tag{9.20}$$

$$\frac{d\Phi_{\nu\alpha}}{d\tau} + p\Phi_{\nu\alpha} + q\frac{d\Phi_\alpha}{d\tau} + pq\Phi_\alpha = -\Phi_{12}. \tag{9.21}$$

Here $\Phi_{21}(\tau) = \langle \delta\chi''(t)\delta\chi'(t+\tau) \rangle$ and $\Phi_{12}(\tau) = \langle \delta\chi'(t)\delta\chi''(t+\tau) \rangle$.

Note that any cross-correlation function describing the correlation between two fluctuating variables $\Phi_{x,y}(\tau)$ can be represented as the sum of its symmetric, $\Phi^{(0)}_{x,y}(\tau) = \Phi^{(0)}_{x,y}(-\tau)$, and antisymmetric, $\Phi^{(1)}_{x,y}(\tau) = -\Phi^{(1)}_{x,y}(-\tau)$, components. Also, as it follows from the definition of these functions, $\Phi_{x,y}(\tau) = \Phi_{y,x}(-\tau)$. These properties yield $\Phi^{(0)}_{y,x}(\tau) = \Phi^{(0)}_{x,y}(\tau)$ and $\Phi^{(1)}_{y,x}(\tau) = -\Phi^{(1)}_{x,y}(\tau)$. As a result, Eqs. (9.20) and (9.21) can be rewritten for the symmetric and antisymmetric cross-correlation functions as

$$\frac{d\Phi^{(0)}_{\alpha\nu}}{d\tau} - p\Phi^{(1)}_{\alpha\nu} + q\frac{d\Phi_\alpha}{d\tau} = -\Phi^{(1)}_{12}, \tag{9.22}$$

$$\frac{d\Phi^{(1)}_{\alpha\nu}}{d\tau} - p\Phi^{(0)}_{\alpha\nu} - pq\Phi_\alpha = \Phi^{(0)}_{12}. \tag{9.23}$$

Also (9.19) can be reduced to

$$\Phi_\nu + 2q\Phi^{(0)}_{\alpha\nu} + q^2\Phi_\alpha = \Phi_{11}. \tag{9.24}$$

The last step that we should make in order to determine the spectral line of the radiation in general terms is to make a Fourier transform of the correlation functions. Their spectral densities can be determined for the case of one variable as $W_x(\Omega) = \frac{1}{2\pi}\int_{-\infty}^{\infty}\Phi_x(\tau)\cos\Omega\tau d\tau$ and for the case of two variables as $W^{(0)}_{x,y}(\Omega) = \frac{1}{2\pi}\int_{-\infty}^{\infty}\Phi^{(0)}_{x,y}(\tau)\cos\Omega\tau d\tau$, $W^{(1)}_{x,y}(\Omega) = \frac{1}{2\pi}\int_{-\infty}^{\infty}\Phi^{(1)}_{x,y}(\tau)\sin\Omega\tau d\tau$. This allows us to rewrite Eqs. (9.18), (9.24), (9.22), and (9.23) as, respectively,

$$W_\alpha(\Omega) = \frac{W^{(0)}_{22}(\Omega)}{p^2+\Omega^2}, \tag{9.25}$$

$$W_\nu(\Omega) + 2qW^{(0)}_{\alpha\nu} + q^2W_\alpha = W^{(0)}_{11}, \tag{9.26}$$

$$W^{(0)}_{\alpha\nu} = -\frac{1}{p^2+\Omega^2}\left(qW^{(0)}_{22} + pW^{(0)}_{12} + \Omega W^{(1)}_{12}\right), \tag{9.27}$$

$$W^{(1)}_{\alpha\nu} = \frac{1}{p^2+\Omega^2}\left(pW^{(1)}_{12} - \Omega W^{(0)}_{12}\right). \tag{9.28}$$

As follows from (9.25)–(9.28), the spectral density of the frequency noise is equal to

$$\begin{aligned} W_\nu(\Omega) = \frac{1}{p^2+\Omega^2}\Bigg\{&\sum_i\left[(p^2+\Omega^2)B_i^2 + q^2A_i^2 + 2qpA_iB_i\right]W_i \\ &+2\sum_{k>i}\sum_i\left[(p^2+\Omega^2)B_iB_k + q^2A_iA_k + qp(B_iA_k+B_kA_i)\right]W^{(0)}_{ik} \\ &+2q\Omega\sum_{k>i}\sum_i(B_iA_k - B_kA_i)W^{(1)}_{ik}\Bigg\}. \end{aligned} \tag{9.29}$$

These equations express all noise spectral densities in terms of the spectral densities of correlation functions for fluctuating susceptibilities. These correlation functions, with the use of Eq. (9.11), can be expressed, in turn, in terms of the spectral densities of fluctuations in technical parameters as

$$W_{11}^{(0)} = \sum_i B_i^2 W_i + 2\sum_{k>i}\sum_i B_i B_k W_{ik}^{(0)}, \tag{9.30}$$

$$W_{22}^{(0)} = \sum_i A_i^2 W_i + 2\sum_{k>i}\sum_i A_i A_k W_{ik}^{(0)}, \tag{9.31}$$

$$W_{12}^{(0)} = \sum_i A_i B_i W_i + \sum_{k>i}\sum_i (B_i A_k + B_k A_i) W_{ik}^{(0)}, \tag{9.32}$$

$$W_{12}^{(1)} = \sum_{k>i}\sum_i (B_i A_k - B_k A_i) W_{ik}^{(1)}. \tag{9.33}$$

Here, on the RHS of these equations we used the spectral densities of fluctuations in the parameters that correspond to the cross-correlation function

$$\Phi_{ik}(\tau) = \left\langle \frac{\delta\xi_i}{\xi_{i0}} \frac{(\delta\xi_k)_\tau}{\xi_{k0}} \right\rangle.$$

Properties of this function are the same as properties of any cross-correlation function, which were discussed above.

These equations complete our derivation of general equations, which was quite straightforward, although rather lengthy. Now we should apply this formalism to analyze the noise in gyrotron oscillators.

Gyrotron Noise Formalism

Let us start from considering the simplest case of a free-running gyrotron oscillator. As was shown in Chapter 3 [see (3.61)], in this device, the susceptibility of an electron beam is equal to $\chi = (I_0/4\pi Q)\hat{\chi}$, where the normalized current parameter I_0 is proportional to the beam current [see (3.57) or (3.65)], and the susceptibility $\hat{\chi}$ depends on three normalized parameters: normalized field amplitude,

$$F = i\frac{eC_s}{mc\omega\gamma_0}\beta_{\perp 0}^{s-4}\frac{s^s}{2^{s-1}s!}L_s, \tag{9.34}$$

normalized cyclotron resonance mismatch,

$$\Delta = \frac{2(\omega - sh_H v_{z0})}{\beta_{\perp 0}^2\omega} = \frac{2}{\beta_{\perp 0}^2}\frac{\omega - s\Omega_0}{\omega}, \tag{9.35}$$

and normalized interaction length,

$$\mu = \left(\beta_{\perp 0}^2/2\beta_{z0}\right)(\omega L/c). \tag{9.36}$$

Below we will denote these parameters as u_1, u_2, and u_3, respectively.

These normalized parameters depend on the orbital and axial velocity components; also the cyclotron resonance mismatch depends on the electron cyclotron frequency, which, in turn, depends on the external magnetic field in the interaction region and on the beam voltage, which determines the electron energy. The dependence of velocity components on technical parameters of the device, in principle, can be analyzed with the use of one of the codes, which describe electron trajectories on the way from an electron gun to the interaction region. This problem, however, can be greatly simplified by using the adiabatic theory of magnetron-type electron guns, which we briefly discussed in Chapter 2. In the framework of this theory, the electron orbital velocity can be determined by (2.4), which can be rewritten as (Baird and Lawson 1986)

$$\beta_{\perp 0} = \frac{(B_0')^{1/2}}{(B_c')^{3/2}} \frac{\lambda \cos\psi}{2\pi R_c \ln(R_a/R_c)} \frac{V_m'}{1+V_b'}. \tag{9.37}$$

In this equation, ψ is the angle between the cathode surface and the axis, R_c is the cathode radius, the anode radius R_a is determined as $R_c + d\cos\psi$, where d is the distance between the cathode and the anode; primed mod-anode and beam voltages are normalized to mc^2/e (correspondingly, $\gamma_0 = 1 + V_b'$), and primed magnetic fields are normalized to $mc\omega/e$. Then, the axial electron velocity can be determined by the standard equation

$$\beta_{z0}^2 = 1 - \frac{1}{\gamma_0^2} - \beta_{\perp 0}^2. \tag{9.38}$$

Also recall that, as follows from Busch's theorem, the radius of the electron guiding center, which determines the coupling impedance of electrons to the resonator field [see (3.59) and (9.34)], depends on the magnetic field compression, as given by (2.5), and hence, fluctuates in fluctuating magnetic fields.

When an electron gun operates in the triode regime, which implies the use of a separate voltage applied to the modulating anode, and there is an additional gun coil for controlling the magnetic field compression and thus the orbital-to-axial velocity ratio, the gyrotron operation is controlled by at least five independent physical parameters. These parameters are the beam current, $I_b = \xi_1$, the modulating anode voltage, $V_m = \xi_2$, the beam voltage, $V_b = \xi_3$, the magnetic field in the interaction region, $B_0 = \xi_4$, and the magnetic field in the gun region near the cathode, $B_c = \xi_5$. So, the transfer coefficients A_i, B_i given by (9.14) can be rewritten for the gyrotron as

$$A_i = \frac{1}{4\pi Q}\left(\beta_i + \sum_{j=1}^{3} d_j'' \alpha_{ij}\right), \tag{9.39}$$

$$B_i = \frac{1}{4\pi Q}\left(\frac{\hat{\chi}'}{\hat{\chi}''}\beta_i + \sum_{j=1}^{3} d_j' \alpha_{ij}\right). \tag{9.40}$$

Here the coefficient $\beta_i = \frac{\xi_{i0}}{I_0}\frac{\partial I_0}{\partial \xi_i}\big|_{\xi_{i0}}$ determines the sensitivity of the normalized current parameter to fluctuations in technical parameters ξ_i. Coefficients $d'_j = \frac{u_{j0}}{\hat{\chi}''}\frac{\partial \hat{\chi}'}{\partial u_j}\big|_{u_{j0}}$ and $d''_j = \frac{u_{j0}}{\hat{\chi}''}\frac{\partial \hat{\chi}''}{\partial u_j}\big|_{u_{j0}}$ determine the sensitivity of the susceptibility to fluctuations in normalized parameters of the gyrotron theory. Also, the coefficients $\alpha_{ij} = \frac{\xi_{i0}}{u_{j0}}\frac{\partial u_j}{\partial \xi_i}\big|_{\xi_{i0}}$ determine the sensitivity of the jth normalized parameter of this theory to the ith technical parameter.

Taking into account the definitions of the normalized parameters (9.34)–(9.36) and using Eqs. (9.37), (9.38), and (2.5), one can readily determine the coefficients α_{ij}. In order to determine the coefficients d'_j and d''_j, one should numerically study the set of gyrotron equations, which determine the susceptibility and its derivatives with respect to various normalized parameters. Such an analysis has been carried out elsewhere (Dumbrajs and Nusinovich 1997). This analysis was also supplemented by the consideration of some typical examples.

Gyroklystron

In the case of gyroklystrons and gyrotron oscillators phase-locked by means of a prebunching cavity, the processes in the output cavity of a gyroklystron or the main cavity of a gyromonotron can be described by the formalism developed above, which is based on the use of Eqs. (9.3) and (9.4). In these equations, however, first, the frequency detuning, $\omega - \omega_s$, designates now the difference between the signal frequency and the cold-cavity frequency. Second, the phase of the field is now a controllable parameter. Finally, it is necessary to take into account the fact that the output cavity is excited by a prebunched electron beam. Corresponding bunching parameters, in the case of using long drift sections, are much more sensitive to various fluctuations in the input cavity than other parameters responsible for operation of the output cavity.

Let us restrict our consideration by a two-cavity device. The boundary condition for the electron phase at the entrance to the second cavity, as follows from (8.31), can be written as

$$\vartheta(0) = \vartheta_0 + q \sin \vartheta_0 - \Theta + \psi_1 - \psi_2, \tag{9.41}$$

where

$$\Theta = \Delta\mu_{dr} = (\omega - s\Omega_0)L_{dr}/v_{z0} \tag{9.42}$$

is the transit angle in the drift region, q is the bunching parameter, and ψ_1 and ψ_2 are the phases of oscillations in the input and output cavities, respectively. (The first phase is determined by the phase of the input signal.) Note that, in principle, any driver has a certain noise, which also causes fluctuations in the phase ψ_1 and bunching parameter q.

9.3 Intrinsic Noise Sources. Shot Noise

In Sec. 9.1 it was already mentioned that in microwave tubes there are several noise sources, which can be classified as the sources of intrinsic noise. These are shot noise, flicker noise, velocity noise, and thermal noise. The thermal noise is caused by the Kirchhoff radiation from microwave circuit walls, which have a finite temperature. Ergakov, Moiseev, and Shaposhnikov (1983) have studied this noise in gyrodevices and showed that this is an extremely low-frequency noise. The velocity noise (also known as Rack velocity noise; see, e.g., Sec. 15.3 in Gilmour 1994) is caused by initial distribution of electrons in velocities. Typically, this spread corresponds to the electron spread in energies below or about 1 eV, while gyrotrons operate at voltages on the order of 60–70 kV and above. Therefore, the contribution from the velocity noise is rather small and this noise does not seem to be one of the key players in the noise figures of gyrodevices, at least in the case of cathodes with temperature-limited emission.

As discussed above, in the case of strong magnetic compression of electrons in the region of increasing magnetic field between the gun and the circuit, some electrons with large initial orbital velocities can be reflected from this magnetic bottle, which is often also called a magnetic mirror. Such mirrored particles move back to the cathode, where, in turn, they are reflected by the electric field existing between the cathode and anode. So, these trapped electrons start to oscillate in the region between the gun and the circuit, and these oscillations can be unstable due to the space charge effect (Bratman 1976, Li and Antonsen 1994). In principle, this region of electron oscillations can also serve as a trap for some ions when an electric field in the potential well caused by trapped electrons exceeds the external electric field caused by the voltage applied between cathode and anode, which attracts ions to the cathode (see Manheimer et al. 2001 for recent studies of these effects in linear-beam TWTs). However, so far the ion trapping in gyrotrons has never been analyzed.

Flicker noise (also known as Johnson noise) is caused by imperfections in the emission process. Detailed analysis of this sort of noise can be found elsewhere (Chapter 2 in Smullin and Haus 1959). At low frequencies (below about 1 kHz; see, e.g., Sec. 15.5.3 in Gilmour 1994), flicker noise dominates over shot noise. (A degree of this domination can be estimated by using a simple method proposed by Y. Y. Lau, Jensen, and Levush 1998.) An electron beam entering the interaction space carries both of these sources of noise. Thus, both of them modulate a microwave signal. Therefore, the natural width of the peak shown in Fig. 9.1 is determined mostly by the flicker noise, while at larger frequency separation from the carrier frequency, the shot noise is important. Clearly, the causes for flicker noise and its characteristics depend on the type of emitter, methods of emitter fabrication, vacuum in the gun region, and other technical factors. Recall that typically the cathodes used in gyrodevices operate in the regime of temperature-limited emission. Therefore, many arguments used for explanation of the origin and spectral dependence of the flicker noise in other devices using thermionic cathodes with temperature-limited emission can be applied to gyrotrons as well.

The rest of this section will be devoted to the shot noise, which is the noise responsible for the pedestal of the radiation line shown in Fig. 9.1. This noise is present at practically all frequencies because it corresponds to the random process of emitting discrete charged particles. The shot noise in gyrotron oscillators, gyroklystrons, and gyro-TWTs was analyzed, respectively, by Ergakov, Moiseev, and Shaposhnikov (1977), Antonsen and Manheimer (1998), and Antonsen et al. (2002).

To account for shot noise effects, let us come back to Eq. (9.5) and assume that, in addition to the stationary beam current

$$I_0 = I_b \int f_0(\vec{r}_{\perp 0}, \vec{v}_0) d\vec{r}_{\perp 0} d\vec{v}_0,$$

there is also a time-dependent fluctuating component of the beam current

$$I_f(t) = I_b \int f_f(\vec{r}_{\perp 0}, \vec{v}_0, t) d\vec{r}_{\perp 0} d\vec{v}_0. \tag{9.43}$$

Then, in addition to the susceptibility given by (9.5), which is determined by the stationary current density, we can calculate in a similar fashion perturbations in the susceptibility caused by the current fluctuations and apply the formalism developed above for calculating corresponding correlation functions.

Let us start from the simplest case of a "bare" shot noise. This implies an assumption that all particles in the electron beam are uncorrelated with one another. Correspondingly, the product of two distribution functions describing the shot noise at two different instants of time can be written as (Ergakov, Moiseev, and Shaposhnikov 1977)

$$f_f(\vec{r}_{\perp 0}, \vec{v}_0, t) f_f(\vec{r}_{\perp 0}, \vec{v}_0, t+\tau) = A(\vec{r}_{\perp 0}, \vec{v}_0)\delta(\tau)\delta(\vec{r}_{\perp 0} - \vec{r}\,'_{\perp 0})\delta(\vec{v}_0 - \vec{v}\,'_0). \tag{9.44}$$

Here the coefficient $A(\vec{r}_{\perp 0}, \vec{v}_0)$ can be determined by using the following arguments. Consider two groups of electrons with close velocities that were emitted from the same element of a cathode at different instants of time. Our assumption means that both groups exist in a small enough element of the phase space $(\Delta\vec{r}_{\perp 0}, \Delta\vec{v}_0)$. Then, as follows from (9.43) and (9.44), the statistical average of the product of these two perturbations in the current, which exist in the same element of the phase space, is equal to

$$\overline{\Delta I_f(t)\Delta I_f(t+\tau)} = I_b^2\delta(\tau)\int\limits_{\Delta\vec{r}_{\perp 0}, \Delta\vec{v}_0} A d\vec{r}_{\perp 0} d\vec{v}_0. \tag{9.45}$$

Since the emission of electrons of any group is uncorrelated, the correlation function for the current of any group is proportional to the DC current in this element of the phase space

$$\Delta I_0 = I_b \int\limits_{\Delta\vec{r}_{\perp 0}, \Delta\vec{v}_0} f_0 d\vec{r}_{\perp 0} d\vec{v}_0 \tag{9.46}$$

and is equal to

$$\overline{\Delta I_f(t)\Delta I_f(t+\tau)} = e\Delta I_0\delta(\tau). \tag{9.47}$$

As follows from (9.45)–(9.47), the constant $A(\vec{r}_{\perp 0}, \vec{v}_0)$ is equal to $(e/I_b)f_0(\vec{r}_{\perp 0}, \vec{v}_0)$.

The fact that electron fluctuations at the entrance to the resonator are uncorrelated results in the delta-correlation of fluctuations in the susceptibility. Therefore, for the gyrotron with a negligibly small electron velocity spread the spectral fluctuations caused by the shot noise can be determined by general expressions (9.25)–(9.29), in which all terms are now proportional to (e/I_b). Ergakov, Moiseev, and Shaposhnikov (1977) carried out a numerical analysis of corresponding characteristics for a gyrotron with the Gaussian axial distribution of the resonator field. In particular, they showed that in the gyrotron with parameters optimal for the orbital efficiency, the natural radiation line width determined by the shot noise is mostly determined by the spectral density of the frequency noise, $W_\nu(\Omega)$ (see (9.25)–(9.28)), and is equal to $0.15(e\omega^2/I_b Q^2)$. (Here the ratio $e\omega/I_b$ can be rewritten as $(mc^3/eI_b)2\pi(r_e/\lambda)$, where $r_e = e^2/mc^2 \approx 2.8 \cdot 10^{-13}$ cm is the classical electron radius.) As follows from this simple formula, for such typical gyrotron parameters as 100 GHz frequency, 10 A beam current and Q-factors on the order of several hundreds, the natural line width is smaller than 1 Hz. The spectral densities of the amplitude and frequency fluctuations, however, become larger when the normalized interaction length μ decreases. Note that this estimate of the natural line width determined by the shot noise agrees with the estimate of such a width for a low-power, 10 GHz klystron. That estimate was done by Malakhov (1968, p. 481), who found that the width of this line should be smaller than 0.1 Hz.

So far, we were considering the effect of the shot noise on the radiation line width of a signal. In principle, the uncorrelated shot noise is the "white" noise, which is equally present at all frequencies and, hence, can exist in the resonator even in the absence of any signal. Antonsen and Manheimer (1998) developed a general theory of the shot noise in a gyroklystron for the case where a drive signal is absent ("carrier-free" conditions). They showed that the noise temperature, which is the noise power per unit frequency dissipated in the cavity, as the function of frequency has a Lorentzian shape

$$T(f) = \frac{T_{\max}}{1 + [2Q_l(f - f_l)/f_l]^2}.$$

Here $T_{\max}$ is the peak noise temperature, and Q_l and f_l are the beam-loaded quality factor and the cavity resonant frequency, respectively. This dependence is due to the fact that the noise amplification is determined by the cavity frequency response when the cavity filling time is much larger than the electron transit time through the resonator, which is a typical case in gyrodevices. All the noise power in this cavity is much smaller than the drive power injected into the input cavity. A rule of thumb proposed by Antonsen and Manheimer suggests the following formula for the signal-to-noise ratio measured in dB:

$$\text{Signal/Noise} = 10 \log_{10} N - G[\text{dB}]. \tag{9.48}$$

Here N is the number of electrons that pass through the cavity in a cavity decay time (a typical value of N is 10^{10}) and G is the gain of the amplifier. This estimate is based on the comparison of the noise power with the beam power and on the assumption that the radiation power is on the order of the beam power. However, as mentioned by these authors, it is more proper to compare the noise power with the drive power coupled into the input cavity because then both signal and noise powers are equally amplified by the beam. Such a comparison yields for the signal-to-noise ratio a value about 80 dB (Calame, Danly, and Garven, 1999).

Antonsen and Manheimer (1998) have also discussed a possibility of significant noise amplification due to the electrostatic cyclotron instability, which may occur in the region of the beam magnetic compression. However, detailed experimental studies of the shot noise in a 35 GHz gyroklystron, which were described by Calame, Danly, and Garven (1999), showed the absence of such noise-enhancing effects, at least in the device under study.

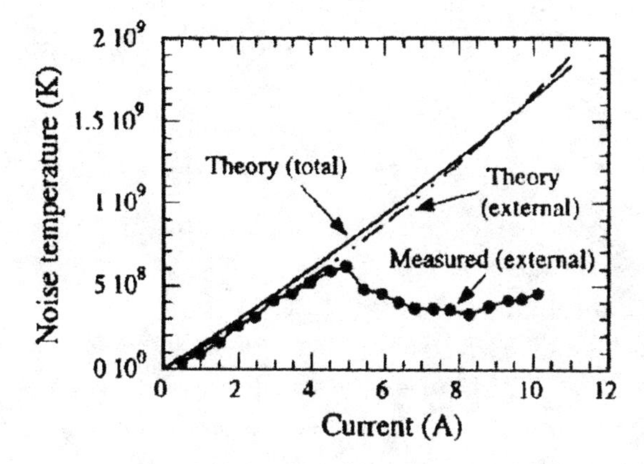

Fig. 9.2. Peak noise temperature measured from the input cavity (circles) and theoretically predicted for the NRL gyroklystron. (Reproduced from Calame, Danly, and Garven 1999)

At the same time, in these experiments it was shown that at high currents the noise temperature is much lower than the temperature predicted by a simple theory of the uncorrelated shot noise. These results shown in Fig. 9.2 agree well with the concept of dielectric shielding of the discrete electron charges by the beam itself, which was discussed by Antonsen and Manheimer (1998). Note that this effect is well known in linear-beam tubes. For the sake of comparison, we present here Fig. 9.3 taken from Gilmour (1994), which demonstrates the dependence of the noise power on the beam current in the

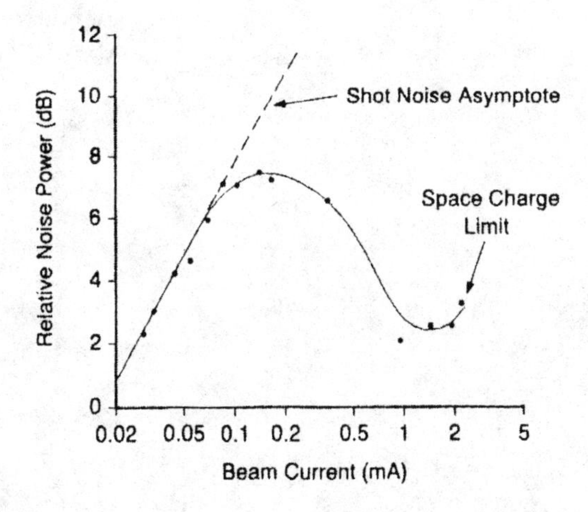

Fig. 9.3. Noise power versus beam current for temperature limited emission. (Reproduced from Gilmour, Jr., 1994)

case of temperature-limited emission, which is quite similar to that shown in Fig. 9.2.

9.4 Extrinsic Noise

Extrinsic or technical noise is the noise caused by fluctuations in such operating parameters as voltages and currents. By "currents" we mean here not only the electron beam current, but also the currents in solenoids creating external magnetic fields and the current in a heater of a thermionic cathode. The theory describing the extrinsic noise was already presented above in Sec. 9.2. Below we shall describe, following Dumbrajs and Nusinovich (1997), only one example showing how to apply this theory to gyrodevices.

Let us consider a free-running gyrotron oscillator immersed in the magnetic fields produced by superconducting solenoids. Also assume that in this device a diode-type electron gun is used, i.e., $V_{mod} = V_b$, and the most important effect is the beam voltage fluctuations.

This allows us to rewrite (9.29) in the following simplified form:

$$W_\nu(\Omega) = \frac{1}{p^2+\Omega^2}\{(p^2+\Omega^2)(B_2+B_3)^2+q^2(A_2+A_3)^2 \\ +2qp(A_2B_2+A_2B_3+A_3B_2+A_3B_3)\}W_3. \tag{9.49}$$

Dumbrajs and Nusinovich (1997) considered a gyrotron driven by a 60 kV electron beam with a 1.5 orbital-to-axial velocity ratio. It was also assumed that the axial structure of the resonator field can be approximated by the Gaussian profile with the normalized length, μ, equal to 15, and the normalized field amplitude F and the cyclotron resonance mismatch Δ are optimally chosen for this length. For this set of parameters, the spectral density of the frequency noise equals

$$W_\nu(\Omega_r) = \frac{1}{3.7+(2Q\Omega_r/\omega)^2}\left[528.7+177.7\left(\frac{2Q\Omega_r}{\omega}\right)^2\right]\frac{W_3}{(4\pi Q)^2}. \tag{9.50}$$

Here subscript r for the frequency Ω indicates that, in contrast to the notations used above where the frequency was normalized to $2\pi\omega_s$ (see notations introduced in (9.9)–(9.10)), in (9.50) the real frequency Ω is present. Typically, the radiation linewidth is much smaller than the width of the resonator curve. As follows from (9.50), in this case the spectral density of the frequency noise is very similar to the spectral density of the correlation function $W_3(\Omega)$ describing the fluctuations in the beam voltage, because the contributions from other terms containing Ω_r in (9.50) are rather small.

The spectral density $W_3(\Omega)$ depends on the kind of high-voltage supply. Consider, for example, the pulse operation and assume that the voltage pulse is formed by a modulator, which is followed by a pulse transformer. In such a circuit, some voltage ripples at the transformer output can be due to parasitic interactions between distributed stray inductance and stray capacitance in the transformer secondary. Assume now that fluctuations in the beam voltage can be treated as a stationary random process that can be described by the correlation function,

$$\Phi(\tau) = \left(\frac{\Delta V}{V_{b0}}\right)^2 \exp(-b\,|\tau|),$$

where $\Delta V = \sqrt{\langle(\delta V_b)^2\rangle}$ is the rms value of the voltage fluctuations, and b is the frequency scale characterizing the width of their spectrum. The spectral density of such a correlation function is (Malakhov 1968)

$$W_3(\Omega_r) = \frac{1}{\pi}\left(\frac{\Delta V}{V_{b0}}\right)^2 \frac{b}{b^2 + \Omega_r^2}. \tag{9.51}$$

As follows from (9.50) and (9.51), when the beam voltage fluctuations $\Delta V / V_{b0}$ are on the order of 10^{-3}, the cavity Q-factor is about 10^3, and the frequency scale b is equal to 1 kHz, the spectral density of the frequency noise is given by

$$W_\nu(\Omega_r) = \frac{2.9 \cdot 10^{-16}}{1 + (\Omega_r/b)^2}(1/\text{Hz}). \tag{9.52}$$

From (9.52) one can easily find that at 1 kHz off the carrier frequency, the frequency noise density is a little higher than 10^{-16} Hz^{-1}, while at 1 MHz off it is about six orders of magnitude lower. This example shows how to use the formalism developed above. Certainly, accurate characterization of the extrinsic noise can be done only based on the measurements of the technical fluctuations in the gyrotron parameters.

9.5 Phase Stability of Gyroamplifiers

For such applications of gyroamplifiers as advanced radars, communication systems, and linear colliders, one of the important characteristics of any device is its phase sensitivity to operating parameters. The phase sensitivity, which is often called a *pushing factor*, is the issue closely related to the technical fluctuations, which we just considered. It shows the degree of the output phase deviation when one of the technical parameters varies. Below, we shall consider these characteristics for the gyroklystron and the gyro-TWT.

Gyroklystrons

Our treatment will be based on a simple theory of the phase-pushing factor in gyroklystrons developed by Park et al. (1991). This theory is developed for the point-gap model of a two-cavity gyroklystron. It is also compared with a similar theory developed for the point-gap model of a two-cavity linear-beam klystron. The theory developed for both devices implies that the most important for variation of the output signal phase is the effect of fluctuations in the beam voltage on the transit angle of electrons through a long drift space. It is assumed that the voltage on the modulating anode is a fixed fraction of the beam voltage. This means that our treatment can be used in the case of diode-type electron guns as well as in the case of triode-type guns with the mod-anode voltage taken of a resistive divider.

Assume that the output signal phase differs from the input signal phase just by the phase determined by the transit angle through the drift region of the length L_{dr}. Then, from (9.41), (9.42) and definitions of the relativistic electron cyclotron frequency and the electron axial velocity, one can readily find (see also Problem 1 in Sec. 9.6) that the deviation of the output phase is equal to

$$\Delta\varphi = \frac{\omega L_{dr}}{v_{z0}} \frac{1}{\gamma_0(\gamma_0+1)} \left[\left(\frac{s\Omega_0}{\omega} \gamma_0^2 - 1 \right) + \alpha^2 \gamma_0 \left(1 - \frac{s\Omega_0}{\omega} \right) \right] \frac{\delta V}{V_{b0}}. \quad (9.53)$$

So, as one could expect, the phase deviation is proportional to the electron transit time, L_{dr}/v_{z0}, and the change in the beam voltage, δV. The first term in square brackets in (9.53) is due to the changes in the electron energy, and the second term is due to the changes in the electron orbital-to-axial velocity ratio. Clearly, this is a simplified version of the approach developed above in Sec. 9.2. Now we can easily identify the changes in the output phase given by (9.7) with the changes in the transit angle, which is a parameter of the gyroklystron theory additional to the normalized parameters of the free-running gyrotron oscillator, which were given by (9.34)–(9.36). The terms in the RHS of (9.53) correspond to the coefficients α_{ij} used in (9.39) and (9.40), which are determined now for the case when the jth normalized parameter of the theory is the transit angle and the ith technical parameter is the beam voltage.

Note that in weakly relativistic gyroklystrons the normalized electron energy $\gamma_0 = 1 + eV_b/mc^2$ is close to one, and the harmonic of the electron cyclotron frequency is close to the operating frequency when the cyclotron resonance condition is fulfilled. Therefore, the phase fluctuation determined by (9.53) in such devices can be rather small. However, at high electron energies the phase sensitivity of relativistic gyroklystrons can be much larger.

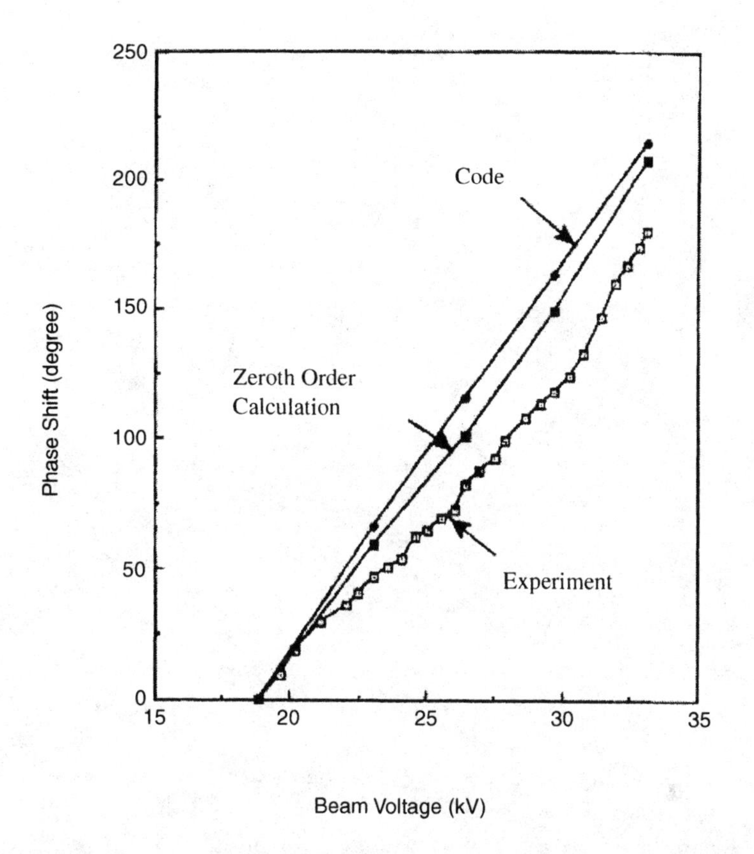

Fig. 9.4. Phase shift versus beam voltage in gyroklystron. (Reproduced from Park et al. 1991)

The analysis carried out by Park et al. (1991) showed that at low voltages the output phase of the gyroklystron radiation is less sensitive to voltage fluctuations than the output phase of the conventional klystron. However, at voltages above 100 kV the klystron radiation phase is more stable. It was also found that, at least at low voltages this simple formalism gives results, which agree reasonably well with experimental data. This agreement is shown in Fig. 9.4. Here predictions of this simple formalism are called "Zeroth Order Calculation" and "code" means results obtained by using the code GKLY (Ganguly, Fliflet, and McCurdy 1985).

Gyro-TWT (Gyro-Traveling-Wave Tubes)

It is also possible to develop a simple theory describing the phase sensitivity of the gyro-TWT (Nusinovich et al. 2001). This theory is focused on the effect of deviation in the technical parameters on the phase difference between the output and input signals, $\varphi_{\text{out}} - \varphi_{\text{in}}$.

In the framework of the small-signal theory, the wave field at the output can be represented as a superposition of three partial waves

$$F(\mu) = \frac{1}{3}F(0)e^{-i\Theta}\sum_{l=1}^{3} e^{i\Gamma_l \mu}. \tag{9.54}$$

Here the coefficient 1/3 originates from the fact that at the entrance an input signal is equally distributed among three partial waves, $\Theta = \Delta\mu$ is the electron transit angle through a waveguide of the normalized length, μ, and Γ_l are the roots of the dispersion equation (7.16). Representing the sum in (9.54) as

$$\sum_{l=1}^{3} e^{i\Gamma_l \mu} = \Phi(\mu) \equiv |\Phi| e^{i\phi}, \tag{9.55}$$

one can determine the phase difference between input and output as

$$\varphi_{\text{out}} - \varphi_{\text{in}} = \phi - \Theta. \tag{9.56}$$

So, the phase-pushing factors are determined by the effect of deviation in such technical parameters as, let say, voltage and magnetic field on normalized parameters of the gyro-TWT theory. These normalized parameters, in turn, determine the propagation constants Γ_l, the normalized cyclotron resonance mismatch Δ, and the normalized interaction length μ. Clearly, as the interaction length increases, the device becomes more sensitive to fluctuations.

Correspondingly, in the large-signal regime, the output phase can be found by integrating Eq. (7.11) for the complex amplitude of the wave. Of course, this equation should be integrated self-consistently with Eqs. (7.9) and (7.10). Then, again, the phase-pushing factors will be determined by the effect of the technical parameters on the normalized parameters present in (7.9)–(7.11) and, in turn, the effect of deviation in these parameters on the output phase of the gyro-TWT.

This simple formalism was applied for the analysis of phase-pushing factors in the two-stage, frequency-doubling gyro-TWT experimentally studied at the University of Maryland. In order to analyze a quite complicated two-stage configuration, some additional assumptions had been made (Nusinovich et al. 2001). Nevertheless, it was found to be a very good agreement between the voltage pushing factor predicted by this simple theory and its experimentally measured value (−0.05°/V versus −0.06°/V). The difference between predictions for the magnetic field phase pushing factor and its experimentally measured value was about two times (5.1°/G versus 2.3°/G). This discrepancy

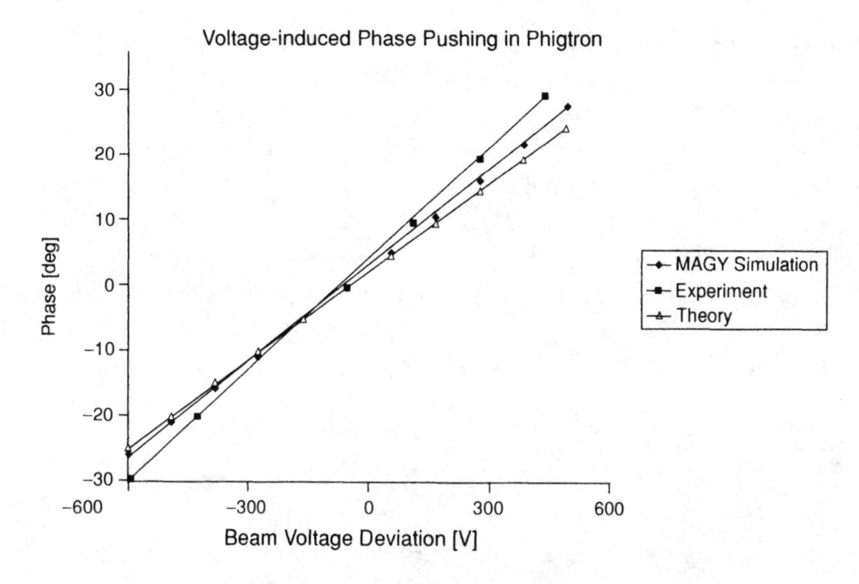

Fig. 9.5. Voltage induced phase pushing factor in the inverted gyrotwystron. (Reproduced from Rodgers et al. 2001)

can, in particular, be explained by the fact that in the experiment the external magnetic field was tapered while in the theory a constant magnetic field profile was assumed.

Later, similar calculations and measurements were carried out in the frequency-doubling inverted gyrotwystron (Rodgers et al. 2001). The principles of operation of this device will be explained later. Here, I would like to reproduce only the beam voltage phase-pushing factor, which was analyzed with the use of this simple theory, calculated by using MAGY and measured experimentally in this device. These results are shown in Fig. 9.5, which demonstrates excellent agreement between the theory and experiment.

9.6 Problems and Solutions

Problems

1. Using the adiabatic theory of magnetron-type electron guns determine the effect of the variation in the beam voltage and external magnetic field on the normalized components of electron velocity, $\beta_\perp$ and β_z. Consider a diode-type electron gun and assume that there is only one solenoid that determines magnetic fields in the gun and interaction regions.

2. Using expressions for the normalized beam current, I_0, normalized field amplitude, F, cyclotron resonance mismatch, Δ, and normalized interaction length, μ, given by (3.65), (9.34), (9.35), and (9.36), respectively, determine the effect of the variation in the beam voltage and external magnetic field on these parameters. In other words, calculate coefficients $\alpha_{j,i} = (\xi_i/u_{j0})(\partial u_j/\partial \xi_i)$ for technical parameters $\xi_1 = V_b$ and $\xi_2 = B_0$ and normalized parameters $u_1 = I_0$, $u_2 = F$, $u_3 = \Delta$ and $u_4 = \mu$. Consider the operation at the fundamental cyclotron resonance and assume that a thin annular electron beam is positioned at the peak of the coupling impedance of the operating mode.

Solutions

1. In a diode-type electron gun $V_b = V_{\text{mod}} = V$. Also, since there is only one solenoid, the magnetic fields in the gun and resonator region are proportional to the same current in the solenoid. Hence, the magnetic compression ratio is fixed. This allows us to rewrite (9.37) for our simple case as $\beta_{\perp 0} = (K/B_0)[V/(1+V)]$, where K is the constant. From this equation it follows that, as the magnetic field and the voltage vary, the normalized orbital electron velocity changes as

$$\frac{\delta\beta_\perp}{\beta_{\perp 0}} = -\frac{\delta B}{B_0} + \frac{1}{\gamma_0}\frac{\delta V}{V_0}.$$

Then, the sensitivity of the electron axial velocity, in accordance with (9.38), can be determined as

$$\frac{\delta\beta_z}{\beta_{z0}} = \alpha^2\frac{\delta B}{B_0} - \frac{\alpha^2\gamma_0 - 1}{\gamma_0(\gamma_0+1)}\frac{\delta V}{V_0}.$$

2. Since it was assumed that the beam coupling is maximal, the effect of the beam voltage and magnetic field variations on the normalized beam current given by (3.65) can be determined as

$$\frac{\delta I_0}{I_0} = -4\frac{\delta\beta_\perp}{\beta_{\perp 0}} - \frac{\gamma_0 - 1}{\gamma_0}\frac{\delta V}{V_0}.$$

Here we use the variation in the electron orbital velocity, which is determined in the solution to Problem 1. Correspondingly, the effect of the voltage and magnetic field variations on the normalized amplitude of the resonator field given by (9.34) can be determined as

$$\frac{\delta F}{F_0} = -3\frac{\delta\beta_\perp}{\beta_{\perp 0}} - \frac{\gamma_0 - 1}{\gamma_0^2}\frac{\delta V}{V_0}.$$

Then, from (9.35) one can readily obtain

$$\frac{\delta\Delta}{\Delta_0} = -2\frac{\delta\beta_\perp}{\beta_{\perp 0}} + \left(\frac{2}{\beta_{\perp 0}^2} - 1\right)\left(\frac{\gamma_0 - 1}{\gamma_0}\frac{\delta V}{V_0} - \frac{\delta B}{B_0}\right)$$

for the cyclotron resonance mismatch, and from (9.36)

$$\frac{\delta\mu}{\mu_0} = 2\frac{\delta\beta_\perp}{\beta_{\perp 0}} - \frac{\delta\beta_z}{\beta_{z0}}.$$

Again, these variations in the electron orbital and axial velocities have already been found in Problem 1. Then, the coefficients characterizing the sensitivity of the jth normalized parameter to the variation in the ith technical parameter in (9.39), (9.40) can be determined as

$$\alpha_{j,i} = (\xi_i/u_{j0})(\partial u_j/\partial \xi_i) = \left(\frac{\delta u_j}{u_{j0}}\right) \Big/ \left(\frac{\delta \xi_i}{\xi_{i0}}\right),$$

where both ratios in the right-hand side directly follow from the formulas given above. Note that, when the beam is not positioned at the peak of the coupling impedance, but we have only one solenoid, the result is the same. This happens because the variation in the magnetic field does not affect the magnetic force lines the electrons follow, and hence does not change the beam location in the cavity. However, when there is an additional gun coil, independent variations of two magnetic fields produced by the main solenoid and the gun coil can shift the beam position, and hence affect the normalized beam current and the normalized field amplitude.

Part III

The Development of Gyrodevices

▪ CHAPTER 10 ▪

Gyrotron Oscillators for Controlled Fusion Experiments

10.1 Historical Introduction

The gyrotron as a device comprising a magnetron-type electron gun with adiabatic compression of a beam of gyrating electrons and a smooth-wall microwave circuit with diffraction output was invented at the Radiophysical Research Institute, Gorki, USSR (Gaponov et al. 1967). The layout of the first gyrotron is shown in Fig. 10.1. According to Petelin (2002), design of this gyrotron was started in 1963 and the tube was tested in September 1964. It delivered 6 W CW power at 3.4 cm wavelength operating at the fundamental cyclotron resonance. After more than doubling the voltage (from 8 kV to 19 kV), it generated 190 W CW power at 1.2 cm wavelength operating at the second cyclotron harmonic (Gaponov et al. 1965).

Then, experiments with gyrotrons operating at the fundamental cyclotron resonance ($s = 1$) and at the second cyclotron harmonic ($s = 2$) at a higher power level were carried out in 1966 (Gaponov et al. 1975). These gyrotrons operated at low-order transverse-electric symmetric modes: TE_{011}-mode ($s = 1$, $\lambda = 2$ cm) and TE_{021}-mode ($s = 2$, $\lambda = 1.2$ cm). CW power in both regimes reached the 4 kW level. In these experiments it was shown that the diffractive output allows for efficient axial mode selection: the only modes, which had high diffractive Q-factors in these axially open resonators, were the modes with one axial variation.

Shortly afterward, a successful attempt was made to demonstrate gyrotron capabilities of mastering the short-millimeter and submillimeter wavelength region at high average power levels. Results of these experiments are shown in Table 10.1 reproduced from Zaytsev et al. (1974). Generation of 1.5 kW CW power at frequencies above 300 GHz reported in this paper was a tremendous success, which greatly impressed the microwave generation community over

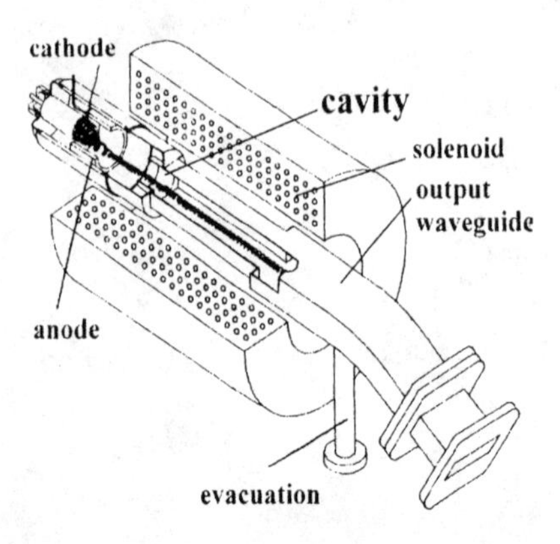

Fig. 10.1. Layout of the first gyrotron.

the world in the mid 1970s (see, e.g., Osepchuk 1978). Note that although a 100 kW-level peak power generation in gyrotrons with pulse solenoids at frequencies up to 600 GHz was reported later (Flyagin, Luchinin, and Nusinovich 1983), this 1.5 kW level of the CW power at submillimeters still remains remarkable. As follows from Table 10.1, in order to handle the ohmic losses of high-millimeter-wave power a step was made for operating in higher-order symmetric modes: in many of these experiments, instead of operating at the TE_{02}-mode the TE_{03}-mode was chosen.

These experiments, however, clearly demonstrated the presence of severe competition between the symmetric TE_{0p}-modes and rotating $TE_{2,p}$-modes, which at radial indices $p \geq 3$ have frequencies very close to the frequencies of symmetric modes. It was found that in many cases it is more convenient and easy to excite the rotating $TE_{2,p}$-mode instead of the symmetric mode because these rotating modes occupy a smaller volume, and therefore their starting currents can be smaller than the starting currents of symmetric modes.

The fact that, among the modes with close frequencies, the cross section occupied by the field of modes with large azimuthal indices is smaller than that of modes with large radial indices has led to the concept of operation in whispering gallery modes (WGMs) mentioned in Sec. 3.5. WGMs are the modes whose azimuthal indices are much larger than the radial ones. The fields of these modes are localized near the wall. More exactly, they occupy the area between the wall of a radius R_w and caustic whose radius, R_c, is equal to $(m/\nu_{m,p})R_w$. Bykov et al. (1975) described results of the first attempts to prove this concept. In these experiments, the operating mode chosen was $TE_{9,1,1}$-mode. It was found, however, that operation at this mode has a

Table 10.1. First Short-Millimeter- and Submillimeter-Wave Experiments in the USSR

Model No.	Operating Mode	Wavelength, mm	Operation	$s = \omega/\Omega$	H_o,kOe	V_b,kV	I_b,A	P, kW	η,%	η_{th},%
1	TE_{021}	2.78	CW	1	40.5	27	1.4	12	31	36
2	TE_{031}	1.91	CW	2	28.9	18	1.4	2.4	9.5	15
	TE_{231}	1.95	pulsed	2	28.5	26	1.8	7	15	20
3	TE_{231}	0.92	CW	2	60.0	27	0.9	1.5	6.2	5.0

Note: η and η_{th} denote experimental and theoretical efficiencies, respectively.

substantial drawback, which is the presence of a very dangerous competing mode $TE_{5,2,1}$ with a very close frequency. Eliminating the excitation of this parasitic mode was possible only by using an inner coaxial insert. Thus, the concept of coaxial cavities was found to be fruitful. The first coaxial gyrotron generating more than 1 MW at the wavelength 6.7 mm in about 100 microsecond pulses was developed in the earlier 1970s. (In these experiments the key role belonged to A. L. Goldenberg and Yu. V. Bykov.) Results of this development were briefly presented by Andronov et al. (1978). However, only later, Gaponov et al. (1981) mentioned that this gyrotron operated in the $TE_{15,1,1}$-mode.

It was also realized that operation at WGMs with $p = 1$ has two more disadvantages. First, to strongly couple a beam of gyrating electrons to such a mode localized near the wall it was necessary to place an electron beam very close to the wall. So, the clearance between the beam and the cavity wall (and, more importantly, cutoff necks) was very small, and it was obvious that with the wavelength shortening this problem will be even more severe. Second, it was found that, since these modes occupy a very small volume, their minimum starting currents are also very small. Correspondingly, the beam current value, which is optimal for efficient interaction, is small as well, and therefore efficient generation typically takes place at rather low power levels.

These reasons have led to the conclusion that it is much more convenient to operate at whispering gallery modes with two radial variations. Indeed, positioning of an electron beam, which provides the maximum coupling to these modes, corresponds to beam radii smaller than the caustic radius of the modes with one radial variation. Thus, the competition with the modes occupying a smaller volume can be eliminated. At the same time, the modes with three and more radial variations have much larger starting currents. Also, the clearance between the beam and the wall in the case of operation at modes with two radial variations is reasonable. All these arguments were successfully proven in the experiments with the first short-pulse (100 microseconds), MW-level, 100 GHz gyrotron (Alikaev et al. 1978), which operated at the $TE_{22,2,1}$-mode. (Note that information about the operating mode in this gyrotron again became available only much later.) A further analysis showed that with the power increase and the wavelength shortening it is necessary to increase the radial index of such rotating modes with large azimuthal indices in order to find a compromise between the mode selection issue and admissible densities of ohmic losses in cavity walls. For providing a reasonable mode selection, the beam should always be located in the inner peak of the electric field strength (see coupling impedances shown in Fig. 3.3), while the radial index should be large enough for combining the required level of radiated microwave power localized between the caustic and cavity wall with the admissible level of the density of ohmic losses.

Shortly after the first demonstration of gyrotron capabilities to generate multi-kW power level at short millimeters, these devices attracted attention of plasma physicists, who carried out experiments with various methods of plasma heating in controlled fusion reactors. In the late 1960s it was well recognized that, parallel with ohmic heating of plasmas, various methods of plasma heating by using RF and microwave power at frequencies resonant to some oscillation frequencies of plasma components are possible. These RF and microwave heating methods include Alfven heating, ion heating, low and upper hybrid heating, and electron cyclotron heating. The last method is based on the initial absorption of a microwave power by electron component of the plasma. Then, in the process of electron collision with plasma ions, the temperature of both components will be equalized. While the RF and microwave sources, which were necessary for experiments on the Alfven, ion, and low hybrid heating were available, the experiments on the electron cyclotron resonance heating (ECRH) required high-power millimeter-wave sources, which did not exist before the gyrotron was invented. To explain this choice of operating wavelength, recall that magnetic fields required for plasma confinement in tokamaks and stellarators are on the order of several tesla. This means that electron cyclotron frequency in these reactors is on the order of 100 GHz. Since the resonant cyclotron absorption of the microwave power by plasma electrons is essentially the process opposite to the electron cyclotron radiation by beam electrons in gyrotrons, it is clear that the frequency of the gyrotron radiation should be close to the electron cyclotron frequency in tokamaks and stellarators. Thus, ECRH experiments required the development of millimeter-wave sources capable of producing a multi-KW power level in long enough pulses. Their pulse duration was dictated by the energy confinement time, which in the tokamak TM-3, for instance, was on the order of milliseconds (Alikaev et al. 1972).

Results of the first experiments with the use of gyrotrons on the tokamak TM-3 at Kurchatov Institute, Moscow (Alikaev et al. 1972), and on the "TUMAN-2"at Ioffe Institute, Leningrad (Golant et al. 1972), were reported in the early 1970s. In these experiments, gyrotrons were used that delivered 80 kW peak power in 0.6 msec pulses at the 4.5 mm wavelength. Although the results were very impressive, the most important immediate message from plasma physicists to gyrotron developers was to significantly extend the gyrotron pulse duration.

Unfortunately, it was impossible (especially in the case of operation at short wavelengths) to solve this problem in a direct way because of the presence of contradicting requirements to the beam collector and output waveguide. Indeed, in the first gyrotrons the output part of the tube was made in a simple way as shown in Fig. 2.2, i.e., a part of an output waveguide was used as a collector for a spent electron beam. This output waveguide starts from the output cross section of the resonator, and the radius of this waveguide should increase gradually in order to avoid conversion of an operating mode into parasitic modes with different radial indices. At the same time, in order to meet requirements to the beam deposition density at the collector, which in long pulse regimes typically should not exceed 0.5 kW/cm^2, the collector area should be large enough. To solve this contradiction, several concepts of quasi-optical, built-in mode converters were proposed by M. I. Petelin. Note that the importance of quasi-optical mode converters became obvious in the process of preparation to the first experiments on ECRH. In these experiments, it was necessary to use linearly polarized wave beams with a reasonably small angular divergence, while the gyrotron output radiation had completely different spatial structure.

10.2 Quasi-Optical Mode Converters

Initially, in the late 1960s, the concept of wave transformers for gyrotron radiation was based on the use of helical corrugations of a waveguide wall (Kovalev, Orlova, and Petelin 1968). A schematic of such a waveguide with corrugated walls is shown in Fig. 10.2. In principle, such a waveguide can be used for transforming any wave with azimuthal index m_1 and axial wavenumber k_{z1} into another wave with azimuthal index m_2 and axial wavenumber k_{z2}. The wall radius of such a waveguide with a spirally corrugated wall can be determined as

$$R_w(z,\varphi) = R_{w0} + l\cos(\bar{k}_z z + \bar{m}\varphi). \tag{10.1}$$

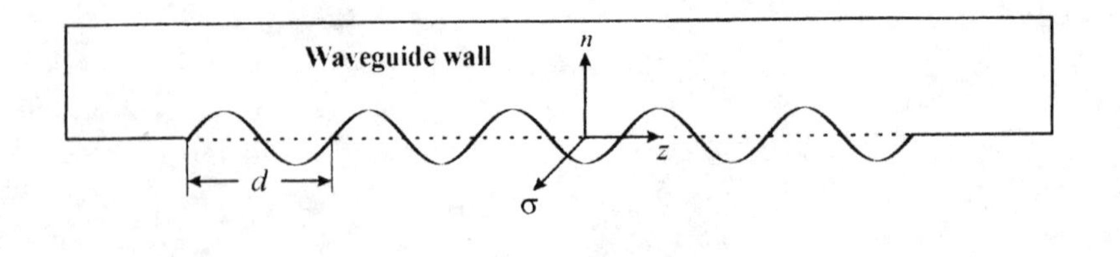

Fig. 10.2. Schematic of a quasi-optical rippled-wall mode converter.

For transforming one wave into another, the corrugation period $d = 2\pi/\bar{k}_z$ and azimuthal index $\bar{m}$ should obey Bragg conditions of two-wave coupling

$$\bar{k}_z = |k_{z1} - k_{z2}|, \tag{10.2}$$

$$\bar{m} = |m_1 - m_2|. \tag{10.3}$$

Kovalev, Orlova, and Petelin (1968) showed that the wave transformation of these two waves due to such corrugation can be described by standard equations for wave excitation by currents located at an inner surface of a waveguide

$$\frac{dC_s}{dz} = -\frac{1}{N_s}\oint \vec{i}_m \cdot \vec{H}_{-s} d\sigma \tag{10.4}$$

[cf. Eq. (6.11)]. In (10.4) the integration is carried out over the contour of the unperturbed waveguide wall. The equivalent surface magnetic current, $\vec{i}_m$, in accordance with Katzenelenbaum et al. (1998), is determined by the waveguide wall profile:

$$\vec{i}_m = \frac{c}{4\pi}\vec{n} \times \left\{\nabla(R_w(\vec{E}\cdot\vec{n})) + i\frac{\omega}{c}R_w(\vec{n}\times\vec{H})\right\}. \tag{10.5}$$

Here $\vec{n}$ is the external normal to the surface of the unperturbed waveguide and electric $\vec{E}$ and magnetic $\vec{H}$ fields of two waves can be represented as

$$\vec{E} = e^{i\omega t}\sum_{s=1}^{2} C_s \vec{E}_s, \qquad \vec{H} = e^{i\omega t}\sum_{s=1}^{2} C_s \vec{H}_s.$$

So, as one can readily check, the right-hand side of (10.4) is not equal to zero, and hence the coupling between two waves occurs only when (10.2) and (10.3) are fulfilled. After making corresponding calculations, one can easily reduce (10.4) to the known equations describing the coupling of two waves [see, e.g., Eqs. (1.57) in Louisell 1960].

It was, however, well understood that selective transformation of the radiating wave into another wave with the use of such a transformer is possible only when relatively low-order waves are coupled. Therefore, for mode transformation in the case of high-order modes, other principles should be used. A new concept of a quasi-optical mode converter was proposed by M. I. Petelin. The starting point for his proposal was the discussion he had with L. I. Pangonis in 1968 (Petelin 2001). By that time Pangonis had studied radiation of waves from an open end of a waveguide and found that the radiation pattern depends on how the waveguide is cut (Pangonis and Persikov 1971). Analyzing this information, Petelin realized that there is a simple way to transform the radiated cylindrical wave into another one by using quasi-optical reflectors.

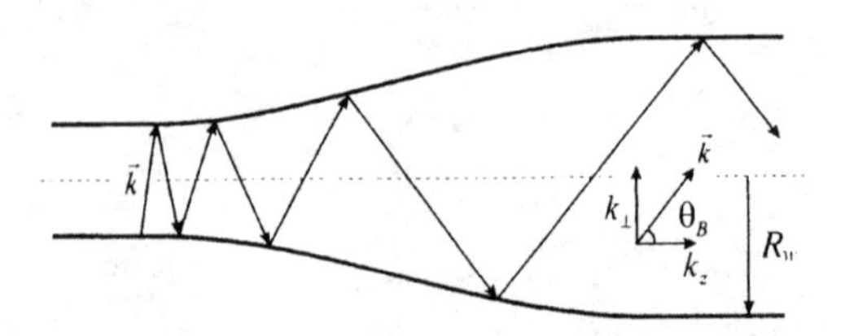

Fig. 10.3. Ray representation of the wave propagation near cutoff in the cavity region and far from cutoff in the output waveguide.

Initially, this method was used for transformation of one symmetric TE-wave into another by means of axially symmetric reflectors (Vlasov, Orlova, and Petelin 1972). (Such reflectors were used later for excitation of input cavities of gyroklystrons.) A little later, a more general method was developed for transformation of any radiating mode into a linearly polarized wave beam. Since this method proved itself as the most useful, we will consider it in more detail.

Principles of such quasi-optical mode converters can be explained in the geometrical-optics approximation as was done elsewhere (Vlasov and Likin 1980, Thumm 1997). In this approximation, any wave of a cylindrical waveguide can be represented as superposition of plane waves propagating at the Brillouin angle, θ_B with respect to the waveguide axis:

$$\sin\theta_B = k_\perp/k = \nu_{m,p}/kR_w. \tag{10.6}$$

This propagation is shown in Fig. 10.3. Here R_w is the waveguide wall radius. So, if the radiation starts from the gyrotron resonator excited near cutoff where $k \approx k_\perp = \nu_{m,p}/R_{\text{res}}$ (here R_{res} is the resonator radius), then $\sin\theta_B \approx R_{\text{res}}/R_w$. In a waveguide cross section, as was shown in Fig. 2.5a, these waves can be represented as a superposition of rays, which propagate between the waveguide wall and the caustic of a radius $R_c = (m/\nu_{m,p})R_w$. (Clearly, symmetric modes (with $m = 0$) have no caustic.) In this transverse plane, an angle between points of two successive reflections is 2α, where $\cos\alpha = R_c/R_w = m/\nu_{m,p}$. The distance such rays pass in the transverse plane between two reflections is equal to

$$L_{\text{tr}} = 2R_w\sqrt{1-(m/\nu_{m,p})^2}. \tag{10.7}$$

Let us consider an unwounded waveguide, shown in Fig. 10.4a. Here thin straight lines show the ray propagation. The dots on one of these lines show the points of reflection of a single ray from the wall. Then, one can readily find that rays that were located at a distance L_{tr} in the transverse plane occupy

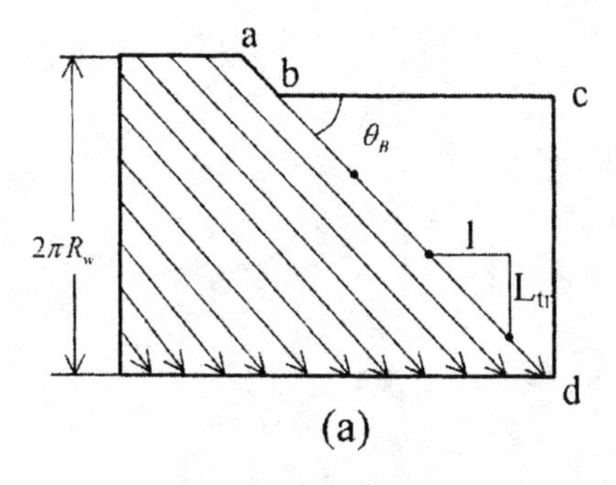

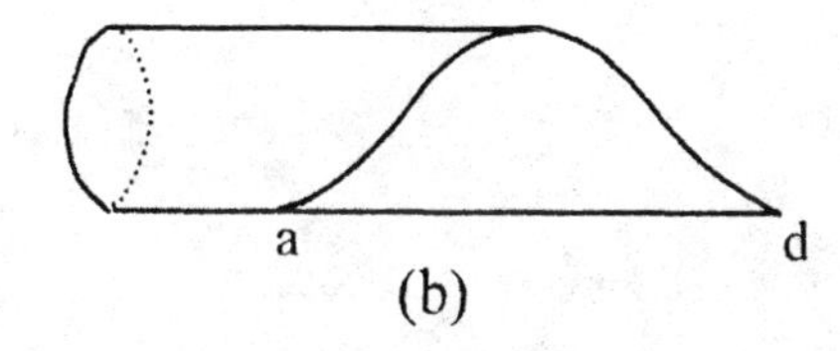

Fig. 10.4. (a) Ray propagation in an unwounded waveguide; (b) helical cut of a waveguide.

the axial distance

$$l = L_{\text{tr}} \cot \theta_B . \tag{10.8}$$

Of course, this distance is not equal to the total distance, which is necessary for the rays to make a full turn. The latter distance, which is often called a Brillouin length, L_B, is larger than l by π/α times. So, using simple formulas given above, one can find that the Brillouin length is equal to

$$L_B = 2\frac{kR_w^2}{\nu_{m,p}}\sqrt{1-(m/\nu_{m,p})^2}\sqrt{1-(\nu_{m,p}/kR_w)^2}\frac{\pi}{\arccos(m/\nu_{m,p})} . \tag{10.9}$$

These results indicate that, if one makes in a cylindrical waveguide a slot of a length L_B with an angular width 2α, then all the wave power will be radiated through this slot. Of course, in the case of symmetric modes this angular width is equal to π. It is also clear that such a slot for radiating the wave power can be made in numerous ways. For instance, in the plane shown in Fig. 10.4a, the whole waveguide area where the rays do not propagate can be deleted, which yields a helical cut shown in Fig. 10.4b. Alternatively, this cut can be made in the direction perpendicular to the ray propagation. Now the question is: "What is the spatial structure of this outgoing radiation, and how can this radiation be transformed into a linearly polarized wave beam?"

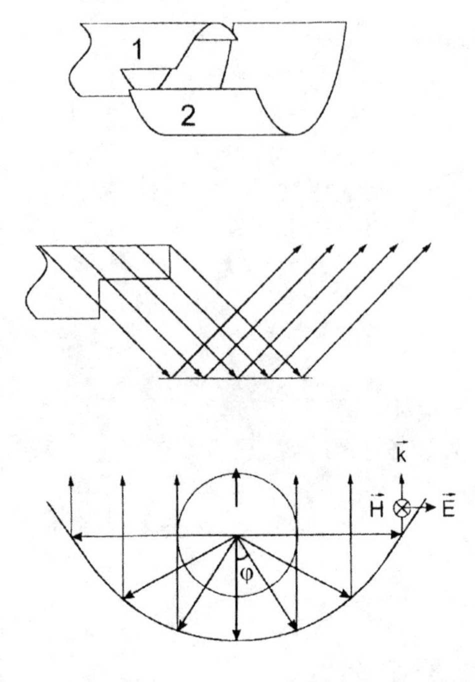

Fig. 10.5. Schematic of a quasi-optical wave transformer for modes with $m = 0, 1$.

Vlasov and Orlova (1974) analyzed the simplest concept of mode converters for symmetric modes and modes with a small azimuthal ($m = 0, 1$) index. A little later, Vlasov, Zagryadskaya, and Petelin (1975) described a concept of a quasi-optical mode converter for modes with a large azimuthal ($m >> 1$) index. Configurations of these mode converters are shown in Figs. 10.5 and 10.6, respectively.

Operation of these converters is based on the fact that radiation of any mode from an open end of an oversized waveguide can be considered as superposition of cylindrical waves limited by the waveguide aperture. So, it is quite obvious that such cylindrical waves can be transformed into a wave beam by a parabolic (or, more exactly, quasi-parabolic) cylindrical reflector, if the focus of this parabolic mirror is located in the region from which these waves are radiated. In the case of symmetric waves and waves with $m = 1$, all rays pass through the waveguide axis, as shown in Fig. 10.5. The only difference between the fields of $TE_{0,p}$-waves and the fields of $TE_{1,p}$-waves is that the fields of symmetric waves do not depend on the incident angle, while the fields of the waves with one azimuthal variation are proportional to $\cos\varphi$ (see Fig. 10.5). So, the focal axis of the parabolic mirror should coincide here with

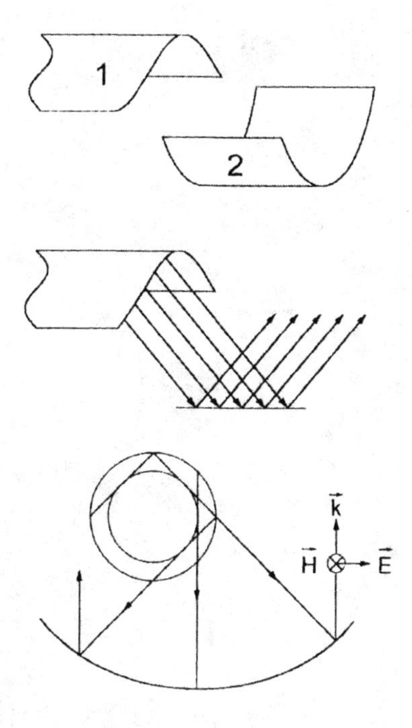

Fig. 10.6. Schematic of a quasi-optical wave transformer for whispering gallery modes with $m \gg 1$.

the waveguide axis. Then, all the rays of a cylindrical wave will be transformed, as is shown in Fig. 10.5, into a set of parallel rays with the linear polarization of the electric field. The length of such a reflector should not be smaller than the Brillouin length. (Strictly speaking, diffraction effects lead to a certain spread of this wave beam, so the length of the mirror should slightly exceed L_B.) Transverse width of this wave beam is equal to

$$L_{\perp} = 4F\sqrt{(\nu_{m,p} - m)/(\nu_{m,p} + m)}, \tag{10.10}$$

where F is the focal length of the mirror.

Let us show how the profile of such a reflector can be described in the case of an arbitrary mode, as was done by Vlasov and Likin (1980). For the case shown in Fig. 10.6, all rays are radiated from the region between the waveguide wall and caustic. Then, after reflection we should get a set of parallel rays. For more detailed consideration, let us consider an arbitrary ray radiated from this area and reflected from the mirror, as shown in Fig. 10.7. Assume that the mirror profile is described by the function $y(x)$, which should be determined. Its derivative is equal to $dy/dx = \tan(\varphi/2)$ where φ is the angle between the

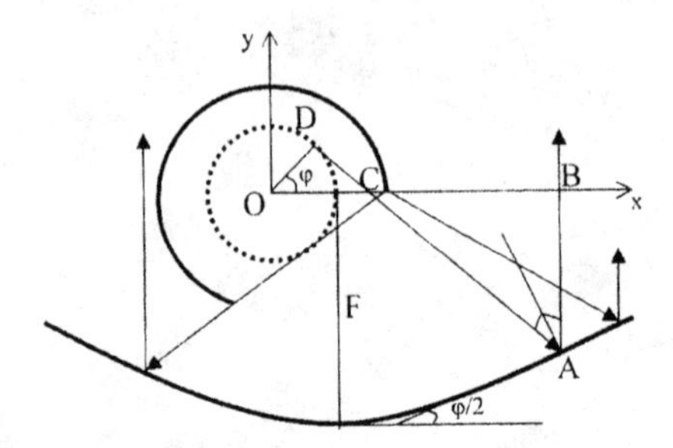

Fig. 10.7. Cross section showing the wave radiation and conversion into a wave beam with a linearly polarized field.

incident and reflected rays, which is also equal to the angle between the point of caustic, which was grazed by this ray, and the horizontal axis (see Fig. 10.7). As follows from the triangle ABC,

$$\tan\varphi = \frac{BC}{AB} = \frac{x - R_C(\frac{1}{\cos\varphi} - 1)}{F - y} = \frac{1}{F - y}\left[x - 2R_C\frac{(dy/dx)^2}{1 - (dy/dx)^2}\right]. \tag{10.11}$$

So, expressing $\tan\varphi$ as $2\tan(\varphi/2)/[1 - \tan^2(\varphi/2)]$, one can easily find that the shape of this mirror is described by the equation

$$(x + 2R_C)\left(\frac{dy}{dx}\right)^2 - 2(y - F)\frac{dy}{dx} - x = 0. \tag{10.12}$$

As shown by Vlasov and Likin (1980), solution of this equation can be given in a parametric form as

$$y = F\left[\tan^2\left(\frac{\varphi}{2}\right) - 1\right] + R_c\left[\frac{\varphi}{2}\tan^2\left(\frac{\varphi}{2}\right) + \tan\left(\frac{\varphi}{2}\right) - \frac{\varphi}{2}\right], \tag{10.13}$$

$$x = 2F\tan\left(\frac{\varphi}{2}\right) + R_c\left[1 + \varphi\tan\left(\frac{\varphi}{2}\right)\right]. \tag{10.14}$$

These equations, in particular, yield (10.11). For symmetric waves $R_C = 0$, and hence Eqs. (10.13) and (10.14) can be reduced to the equation of parabola

$$x^2 = 4F(y + F). \tag{10.15}$$

As one can readily find, the transverse width of such a wave beam is equal to $4F$, as follows from (10.10) for $m = 0$.

Note that the efficiency of wave conversion with the use of such simple quasi-optical converters is relatively low (about 80–90%). Therefore, later various more efficient modifications were proposed, which are described elsewhere (Denisov et al. 1992, Thumm 1997). Of course, to correctly describe the field transformation and wave conversion in quasi-optical mode converters discussed above, one should take into account diffraction effects, i.e., use

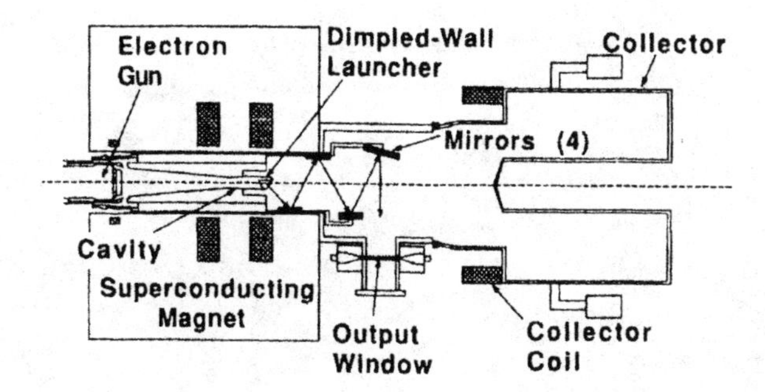

Fig. 10.8. Schematic diagram of internal-converter gyrotron. (Reproduced from Felch et al. 1996)

quasi-optical theory (or even more accurate theory) instead of the geometrical-optics approximation.

After transforming a gyrotron radiation into a wave beam with the use of a quasi-parabolic mirror, this radiation passes through a set of phase correctors before it reaches an output window. Typically, an output part of a gyrotron contains, as is shown in Fig. 10.8, from two to four phase correcting mirrors (Felch et al. 1996). The last mirror directs the wave beam onto the output window.

10.3 Output Windows

An output window, through which a microwave radiation passes from an inner vacuum part of a tube to the air, is one of the key components of any high-power microwave tube. Several issues are important for successful operation of output windows at high-power levels in long pulses or in CW regimes. The most important of them are caused by the dielectric characteristics of window materials, i.e., real and imaginary parts of the dielectric constant, which are known as *dielectric permittivity* and *loss tangent,* respectively. (Mechanical characteristics will be discussed later.)

The permittivity ε_r' determines the radiation wavelength in a given dielectric, $\lambda_d = \lambda/\sqrt{\varepsilon_r'}$ (here λ is the wavelength in a free space). To avoid reflections of a wave from the window, the window thickness d should be equal to an integer number of such half-wavelengths:

$$d = N(\lambda_d/2). \tag{10.16}$$

At first glance, this condition can be easily fulfilled. However, first, dielectric constants are temperature dependent, and also the dielectric itself experiences thermal expansion. So, a window whose thickness was matched at room temperature, being heated in the process of microwave power absorption, can reflect some wave power back into the tube. The effect of the reflected power on the gyrotron operation was studied by Antonsen et al. (1992) and Glyavin et al. (1999). Clearly, for providing stable gyrotron operation, it is desirable to avoid any reflection. One obvious way to avoid reflections is to position a window at the Brewster angle θ_B, which is determined by the equation $\tan\theta_B = \sqrt{\varepsilon_r'}$. Then, a linearly polarized wave with the electric field oriented in the incidence plane will pass through a boundary between the air and the dielectric without any reflection. Operation of a gyrotron with a Brewster window was described by Dammertz et al. (1999).

Another problem associated with reflections from the window occurs when tubes are designed for operation at more than one frequency. Clearly, Eq. (10.16) can be fulfilled only for several frequencies but not for a wide spectrum. In addition to the concept of Brewster angle, such concepts as a multilayer window with variation of permittivity for "anti-reflection coating" (this concept is essentially the same as the stealth technology known in military aviation) and "moth-eye-type" window configuration were considered (Thumm 1998). Note that initially the concept of night-moth-eye window was proposed for the use in gyrotrons by Ma, Blanco, and Robinson (1983); later this idea was advanced by Petelin and Kasparek (1991). Also note that a wave beam passing through a window has a finite cross section. Therefore, it cannot be considered as just one plane wave, but should be considered as a superposition of plane waves. In this regard, we should mention the paper by Golubyatnikova and Petelin (1994), where some conclusions about minimizing the reflection coefficient for such a wave beam were made.

So far we have been concerned with reflections from the dielectric window. However, the most important issue for long-pulse and CW operation is the window heating caused by the losses of microwave power in the window. This factor is determined by the above mentioned loss tangent, $\tan\delta$, as well as the microwave power deposition profile. This factor also depends on the thermal conductivity, heat capacity, and other thermophysical and mechanical characteristics (including resistance against thermal crack formation) of window materials. For window operation, the methods of window cooling are also important.

In the first long-pulse experiments, such known materials as alumina and BeO were used. These windows were fabricated in an edge-cooled configuration. However, first attempts to use such windows in CW high-power gyrotrons resulted in a thermal failure of the window. This happened when a 28 GHz gyrotron built at Varian operated at more than a 100 kW power level (Jory

1978). Therefore, in the next series of experiments with 28 GHz and 60 GHz CW gyrotrons, Varian developers successfully used BeO double-disk window configurations with a face cooling of disk surfaces with a low-loss dielectric fluid (such as, for instance, FC-75) (Felch et al. 1984). During the late 1980s and the 1990s, the gyrotron development for electron cyclotron plasma heating and current drive was focused on the power increase and extension of pulse duration (up to CW operation) as well as on the frequency increase. Therefore, it was a time of active characterization of frequency dependencies of loss tangent of various dielectrics (Sushilin, Fix, and Parshin 1989, Parshin et al. 1995). These dependencies were measured at various temperatures, including cryogenic temperatures (see, e.g., tables presented by Thumm 1998), because it is known that, for instance, at the temperature of liquid nitrogen the loss tangent can be much smaller than at room temperature.

Note that during the initial phase of the window performance, when a window operates in the nonstationary regime, not only thermal conductivity but also the radial profile of the microwave beam is important for the window heating and mechanical stresses. Therefore, it makes sense to treat this problem self-consistently. Consider, for instance, the windows with edge cooling. In such windows, a wave beam with a Gaussian radial profile deposits more power in the noncooled center than at the cooled periphery. Thus, in long pulses the central part of such a window can be overheated. To avoid this, it is necessary to optimize the shape of wave reflectors in such a way that the radial profile of a wave beam passing through the window will correspond to the temperature runaway from the center to the window periphery, where the window is cooled. Note that optimization of the wave beam configuration should also be done taking into account a possible multipactor discharge at the window surface. The term "multipactor" means here a resonance multiplication of the secondary emission, which may occur at the surface of dielectric under certain resonance conditions (Vaughan 1988). Such a discharge was, in particular, a serious issue in the first Russian short-pulse, 1 MW gyrotron operating at the whispering gallery mode ($TE_{15,1}$) without mode conversion, because the field of this mode was localized near the wall.

During the 1990s various materials and window configurations were tested. As one of the most impressive results, let us mention a long-pulsed 0.5 MW, 118 GHz gyrotron, which was developed at Thomson Tubes Electronique (France) in collaboration with researchers from Lausanne (Switzerland) and Karlsruhe (Germany) (Giguet et al. 1997). In this tube a liquid-nitrogen, edge-cooled, single-disk sapphire window was successfully used. Later on, however, it was well recognized that the most promising material for MW-level, CW (or quasi-CW) gyrotrons is CVD diamond (here CVD is an abbreviation for chemical vapor deposition). This material is characterized by a combination of a very small loss tangent and very high thermal conductivity (corresponding data can be found in Thumm 1998; see also Heidinger et al. 1997). Therefore, all the latest achievements in gyrotron performance throughout the world were demonstrated with the use of CVD diamond windows.

10.4 Depressed Collectors

Depressed collectors are widely used in vacuum microwave tubes for enhancement of overall efficiency. This issue is especially important for TWTs used for satellite communications where the problem of heat dissipation at the collector is extremely severe. (A very impressive example of the importance of this problem for TWTs used in satellite communication systems is given by Abrams et al. 2001.)

By "depressed collectors" I mean collectors whose potential is lower than the potential of the beam in the interaction region. Typically, electrons in the interaction region do not give up all their kinetic energy to the microwave field. Therefore, the potential of a beam can be reduced at the collector, which will slow down electrons and reduce the beam power deposited at the collector. The efficiency of this method can be easily evaluated by using the following simple procedure. Consider an electron beam that has at the entrance to the interaction region initial normalized energy $\gamma_0 = 1 + eV_b/mc^2$. Here V_b is the beam voltage determined by the potential difference between the cathode and resonator/waveguide. At the exit the average normalized energy of this beam after the interaction is $\langle\gamma(L)\rangle$. Here L designates the coordinate of the exit cross section, and angular brackets mean electron averaging over entrance phases. (Ignore, for the sake of simplicity, all other possible distributions.) So, the interaction efficiency in this case is equal to

$$\eta_{\text{int}} = \frac{\gamma_0 - \langle\gamma(L)\rangle}{\gamma_0 - 1}. \qquad (10.17)$$

When the collector potential is lower than the resonator/waveguide potential, the overall or wall-plug efficiency, η_w, is determined by the ratio of the electron energy given up to the microwave field to the electron energy deposited to the collector:

$$\eta_w = \frac{\gamma_0 - \langle\gamma(L)\rangle}{e(V_b - V_{b-c})/mc^2}. \qquad (10.18)$$

Here V_{b-c} is the potential difference between the resonator and collector. [So, when resonator and collector are at the same potential, $V_{b-c} = 0$ and (10.18) reduces to (10.17).] Thus, one can introduce collector efficiency as the ratio of V_{b-c} to the energy of a spent beam,

$$\eta_{\text{coll}} = \frac{eV_{b-c}/mc^2}{\langle\gamma(L)\rangle - 1}. \tag{10.19}$$

In the ideal case of deceleration of all electrons as one macrobunch, the limiting value of this efficiency is 100%. However, in real devices, there is always a certain spread in electron energies of a spent beam; therefore, the collector efficiency is smaller than this limiting number, because the voltage depression in the collector region should not stop the slowest electrons.

Rewriting (10.18) as

$$\eta_w = \frac{\gamma_0 - \langle\gamma(L)\rangle}{\gamma_0 - 1 - eV_{b-c}/mc^2}$$

and using definitions of the interaction and collector efficiencies given by (10.17) and (10.19), respectively, one can easily obtain the following formula for the overall efficiency,

$$\eta_w = \frac{\eta_{\text{int}}}{1 - \eta_{\text{coll}}(1 - \eta_{\text{int}})}. \tag{10.20}$$

This is a general definition of the overall efficiency of any device with a depressed collector in which the definition of the collector efficiency given by (10.19) can easily be generalized for the case of multistage depressed collectors.

In the case of gyrotrons, the situation is more specific than in linear-beam devices in the sense that in gyrotrons the electrons, in the process of interaction, lose primarily the energy of their gyration while the energy associated with their axial motion remains practically unchanged. After exiting from the interaction region, these electrons move in the decreasing external magnetic field where they experience adiabatic decompression, i.e., their transverse momentum transforms into the axial one. Corresponding changes in electron beam location in the electron momentum phase space are shown schematically in Fig. 10.9. So, the potential depression at the collector should be small enough in order to avoid reflection of slow particles back to the interaction region.

Presently, numerous codes are developed for optimizing performance of single-stage and multistage collectors (Singh et al. 1999). There are also numerous experimental evidences starting from Sakamoto et al. (1994) that the use of even single-stage depressed collectors can improve the gyrotron efficiency from the typical 30–35% to 50% and more. Designs of two-stage depressed collectors predict overall efficiencies up to 65–70%.

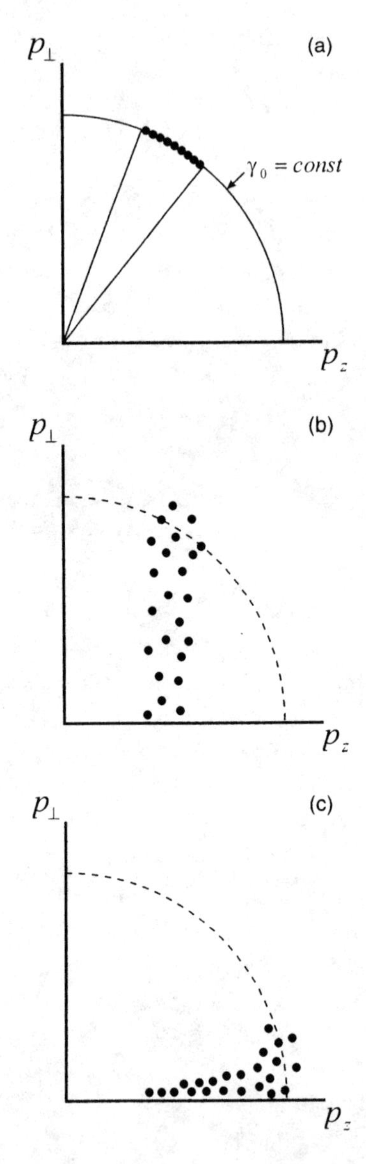

Fig. 10.9. Initially monoenergetic electrons with a spread in the orbital-to-axial velocity ratio in the momentum space at the entrance to the interaction region (a), at the exit from the interaction region (b), and after magnetic decompression (c).

10.5 Experimental Results

First steps in the development of gyrotrons for ECRH experiments in controlled fusion reactors were already discussed in Sec. 10.1. Present-day gyrotrons operate at about 1 MW level in multisecond pulses at frequencies above 100 GHz. Practically all of them use quasi-optical built-in mode converters, single-stage depressed collectors, and CVD-diamond windows. In some tubes an additional window is added for extracting the stray radiation from the gyrotron output part. In some tubes, a dual-window concept is used, which allows one to extract about 0.5 MW power through each window in a 1-MW tube. The state-of-the art in this development is briefly summarized in Table 10.2. Let us call the reader's attention to the increase in the indices of operating modes with the frequency shown in this table. This tendency is in line with the general dependence of the ohmic loss density on the wavelength and resonator radius, which is given in Chapter 4 by (4.19), and is also discussed in Sec. 10.1. (In more detail, the choice of modes for quasi-CW gyrotrons is discussed by Kreischer et al. 1985 and Flyagin and Nusinovich 1988.) Note that, as was mentioned by Myasnikov et al. (2002), in Russian gyrotrons with a BN window the effect we discussed in Sec. 10.3 was observed. Namely, it was found that the thermal expansion of this window caused reflections of the microwave power that limited pulse duration.

Practically all gyrotrons whose parameters are given in Table 10.2 operate at frequencies from 110 GHz to 170 GHz. This choice of frequencies is dictated by the needs of large fusion reactors. For instance, the International Thermonuclear Experimental Reactor (ITER) requires 170 GHz frequency (Makowski 1996). U.S. tokamak "Doublet-IIID" (Felch et al. 2002) and Japanese tokamak JT-60U (Sakamoto et al. 2002) need 110 GHz gyrotrons. German large-scale stellarator "Wendelstein 7-X" requires 70 GHz and 140 GHz gyrotrons (for the first and second harmonic experiments, respectively) (Erckmann et al. 1999), and French tokamak Tore Supra and tokamak TCV (Lausanne, Switzerland) need 118 GHz tubes (Giguet et al. 1997). It is necessary to emphasize that presently gyrotrons are used not only for the electron cyclotron heating experiments, but also for current drive experiments (Erckmann et al. 1999, Lohr et al. 2000), suppression of neoclassical tearing modes (Gantenbein et al. 2000), tokamak start-up, wall conditioning, etc. The ability to localize microwave power deposition in any desired area makes gyrotrons unique sources for many applications in the fusion reactors. It is worthy to mention that gyrotrons can also be used for active plasma diagnostics (see, e.g., Woskoboinikow, Cohn, and Temkin 1983, Ogawa et al. 1997, Lubyako et al. 1998, and references therein). In some of the applications mentioned above, it is preferable to use frequency-tunable gyrotrons. This is why some possibilities of the development of such tubes are actively studied both theoretically (Dumbrajs and Nusinovich 1992, Dumbrajs et al. 2000) and experimentally (Pyosczyk et al. 2000).

Table 10.2. State-of-the Art in Gyrotron Development for Experiments with Plasma in Large Reactors

Frequency, GHz	Reactors	Company	Operating Mode	Power, MW	Pulse Duration (sec)	Depressed Collector
110	Doublet III-D Tokamak (USA)	CPI (w/partners)	$TE_{22,6}$	1.0	5	No
	JT-60U Tokamak (Japan)	Toshiba Co. and JAERI	$TE_{22,6}$	1.0	5	Not available
				1.2	4.1	
140	Wendelstein VII-X Stellarator (Germany)	Euroatom Team	$TE_{28,8}$	0.9	180	Yes
				0.92	55	
		CPI	$TE_{28,7}$	0.5	700	Yes
				0.9	<1	
		GYCOM	$TE_{22,8}$	0.92	3.5	Not available
170	ITER	Toshiba Co. and JAERI	$TE_{31,8}$	0.45	47	Yes
				0.9	9.2	
		GYCOM	$TE_{25,10}$	0.54	40	Yes

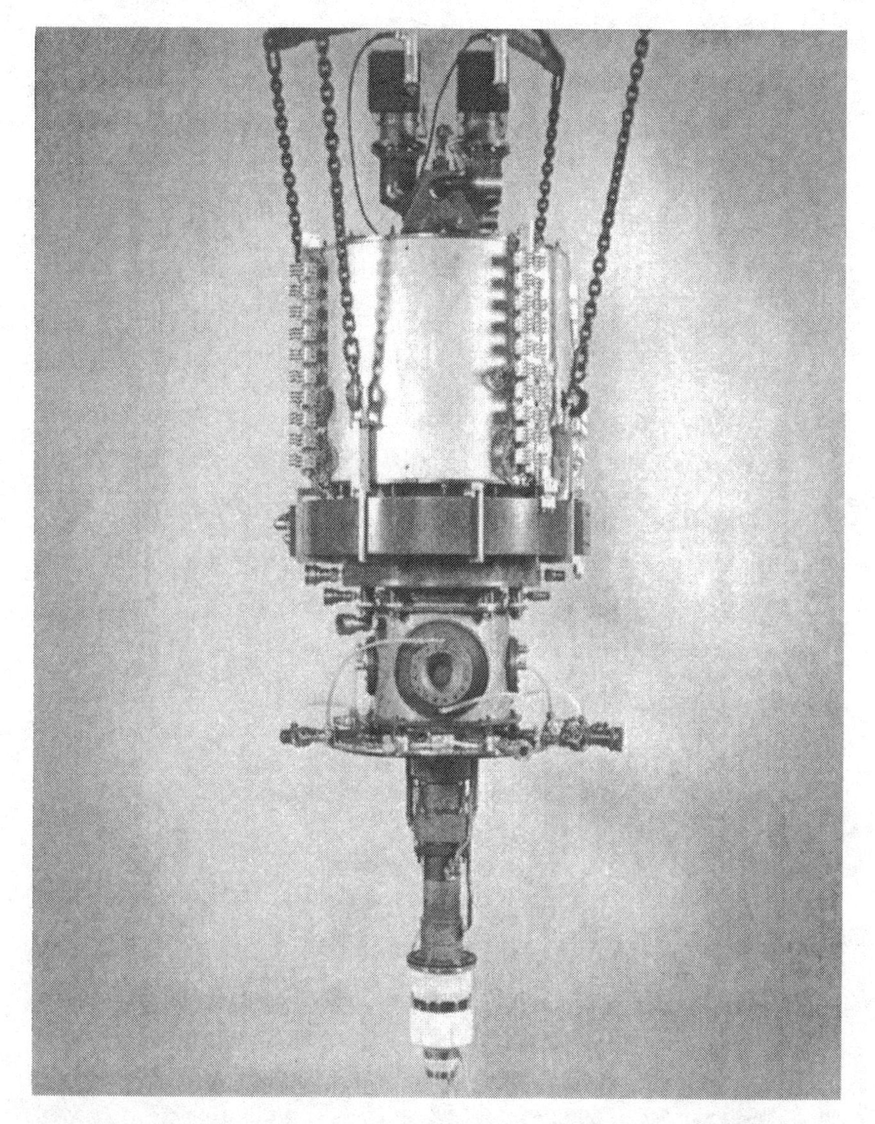

Fig. 10.10. Photo of a 110 GHz, 1 MW, CPI gyrotron with a CVD diamond window. (Courtesy of CPI)

As follows from (4.19), to increase the output power and operating frequency further, one should operate at higher-order modes. Selection of such modes is possible in coaxial cavities, which also allow one to compensate for the voltage depression caused by the beam space charge fields. (Such a depression can be significant when an electron beam is positioned at the inner

Fig. 10.11. Photo of a 1 MW, 140 GHz, CPI gyrotron with a CVD diamond window and a single-stage depressed collector. (Courtesy of CPI)

peak of the radial profile of a rotating mode with a large radial index.) In coaxial cavities, support of an inner conductor can be easily provided when an electron gun has an inverted geometry (Lygin et al. 1995). Such inverted guns were successfully used in many gyrotrons with coaxial cavities starting from one mentioned by Alikaev et al. (1978).

Let us note that just from comparison of the data on first gyrotron experiments with the latest achievements, one may conclude that gyrotron development already reached its saturation stage, and there is no reason to expect further progress from it. This statement would be incorrect. In order to put this development in proper perspective, let us mention that first experimental results on generating over 1 megajoule microwave energy per pulse in 110 GHz gyrotrons were reported by the Russian and U.S. teams in 1992 (Flyagin, Goldenberg, and Zapevalov 1992 and Felch et al. 1992, respectively). Present-day gyrotrons deliver multimegajoule pulses at this and higher frequencies. Three of such gyrotrons are shown in Figs. 10.10, 10.11, and 10.12. The gyrotron shown in Fig. 10.10 is a 110-GHz tube developed at CPI, which delivers 1 MW power in 5-second pulses. A 1 MW, 140 GHz, CPI gyrotron with a single-stage depressed collector is shown in Fig. 10.11 (This photo is

Fig. 10.12. Photo of a 170 GHz, JAERI gyrotron with a CVD diamond window and a single-stage depressed collector. (Courtesy of JAERI)

reproduced from Felch and Temkin, 2003.) In Fig. 10.12, a 170 GHz gyrotron developed at JAERI is shown. Its overall length is about 3 meters; other parameters were given in Table 10.2. (The tube developed by JAERI at 110 GHz looks very much like this one.)

There is no doubt that in the near future further progress in the development of windows and other gyrotron components will result in real CW operation at the megawatt level.

▪ CHAPTER 11 ▪

Gyroklystrons

11.1 Historical Introduction

In pre-gyrotron time, J. M. Wachtel and J. L. Hirshfield (1966) did the first experimental study of a CRM-klystron. Once the concept of gyrodevices had been clearly understood, the gyroklystron was invented (Antakov et al. 1975, Gaponov et al. 1975; note that in both cases the priority is dated by 1967) simultaneously with the gyrotron oscillator. Initially, in the late 1960s, gyroklystron experiments were carried out in the X-band. However, their results were published much later (Andronov et al. 1978). The most impressive result of these experiments was an experimentally demonstrated 70% electronic efficiency. This record level of efficiency was realized due to operation with a very high orbital-to-axial velocity ratio (about 3), which was possible because of operation at low currents. (The maximum efficiency was obtained at 0.3 A, the operating voltage was in the range from 20 kV to 30 kV; thus the output power was about several kW.) At such high velocity ratios, it follows from (3.37) that the electronic efficiency is close to the orbital efficiency, whose maximum, in gyroklystrons, may exceed 80%, as was discussed in Chapter 8.

Beginning in the 1970s, millimeter-wave (Ka-band and W-band) gyroklystrons operating at moderate voltages (below 80 kV) have been actively developed by a group led by A. V. Gaponov for radar applications. As a rule, some results of this development were published much later, in the 1990s (Antakov et al. 1990, 1993; Antakov, Sokolov, and Zasypkin 1993; Zasypkin et al. 1995). Also a Ka-band, 1-MW radar was described by Tolkachev et al. (2000) much later than the system was built and tested. In this radar, two gyroklystrons developed by Gaponov's group in collaboration with the industrial company Toriy were used.

In the United States, the development of gyroklystrons began, along with the development of gyrotron oscillators, in the late 1970s at Varian (Jory et al. 1977, Jory 1978). The first tube operated at 28 GHz frequency and was designed for delivering 200 kW output power. However, due to excitation of spurious oscillations, the maximum power achieved in the regime of amplification was 65 kW. In parallel with this tube operating at the fundamental cyclotron resonance, a 10 GHz, 100 kW gyroklystron designed for operation at the second harmonic was developed at Varian. Symons and Jory (1981) described some details of this activity. Shortly thereafter, the gyroklystron development was started at the Naval Research Laboratory (NRL) in the United States. In the first experiments done at 4.5 GHz frequency the maximum power was over 50 kW and the efficiency was 25% (Bollen et al. 1985). The aim of this program was to develop gyroklystrons for radar applications.

In general, an interest in the wavelength shortening in radars is obvious, since it allows one to better focus a radar wave beam, and hence, to improve an angular resolution and imaging. It also allows one to use a wider instantaneous bandwidth. Unfortunately, as the wavelength shortens, the atmospheric attenuation of radiation increases even in such "windows" as Ka-band and W-band. The attenuation also increases with the decrease in the elevation angle. It should be noted that, in spite of recent progress in the development of solid-state phased arrays (see, e.g., Sarcione et al. 1996), vacuum electron tubes have many advantages, which make them preferable, especially at short wavelengths. In particular, the use of millimeter-wave gyroklystrons in radar systems greatly reduces their cost. A cost analysis presented by Tolkachev (2002) shows that at millimeter wavelengths (and, especially, at short millimeters) the cost of phased arrays based on the use of gyroamplifiers is much smaller than the cost of other phased arrays.

Simultaneous to the development of gyroklystrons for radar applications, starting from the mid-1980s, a program for the development of gyroklystrons driven by relativistic electron beams was started at the University of Maryland (Chu et al. 1985). This program, sponsored by the Division of High Energy Physics of the U.S. Department of Energy, is focused on the development of high-power, millimeter-wave sources, which can be used as RF drivers in future linear colliders. The tendency of wavelength shortening of RF drivers is motivated by the fact that this shortening allows one to increase the accelerating gradient, and hence reduce the distance required for accelerating charged particles up to the desired high-energy level. In the linear accelerators of the next generation, these energies will be in the range of several TeV's. The state-of-the art in this development as well as in the development of gyroklystrons for radar applications is presented below.

11.2 Gyroklystrons for Radar Applications

As was mentioned above, development of gyroklystrons for radar applications is focused on two frequency ranges: Ka-band (34–35 GHz) and W-band (93–95 GHz), which are the atmospheric windows in the millimeter-wave attenuation. A Russian Ka-band gyroklystron developed jointly by the Institute of Applied Physics (Nizhny Novgorod) and industrial company Toriy (Moscow) delivers 750 kW output power in up to 100 microsecond pulses (Antakov et al. 1993). This is the highest power level for millimeter-wave amplifiers driven by moderately relativistic electron beams. The operating voltage in this tube is in the range from 50 to 65 kV. The tube consists of only two cavities, thus the gain is relatively low: 30 dB and 22 dB in the small-signal and large-signal regimes, respectively. The operating mode in both cavities was the TE_{02}-mode. The Q-factor of the input cavity was about 100, while the Q-factor of the output cavity was about 320. The maximum power was obtained with an efficiency of 24%. The maximum efficiency of 32% was realized at the 300 kW output power level. At the 600 kW power level, the instantaneous bandwidth at the −3 dB level was 0.61%. This bandwidth was limited by the Q-factor of the output cavity.

Two such gyroklystrons were used in the above-mentioned 1-MW, Ka-band radar "Ruza" located in Kazakhstan. A photo of this radar is shown in Fig. 11.1, reproduced from Tolkachev et al. (2000). Fig. 11.2, reproduced from the same paper, shows a mechanically steering, phased-array antenna of this radar.

In the second half of the 1990s a detailed study of two-cavity and three-cavity Ka-band gyroklystrons was carried out at NRL (Calame et al. 1999). The three-cavity device performance was extensively characterized in a wide range of operating parameters. In this device all cavities operated at TE_{01}-modes with Q-factors in the range from 150 to about 200. All cavities were heavily stagger tuned to increase the bandwidth. The signal was directed into the input cavity via a passive TE_{41} coaxial resonator, which surrounds this cavity, having four axial slots placed every 90 degrees in azimuth. By varying the external magnetic field it was possible either to produce the maximum power of 245 kW with the bandwidth 0.63% at a low field or to increase the bandwidth up to almost 1% while lowering the power to 200 kW at a high field. One of the most important conclusions of this study was excellent agreement

Fig. 11.1. Photo of a Russian 1MW, Ka-band radar "Ruza." (Reproduced from Tolkachev et al. 2000)

between experimental results and numerical calculations done with the use of codes MAGY and MAGYKL. An example of this agreement is shown in Fig. 11.3. As one can see in Fig. 11.3, the device operated with an efficiency exceeding 30%. The saturated gain in this regime of operation exceeds 30 dB and the power exceeds 200 kW.

First W-band gyroklystrons have been developed at the Institute of Applied Physics (Nizhny Novgorod, Russia) for operation in the pulsed and CW

Fig. 11.2. Photo of a mechanically steering phased-array antenna of the "Ruza" radar. (Reproduced from Tolkachev et al. 2000)

regimes (Antakov et al. 1993, Antakov, Sokolov, and Zasypkin 1993). Both tubes were four-cavity gyroklystrons, in which all cavities operated at the TE_{01}-mode. The pulsed device had cavities with the Q-factors of 150 (for the input cavity), 250 (for intermediate cavities), and 300 (for output cavity). As a result, the bandwidth limited by these Q-factors was about 0.3%. The maximum power and efficiency were equal to 65 kW and 34%, respectively. The

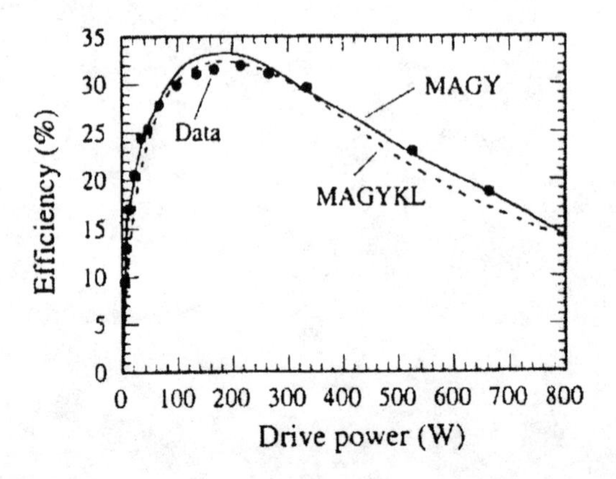

Fig. 11.3. Comparison of experimental data and simulation results for a Ka-band, NRL gyroklystron. (Reproduced from Calame et al. 1999)

small-signal gain achieved 40 dB, while the saturated gain was about 33 dB. The duty cycle of operation was rather low.

In the device intended for CW operation, the maximum output power of 2.5 kW was produced with an efficiency of 25%. (This device was driven by a 22 kV electron beam with the current about 0.5 A.) The large-signal gain and the −3 dB bandwidth were equal to 30 dB and 0.35%, respectively.

Later, remarkable progress in the development of W-band gyroklystrons capable of high average power operation was demonstrated at NRL. First, the pulsed performance of a four-cavity device operating at the TE_{01}-mode in all cavities was studied. In this device, a bandwidth greater than 460 MHz, peak output power of 67 kW, 28% efficiency and about 30 dB large-signal gain were realized (Blank et al. 1997). So, the bandwidth achieved in this device was a significant increase over the bandwidth demonstrated previously in W-band gyroklystrons. (A further bandwidth enhancement can be realized in other versions of gyroamplifiers, such as gyrotwystrons and gyro-TWTs, which will be considered below.)

The device was then redesigned for CW operation. Again, all cavities operated at the TE_{01}-mode with Q-factors in the range from 130 to 175. These efforts resulted in the development of a four-cavity W-band gyroklystron delivering over 10 kW average power with 11% duty cycle (Blank et al. 1999). This world record performance was realized with 33% efficiency. (Note that the best W-band coupled-cavity traveling wave tube delivers about 0.5 kW CW power only; see James and Kreutzer 1995.) The instantaneous bandwidth of this gyroklystron was 600 MHz. This was a result of joint activity between researchers from NRL, the industrial companies Litton and CPI, and the University of Maryland. A photo of this tube and cryomagnet for it are shown in Fig. 11.4.

Regarding a high average power operation of gyroklystrons, it is necessary to mention that one of the serious problems in such devices is the thermal loading of the penultimate cavity. Indeed, if we assume that each stage of this device yields about 10 dB gain, then 10 kW average output power corresponds to about 1 kW power in the penultimate cavity. This power should be somehow dissipated in this cavity that, once we take into account that diffraction losses in the axial direction are extremely undesirable, becomes a serious problem.

Presently, this gyroklystron is installed in the "WARLOC" W-band radar developed and tested at NRL (Ngo et al. 2002). A photo of this radar is shown in Fig. 11.5.

11.3 Gyroklystrons for Charged Particle Accelerators

Present linear accelerators are driven by high-power klystrons. For instance, Stanford Linear Collider is driven by 65 MW, S-band (2.856 GHz frequency) linear-beam klystrons. For the Next Linear Collider, which is planned to be built at Stanford, 75 MW linear-beam klystrons operating at four times the present frequency, i.e., at 11.424 GHz, have been developed at the Stanford Linear Accelerator Center (SLAC) (Phillips and Sprehn 1999). P. Wilson (1994) recently discussed the needs of future linear colliders in high-power microwave sources.

For a further increase in the operating frequency two kinds of high-power microwave sources alternative to linear-beam klystrons are currently under development. Those sources are magnicons (see Nezhevenko et al. 1998) and gyroklystrons. Initially, gyroklystrons were designed for operation at the fundamental cyclotron resonance. The operating frequency was about 9.85 GHz, and the operating mode in all cylindrical cavities was the TE_{01}-mode. The tubes were driven by a 425 kV, 200 A electron beam with a pulse duration of about 1.5 microseconds. In experiments with two-cavity gyroklystrons, 27 MW output power with the 36 dB gain and 32% efficiency was demonstrated (Lawson et al. 1991). Then, in three-cavity devices with a tunable penultimate cavity the gain was increased up to 50 dB, while the maximum power was about 27 MW and the maximum efficiency was about 37% (Tantawi et al. 1992).

In order to increase the frequency of outgoing radiation, a two-cavity gyroklystron, with the same input cavity operating at the fundamental cyclotron resonance but the output cavity operating in the frequency-doubling regime, was designed and tested. This output cavity was designed for operation at the

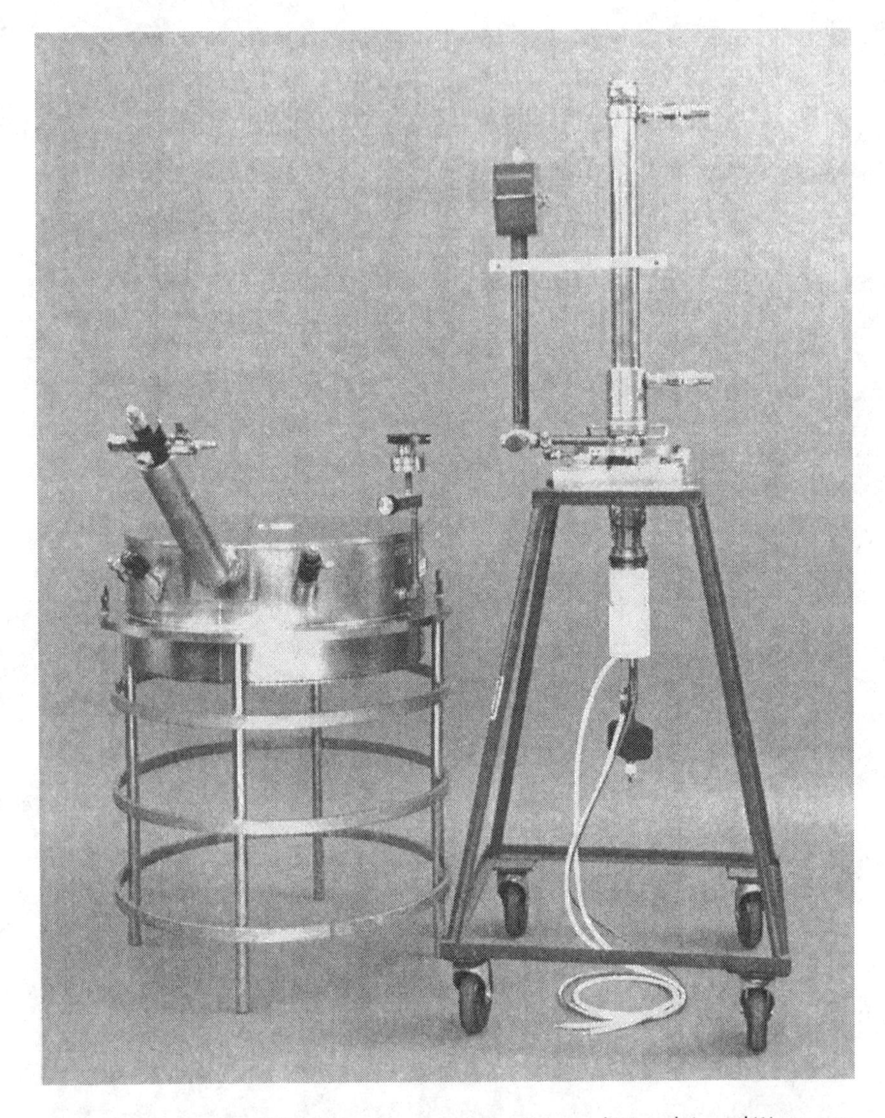

Fig. 11.4. Photo of a W-band gyroklystron delivering 100 kW peak/10.5 kW average power. On the left, a liquid He cooled cryomagnet is shown, which was later replaced by a liquid-He-free cryomagnet. (Courtesy of CPI)

TE_{02}-mode. The drift space of this device was heavily loaded with absorbers in order to eliminate the propagation of the output cavity field back into the input cavity. As a result, peak power exceeding 21 MW with efficiency near 21% and a large-signal gain above 25 dB was produced at a frequency of about 20 GHz (Lawson et al. 1993). Note that the output power in the devices operating in the regime of frequency multiplication scales as $P_{out} \propto P_{in}^{s}$, where s is the

Fig. 11.5. Photo of an NRL WARLOC radar system located at the NRL Chesapeake Beach Detachment site. (Reproduced from Danly et al. 2002)

frequency multiplication factor. Therefore, there is no sense to define the small-signal gain of such devices.

A desire to extend the device performance above the 100 MW power level has led to the concept of gyroklystrons operating in higher-order modes. It was chosen to operate at the same voltage below 500 kV—because this is the operating voltage level of klystrons developed at SLAC—but to increase the beam current. It was found that in order to triple the output power by tripling the beam current, it is necessary to approximately triple the beam radius. Since it is impossible to provide a transport of such a large beam through the cylindrical drift section and simultaneously eliminate propagation of the microwave radiation in it, for cavities' isolation the decision was made to use coaxial configurations instead of cylindrical ones. Thus, the concept of a coaxial gyroklystron was chosen.

The first experiment with the coaxial gyroklystron was carried out at the fundamental cyclotron resonance. A three-cavity device was designed for operation in the TE_{011}-mode in all cavities at the frequency close to 8.6 GHz.

The output power achieved in these experiments was in the 75–85 MW range in pulses longer than one microsecond. The maximum efficiency was about 32% and the gain was near 30 dB (Lawson et al. 1998). Presently, at the University of Maryland a frequency-doubling 70–80 MW, Ku-band (17.12 GHz) gyroklystron is under development (Lawson et al. 2002). Also, Ka-band (Arjona and Lawson 2000) and W-band (Neilson et al. 2002) relativistic gyroklystrons have been designed. Note that in the case of coaxial configuration, it is preferable to support the inner coaxial conductor axially rather than with the use of radial pins described by Lawson et al. (1998). Such a concept of relativistic gyroklystrons is currently under consideration (Read et al. 2002).

In addition to the gyroklystron applications considered above, these devices can also be used for low hybrid plasma heating in controlled fusion reactors. This method of heating requires phase-controlled high-power microwave sources at frequencies of 5–10 GHz. The use of gyroklystrons for this application was discussed by Antakov et al. (1990).

▪ CHAPTER 12 ▪

Gyro-Traveling-Wave Tubes

12.1 Historical Introduction

Pantell (1959) and Gaponov and his group (1959) carried out the first experimental studies of the interaction of gyrating electrons with fast traveling waves. At that time, R. H. Pantell reported the results of his experiments in which the radiation at frequencies between 2.5 and 4.0 GHz was obtained due to the interaction between the cyclotron wave and the backward TE_{11}-wave of a rectangular S-band waveguide. The power achieved was about 0.4 W and the corresponding efficiency was 0.5%. That device operated as an oscillator whose starting current was about 100 mA. So, according to our nomenclature discussed in Chapter 1, it should be considered as a CRM-BWO rather than a CRM-TWT.

A. V. Gaponov (Gaponov et al. 1959c) reported the first results of the studies of a CRM-TWT. In his unpublished experiment, in which a beam of electrons gyrating in a constant magnetic field amplified the TE_{10}-wave of a rectangular waveguide, a so-called electronic gain has been demonstrated. The term "electronic gain" needs some explanations. In that experiment the output power was lower than the input power. However, as was discussed in Chapter 6, due to insertion losses (see (6.34) and (6.35)), only one-third of the input power is coupled to the growing wave at the waveguide entrance. Therefore the fact that the output power exceeded $P_{in}/3$ allowed researchers to conclude that in this experiment an amplification of a forward EM wave by a beam of electrons gyrating in a constant external magnetic field has been observed. This and later experiments were, however, quite discouraging in the sense that the efficiency observed was much lower than that predicted by the theory. It soon became clear that the axial velocity spread inherent in the beams of electrons gyrating in a constant external magnetic field causes significant Doppler spread of resonant frequencies in the case of electron interaction with traveling waves (Flyagin et al. 1977, Andronov et al. 1978). Therefore, in the mid-1960s, much better progress was demonstrated in the experiments with trochoidal electron beams. (This issue was already discussed in Sec. 2.1.)

Later, development of gyrodevices in the USSR was focused on the gyrotron oscillators and gyroklystrons, although theoretical studies of gyro-TWTs have been continued. The development of gyro-traveling-wave tubes (gyro-TWTs) was actively started in the United States in the late 1970s, first, by the researchers from the Naval Research Laboratory (NRL) and a little later by Varian. After detailed theoretical studies, researchers at NRL demonstrated successful experimental operation of a wide-band, Ka-band gyro-TWT (Barnett et al. 1981).

The first Varian gyro-TWT operated at 5 GHz at more than 100 kW peak power level (Ferguson, Valier, and Symons 1981). Symons and Jory (1981) described the first stage of these experiments in detail in a review paper. Also, Hirshfield and Granatstein (1977) described in detail the first experiments on amplification of forward EM waves by gyrating beams of relativistic electrons, which took place in the 1970s. (This issue will be discussed in more detail in the next chapter, in the section on CARMs.) In the early 1980s, a W-band gyro-TWT had been developed at Varian (Eckstein, Latshaw, and Stone 1983). Parameters of this tube can be found in Table 6.2 in Barker and Schamiloglu (2001).

Successful development of gyro-TWTs has been continued in many laboratories for over 20 years. Some results of this development are presented below.

12.2 Large-Bandwidth Gyro-TWTs

It is well known that, although it is always desirable to have all performance characteristics of any device superior, in real life a tendency to improve one performance characteristic causes degradation of another. Therefore, there is always a trade-off between tendencies to improve various performance characteristics, and priorities in improving these characteristics are dictated by a given application of a device. Although at present there are no systems in which gyro-TWTs are used, one can easily envision that the focus of the gyro-TWT development could be made either on achieving the largest possible bandwidth (in the case of their use in communication systems), or on gain enhancement. The latter can be the case of the gyro-TWT used in radar systems where a moderate bandwidth exceeding the bandwidth of gyroklystrons should be accompanied with a high enough gain for realizing the maximum gain-bandwidth product. In this section we will consider large-bandwidth gyro-TWTs.

The largest bandwidth obtained in gyro-TWTs with the uniform interaction waveguide and external magnetic field was reported by Chu et al. (1990). In that Ka-band experiment an orbital-to-axial velocity ratio of electrons was rather low (about 0.8), which resulted in the stabilization of operation and allowed researchers to realize the bandwidth in excess of 10% in the large-signal regime.

The first time an understanding of the fact that simultaneous tapering of the waveguide and external magnetic field may result in significant enlargement of the bandwidth of gyro-TWTs was possibly expressed in a general form by Bratman et al. (1973). Nevertheless, a simple concept of a single-stage, wide-band tapered gyro-TWT was first proposed and analyzed by Y. Y. Lau and K. R. Chu (Lau and Chu 1981). These authors proposed to use an up-tapered waveguide as an interaction circuit and to taper an external magnetic field in such a way that a grazing condition for the waveguide mode and the cyclotron wave, which was discussed in Chapter 6, is locally fulfilled everywhere along the waveguide. A schematic drawing of such a circuit is shown in Fig. 12.1. As is seen in the figure, the authors proposed to introduce a signal in the reverse direction as a backward wave, which should be reflected from the cutoff cross section. Then, such a reflected wave starts the resonance

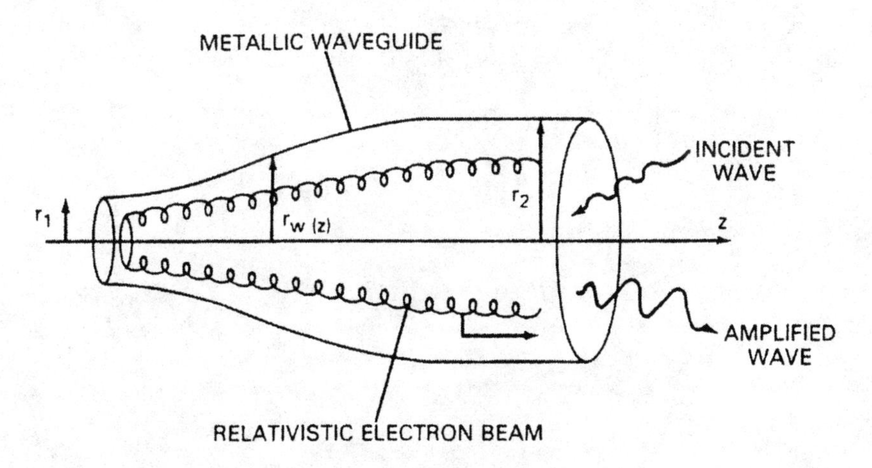

Fig. 12.1. Schematic drawing of a tapered wide-band gyro-TWT. (Reproduced from Lau and Chu 1981)

interaction with electrons as a forward wave. (An alternative side-wall injection scheme, which is more suitable for high gain operation, was discussed by Chu et al. 1981.)

This concept was immediately verified in the proof-of-principle experiment (Barnett et al. 1981), where a 13% bandwidth in the range of frequencies from 32.5 GHz to 38.0 GHz was achieved at the multi-kW peak power level with the gain close to 20 dB and efficiency about 13%. In this experiment the TE_{01} circular electric mode was used. The waveguide was linearly up-tapered from a radius of 4.57 mm to 6.22 mm at a distance of 33.3 cm. In addition to the main superconducting solenoid, two independent trim coils were used to properly taper the profiled magnetic field. It is worthy of mentioning that this tapering of the magnetic field has led to substantial enhancement of the gain, which was about a few dB only in the constant magnetic field. It was also shown that the operation of tapered gyro-TWTs is more stable than that in gyro-TWTs with constant parameters of the circuit because the length of the interaction space is shorter, and therefore the start-oscillation current of the absolute instability is higher. Some improvements in the design allowed researchers from NRL to increase the instantaneous bandwidth of the Ka-band tapered gyro-TWT to 33% with a small-signal gain in excess of 20 dB and efficiency of about 10% (Park et al. 1994). However, window reflections in combination with the use of the signal injection scheme shown in Fig. 12.1 resulted in a round-trip pass for the amplified radiation at any frequency that caused severe limitations on the achievable gain.

The next conceptual step in the development of large-bandwidth gyro-TWTs was made at NRL when the concept of a two-stage tapered gyro-TWT shown schematically in Fig. 12.2 was analyzed theoretically (Ganguly and Ahn

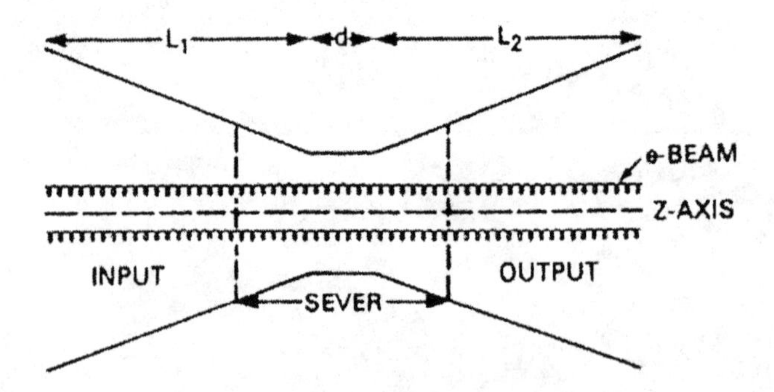

Fig. 12.2. Schematic of a tapered, two-stage gyro-TWT. (Reproduced from Ganguly and Ahn 1982)

1982) and verified experimentally (Park et al. 1995). To some extent, this work was also initiated by the paper (Moiseev 1977) in which the bandwidth of such a device was studied analytically. (Later, Nusinovich and Walter 1999 explained this analytical approach in more detail.)

Such a Ka-band gyro-TWT operated with the small-signal gain of 30 dB and saturated gain of 25 dB over a 20% bandwidth with the efficiency of 16% and output power of about 8 kW. The device was driven by a 1.5 A, 33 kV electron beam with a relatively small orbital-to-axial velocity ratio (about 0.7) and low axial velocity spread (about 4%). Low voltage allowed researchers to operate reasonably close to cutoff, which made the operation relatively insensitive to the velocity spread. A small orbital-to-axial velocity ratio made the device operation more stable with respect to spurious self-excitations.

Recently, a new concept of large-bandwidth gyro-TWTs has been actively studied by a joint group of researchers from the Institute of Applied Physics (Nizhny Novgorod, Russia) and the University of Strathclyde (Glasgow, U.K.) (Denisov et al. 1998, Cooke and Denisov 1998). This concept is based on the use of a cylindrical waveguide having a helically grooved wall. Such a helical corrugation of the inner surface of the waveguide is designed to selectively couple one wave excited at a frequency close to cutoff to another, lower-order wave that propagates at a small Brillouin angle with respect to the waveguide axis. This coupling of two waves due to the periodic waveguide wall perturbations is essentially the same as one discussed in Sec. 10.2: again, the periodic corrugation given by (10.1) couples the waves whose azimuthal indices and axial wave numbers obey Bragg resonance conditions given by (10.2) and (10.3). In a gyro-TWT with such a helically grooved waveguide, electrons can interact with the wave excited near cutoff frequency, which makes the device operation relatively insensitive to the axial velocity spread. Also, a group velocity of a normal wave of such a structure can be more or less constant over a wide frequency band.

Note that initially such a concept of using a helix corrugation for transforming a wave excited near cutoff into a traveling wave was proposed by M. I. Petelin and I. M. Orlova in the late 1960s (Orlova and Petelin 1968) with regard to gyrotron oscillators. (Later, Goldenberg, Nusinovich, and Pavelyev 1980 continued the treatment of this concept.) The driving idea was to directly transform the operating mode into a wave suitable for transmitting with small losses through long waveguides. It was also taken into account that such wave coupling should reduce the diffractive Q-factor of the mode, and hence lower the ohmic losses in the cavity walls, which is important for high-power tubes operating in the CW regime. However, just the fact that the Q-factor of

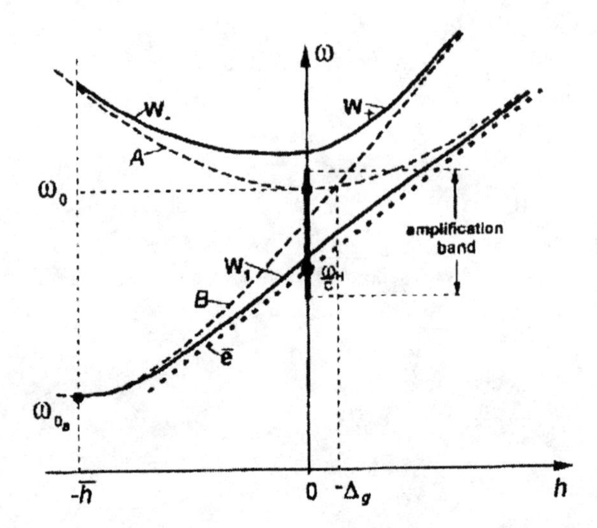

Fig. 12.3. Dispersion diagram for a helically corrugated waveguide. Dashed curves labeled by A and B show dispersion curves for the near-cutoff and traveling waves of a smooth waveguide, respectively. These waves are coupled due to the helical corrugation of the inner surface of a waveguide. Solid lines labeled by W_1 and $W_{+/-}$ show the operating and spurious waves of the corrugated waveguide, respectively. A straight dotted line labeled ē shows the unperturbed electron cyclotron wave, which is located close to the operating wave W_1 over a large frequency range. (Reproduced from Bratman et al. 2000)

the operating mode will be reduced while Q-factors of parasitic modes will remain unchanged made this mode less competitive. Thus, the use of this idea in high-power gyrotrons with a dense spectrum of competing modes looked unrealistic; therefore, this concept had not been studied for a long time until it was revived with regard to gyro-TWTs.

Recently, Bratman et al. (2000) verified the bandwidth capabilities of such a gyro-TWT with a helically corrugated waveguide experimentally. In this experiment, the operating mode was formed by a coupling of the $TE_{2,1}$-wave to the counterrotating $TE_{1,1}$-wave. A corresponding dispersion diagram is shown in Fig. 12.3. The upper boundary of the amplification band shown in this figure is determined by the presence of parasitic waves at higher frequencies. The X-band device was driven by a short-pulse, 110A, 185 keV electron beam with the orbital-to-axial velocity ratio of 1.2. A saturated gain of 37 dB was obtained over a 21% bandwidth centered on 9.4 GHz frequency. The output peak power was 1.1 MW. The efficiency was close to 30% in spite of the fact that the orbital velocity spread was close to 15%. (Note that from publication it is not clear whether the authors mean the RMS spread value.) These results clearly demonstrate the potential of this new concept.

12.3 High-Gain Gyro-TWTs

As we discussed in Chapter 6 (Sec. 6.3), in principle, the gain of the gyro-TWT can be increased proportionally to the interaction length. However, in real devices, the lengthening of the interaction space makes the amplifier operation unstable due to parasitic self-excitation. In Sec. 6.3 we discussed a possibility to avoid this excitation by developing multistage gyro-TWTs in which each stage is short enough for providing stable amplification. We also mentioned there another way to stabilize the gyro-TWT operation, based on the use of lossy materials in an input part of the waveguide. This method, known in linear-beam TWTs, was successfully used in the gyro-TWT experiments carried out at the National Tsing Hua University, Taiwan (Chu et al. 1999). Since in these experiments the highest gain (up to 70 dB) has been demonstrated, we will discuss these experiments below in more detail.

The use of distributed losses as an effective means for stabilizing the gyro-TWT operation with respect to both reflective oscillations caused by the wave reflections from the input/output couplers, and the absolute instability caused by the excitation of backward waves, was proven in previous experiments (Chu et al. 1995). In those experiments a Ka-band gyro-TWT operating at the TE_{11}-wave at the fundamental cyclotron resonance was studied. There, the concept of a severed gyro-TWT was tested as well. It was shown that the wall losses distributed over approximately the first half of the waveguide is a much more efficient means to increase the starting current of the parasitic TE_{21}-wave—which can be excited as a backward wave at the second cyclotron harmonic—than the use of a sever. These distributed losses produced about a 20 dB circuit loss at the band center frequency. Thus, the wave growth in the lossy section was about 13 dB, while the copper output section added about 20 dB to the gain. So, the overall gain was about 33 dB and the efficiency was about 20%.

A careful theoretical analysis of all kinds of instabilities (Chu et al. 1999) allowed researchers to substantially improve the device performance. In a new set of experiments, the length of the graphite-coated lossy section was increased from less than 10 cm to about 20 cm with a much higher cold circuit loss of about 100 dB. Simultaneously, the length of the conducting wall section was shortened from more than 7 cm to 4 cm. Such a Ka-band gyro-TWT driven by a 100 kV, 3.5 A electron beam with the orbital-to-axial velocity ratio in the range from 0.8 to 1.1 demonstrated an excellent performance with the 70 dB saturated gain, about 100 kW output power and 26.5% interaction efficiency. Saturated output power over 80 kW with a gain above 65 dB was observed over a bandwidth of 1 GHz. The full-width half-maximum bandwidth was about 3 GHz, which is approximately 8.6% of the center frequency. Note that such an ultra-high gain permits the use of solid-state sources as drivers for high-power amplifiers.

The success of this experiment sparked an interest in this concept at NRL, where presently a similar Ka-band gyro-TWT is under development (Nguyen et al. 2001, Garven et al. 2002). To accommodate high average power operation, in this tube a diffractive loading scheme instead of lossy ceramic will be employed. Nguyen et al. (2001) showed that the peak power/bandwidth characteristics can be significantly altered by changing the electron orbital-to-axial velocity ratio. In experiments (Garven et al. 2002), the peak power of 137 kW was obtained with the 47.0 dB gain, 17% efficiency and 3.3% bandwidth. This device is intended to serve as the transmitter power amplifier in radar systems used for a variety of applications such as precision tracking and high-resolution imaging.

It is worthwhile to mention another experiment that is of great interest to the developers of gyro-TWTs, although it was not focused on the high-gain operation. In that experiment, described by Wang et al. (1995), the tube operated in the TE_{21}-wave excited at the second cyclotron harmonic. This wave seems to be very convenient for the second harmonic operation because, when the beam interacts with this wave at the grazing condition at the second harmonic, there is no intersection of the fundamental resonance cyclotron mode with any wave of a waveguide. This statement was illustrated by Fig. 6.7, reproduced from Wang et al. (1995). The fact that the beam coupling to the waves in the case of interaction at harmonics is smaller than in the case of the fundamental resonance interaction allowed researchers (Wang et al. 1995) to realize stable amplification at the 200 kW output level with an efficiency close to 13%. The measured saturated gain and bandwidth in this Ku-band (16 GHz) gyro-TWT were 16 dB and 2.1%, respectively.

▪ CHAPTER 13 ▪

Other Types of Gyrodevices

13.1 Gyro-Backward-Wave Oscillator

General Remarks

The gyro-backward-wave oscillator (gyro-BWO) was already mentioned in Chapter 2 as one of the main members of the family of gyrodevices. In accordance with the discussion given in Chapter 6, it is very easy to make a transition from the gyrotron operation (at frequencies near cutoff) to the gyro-BWO operation, which is the case of exciting a wave propagating toward the cathode. To do this, it is enough to increase an external magnetic field. Then, in the dispersion diagram shown in Figs. 6.5 and 6.7 the cyclotron wave, $\omega = s\Omega_0 + k_z v_{z0}$, will intersect the waveguide mode dispersion curve, $\omega = (\omega_{\text{cut}}^2 + c^2 k_z^2)^{1/2}$, at negative k_z's, which means excitation of a wave propagating in the opposite direction. There are certain similarities and differences between gyro-BWOs and linear-beam BWOs. Although briefly discussed above, it is expedient to reiterate. The most important difference between these two kinds of devices is not the obvious fact that in one case a beam of electrons is propagating linearly, while in the other case electrons gyrate in the external magnetic field. Most important is the fact that in linear-beam BWOs the electrons interact with a specific space harmonic of a wave of a slow-wave structure whose phase and group velocities are oriented in opposite directions. Indeed, the phase velocity, $v_{\text{ph}} = \omega/k_z$, of this harmonic is close to the electron velocity, i.e., it is positive, while the group velocity, $v_{\text{gr}} = d\omega/dk_z$, is negative, which means that the electromagnetic energy of the wave propagates in an opposite direction. On the contrary, in gyro-BWOs the gyrating electrons interact with a fast wave whose phase and group velocities are both negative. This is why in the first Russian papers on gyro-BWOs these devices were called gyrotrons with oppositely propagating waves in order to emphasize this difference.

At the same time, these two kinds of devices have very important common features. In both of them, electrons that propagate in the forward direction and a wave propagating backward form a feedback loop. Correspondingly, a characteristic time of the signal propagation via this loop in a waveguide of length L can be determined as $T = L/v_z + L/v_{gr}$. The presence of such a loop is the condition required for the appearance of oscillations in any oscillator when the beam current exceeds a certain threshold. Note that, in addition to this feedback loop that exists even in the devices without any reflections at the ends, in the case of nonzero reflections a part of the wave propagating backward is reflected at the entrance back to the interaction space where it propagates as a forward wave, which, in turn, is partially reflected back from the exit. This reflection leads to the formation of the second feedback loop formed by forward and backward waves. This can also be treated as the formation of a standing-wave pattern of a cavity instead of a traveling wave propagating in a well-matched waveguide.

A similarity between these two kinds of devices is also enhanced by similarity of the dispersion equations describing them. As was shown in Sec. 6.2, under certain conditions the dispersion equation of the gyro-BWO can be reduced to that of the linear-beam BWO in the same fashion as the gyro-TWT dispersion equation can be reduced to that of the TWT. Also, equations describing the large-signal regimes of the gyro-BWO in the low-current limit considered in Sec. 7.3 can be reduced to equations describing the large-signal operation of the BWO. To make a transition from the gyro-TWT equations derived in Sec. 7.3 to the gyro-BWO equations, it is enough to change the sign of the source term in the right-hand side of (7.23). Of course, the boundary conditions for the wave should also be properly modified. While in the reflectionless gyro-TWT the boundary condition for the forward wave at the entrance is the initial value of the wave amplitude, which is determined by the input signal, $F_f(\varsigma_{in}) = F_0$, in the gyro-BWO the situation is more complicated. In the simplest case of the absence of reflections, the wave amplitude is determined not at the entrance, but at the exit from the interaction space. This condition, $F_b(\varsigma_{out}) = 0$, indicates that no backward waves are coming from the output waveguide to the interaction space. This condition is also known in linear-beam BWOs (Johnson 1955). Usually, in order to prevent an electron gun region from microwaves, there is an iris or cutoff narrowing of a waveguide from the cathode side. Therefore, a backward wave excited by the beam, being reflected from the entrance, propagates through the interaction region as a nonsynchronous forward wave without interaction with electrons and then radiates through the exit with no reflections. In the presence of reflections at the exit, of course, the boundary condition for the backward wave should take these reflections into account. This means that instead of the boundary condition given above ($F_b(\varsigma_{out}) = 0$), one should write for the forward wave at the entrance $F_f(\varsigma_{in}) = R_{in} F_b(\varsigma_{in})$ and for the backward wave at the exit $F_b(\varsigma_{out}) = R_{out} F_f(\varsigma_{out})$, where in the absence of interaction of the forward wave with electrons $F_f(\varsigma_{out}) = F_f(\varsigma_{in})$. From these simple relations one can easily derive the self-excitation condition, similar to one formulated earlier in Chapter 6. (See Levush et al. 1992 for more details.)

Efficiency

An axial structure of the RF field in backward-wave oscillators is unfavorable for the efficiency because electrons are modulated near the entrance by a large amplitude field while electron bunches are decelerated near the exit by the field of a small amplitude. This statement is quite obvious for the BWO without reflections where the field has a maximum just at the entrance. This axial structure and the axial structure of the RF field in the gyro-BWO without reflections are shown in Fig. 13.1. In the gyro-BWO, the field maximum is slightly shifted from the entrance toward the center as shown in this figure. This shift can be explained by the presence of additional weak effects associated with the M-type electron bunching (Yulpatov 1965). Because of such

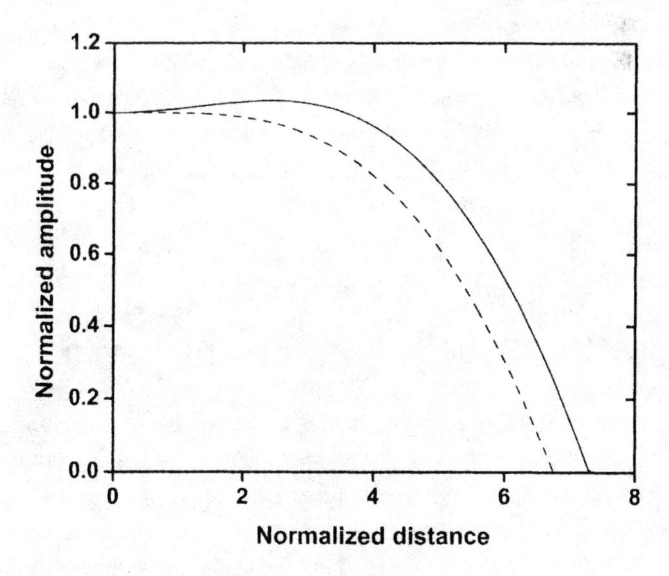

Fig. 13.1. Axial profile of the wave envelope in the linear-beam BWO (dashed line) and gyro-BWO (solid line).

unfavorable structure, the efficiency of such BWOs and gyro-BWOs typically does not exceed 20%. More exactly, for the gyro-BWO driven by a weakly relativistic electron beam it was shown (Yulpatov 1965) that the orbital efficiency does not exceed 20%. Later Ginzburg, Zarnitsyna, and Nusinovich (1979) confirmed that the orbital efficiency of the relativistic gyro-BWO also does not exceed 20%.

These values of efficiency were calculated for devices with constant parameters of the interaction space. The tapering of these parameters, however, can drastically enhance the efficiency. For instance, Ganguly and Ahn (1989) showed for two specific designs of Ka-band gyro-BWOs that the linear tapering of the external magnetic field allows one to increase the electronic efficiency from 10–15% in a uniform magnetic field to 25–30% in a properly up-tapered magnetic field. Later, Nusinovich and Dumbrajs (1996) considered a more general model of the gyro-BWO. They assumed that the efficiency can be maximized due to linear tapering of the cyclotron resonance mismatch, which is proportional to $\omega - k_z v_{z0} - s\Omega_0$ and hence can be varied either due to the tapering of the external magnetic field or due to the tapering of the waveguide radius. They proved in their treatment that such a linear tapering could enhance the orbital efficiency of the gyro-BWO from less than 20% to about 55%, which is consistent with results by Ganguly and Ahn (1989).

Nonstationary Processes

The gyro-BWO (as well as the linear-beam BWO) can serve as an example of an active system with an internally distributed feedback. Such systems can be prone to so-called automodulation instability, which we discussed in Chapter 5 in regard to gyromonotrons. To explain the origin of this instability in simple terms, let us consider a (gyro-) BWO without reflections at the output and assume that the electron velocity is on the order of the group velocity of the wave. This assumption allows us to assume that the FR field amplitude can vary in time during the electron transit time through the interaction region, $T_{tr} = L/v_z$. Therefore, when the beam current greatly exceeds the starting current, the oscillations in a device grow rapidly in the time comparable to T_{tr}. Assume also that the efficiency as the function of the field amplitude, A, has maximum at $A = A_{opt}$ and then, at $A > A_{opt}$, decreases. At high currents, the power extracted by the RF field from the beam can grow faster than the power radiated from the interaction space. Therefore, the field amplitude during this field rise can exceed the optimal value. This will lower the efficiency and RF power that can return a system, after a certain time delay on the order of

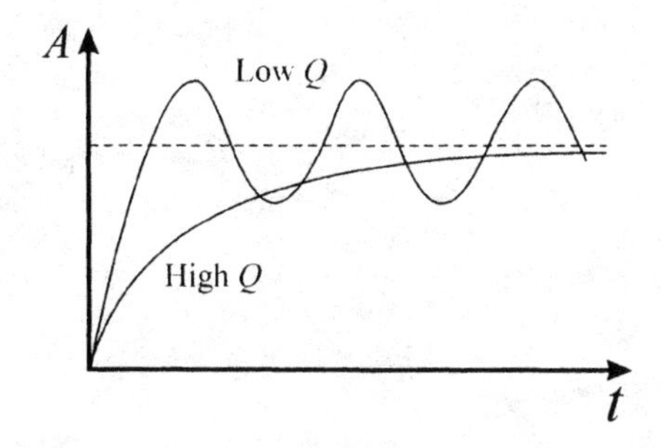

Fig. 13.2. Temporal evolution of the field amplitude in oscillators with high-Q and low-Q resonators.

$T = L/v_z + L/v_{gr}$, to the amplitudes below A_{opt}. So, instead of a slow process of the onset of oscillations with the amplitude close to A_{opt}, which would take place in high-Q systems, in our oscillator with a low inertia the field amplitude can bounce back and forth with the period on the order of T about its optimal value. This statement is qualitatively illustrated by Fig. 13.2. A long time ago, a similar effect was studied in active systems with delayed effects by V. M. Bovsheverov (1936). He showed that with the increase in the beam current (or equivalent parameter of a system under study), the system exhibits a sequence of period-doubling transitions, which results at even higher currents in chaotic oscillations. Much later, Ginzburg, Kuznetsov, and Fedoseeva (1978) developed the nonstationary theory of transients in the BWO. This theory, which was soon confirmed experimentally by Bezruchko, Kuznetsov, and Trubetskov (1979), agrees very well with the above-mentioned studies of other active systems done much earlier. In particular, it was shown that, when the beam current exceeds the optimal value (typically, the optimal current is about 3–4 times higher than the starting current), the stationary oscillations with constant amplitude are replaced by automodulation processes. Later on, Ginzburg, Nusinovich, and Zavolsky (1986) showed that similar transitions also exist in gyro-BWOs and gyrotrons with low-Q resonators.

Recently, nonstationary processes in tapered gyro-BWOs were actively analyzed both theoretically and experimentally. In the experiments by Kou et al. (1993) (see also Chen, Chu, and Chang 2000), it was found that oscillations in the tapered gyro-BWO remain stable even when the beam current exceeds the starting current by more than two orders of magnitude. To explain this interesting result, which, at first glance, contradicts the theory of transients in the BWO developed by Ginzburg, Kuznetsov, and Fedoseeva (1978), numerical

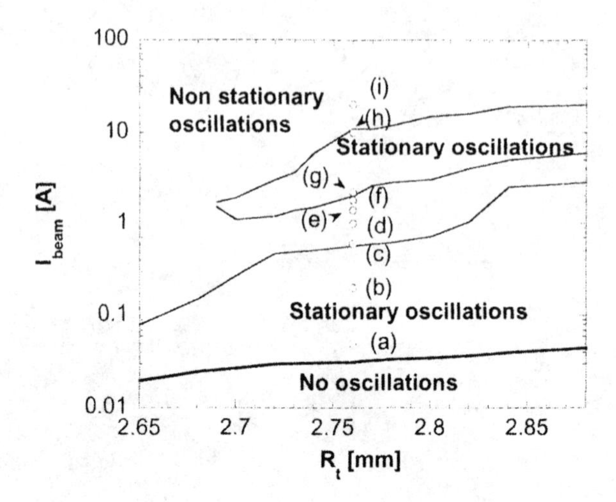

Fig. 13.3. The gyro-BWO map showing the regions of stationary and nonstationary oscillations in the parameters' plane, beam current, I_b, versus uptaper radius, R_t. (Reproduced from Nusinovich, Vlasov, and Antonsen, Jr. 2001)

simulations have been done by Nusinovich, Vlasov, and Antonsen (2001) with the use of the nonstationary self-consistent code MAGY. Some results of these simulations are shown in Figs. 13.3 and 13.4. Here, in Fig. 13.3, the regions of stationary and nonstationary oscillations are shown in the plane of parameters "beam current versus radius of the input up-taper." The value of this radius, equal to 2.65 mm, corresponds to the radius of a uniform waveguide, which follows the taper. So this limiting value designates the case of the uniform gyro-BWO. As one can see in Fig. 13.3, the operation of the nontapered gyro-BWO is essentially the same as that of the linear-beam BWO: as the beam current increases, first the oscillations with constant amplitude appear, and then, at higher currents, these oscillations become unstable. However, when the input part of the gyro-BWO is tapered strongly enough, there is also another zone of stationary oscillations at currents above 1 A. This zone includes the operating point of the experiments by Kou et al. (1993), which was $I_b = 5A$, $R_t = 2.76$ mm. In this zone the RF field mainly occupies the region of the input taper and the beginning of the regular waveguide. This zone lasts up to currents of about 10A while the initial starting current for the first zone, in which the RF field occupies all the waveguide, is less than 0.1A. The sequence of these events in the tapered gyro-BWO is shown in Fig. 13.4, where the snapshots denoted by small letters (a), (b), (c), etc. correspond to the circles shown

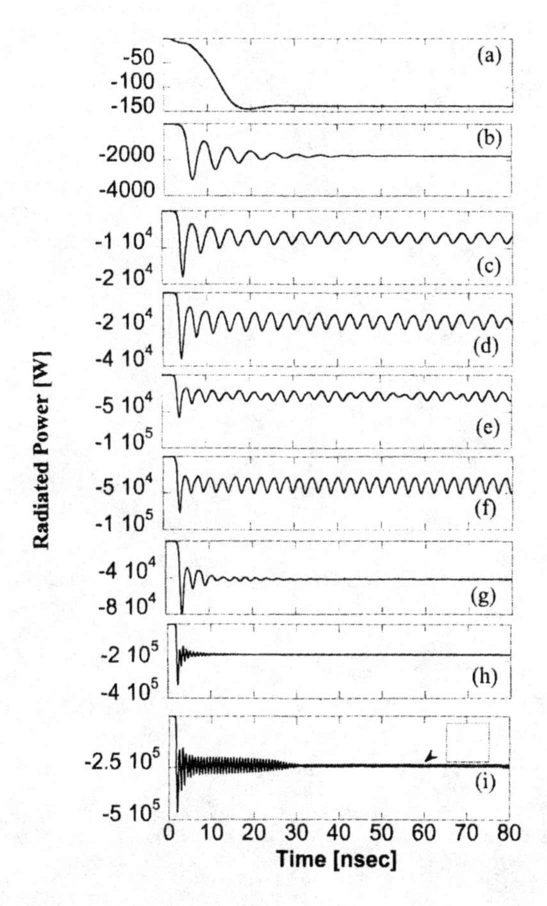

Fig. 13.4. Nonstationary processes in the gyro-BWO with $R_t = 2.76$ mm at different currents. (Reproduced from Nusinovich, Vlasov, and Antonsen, Jr., 2001)

in Fig. 13.3. These results clearly indicate that one can expect various interesting nonstationary and nonlinear phenomena in such active systems with internal feedback and tapered parameters of the interaction region.

Frequency Tuning

Gyro-BWOs are known as devices in which the radiation frequency can easily be tuned by varying either the operating voltage or the external magnetic field. Let us mention in this regard a Ka-band gyro-BWO experiment carried out by S. Y. Park et al. (1990). In this device operating in the fundamental TE_{10}-mode of a rectangular waveguide, it was possible to tune the operating frequency by

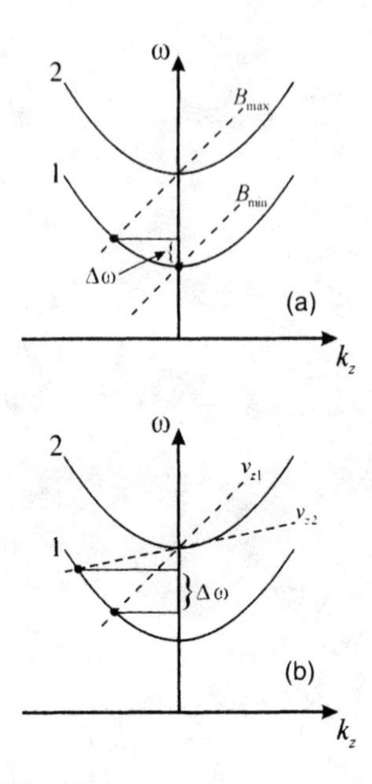

Fig. 13.5. Dispersion diagram illustrating the frequency tunability of the gyro-BWO in the presence of parasitic modes: (a) variable magnetic field; (b) variable axial velocity.

about 13% by varying the external magnetic field by 20%. Also, by changing the beam voltage it was possible to tune the frequency by about 3%.

The general statement about excellent frequency tunability of the gyro-BWO is, however, correct without limitations only for gyro-BWOs operating in low-order modes. In the case of gyrodevices operating in high-order modes, possibilities to tune the frequency of a mode operating in the gyro-BWO regime are restricted by the presence of some modes with higher cutoff frequencies that can be excited near cutoff.

In order to evaluate the frequency tuning of such an operation let us make some simple estimates. Consider a dispersion diagram as shown in Fig. 13.5 where the wave denoted by 1 designates the operating wave while the wave 2 is a higher-frequency parasite that can be excited near cutoff. Let us estimate first the tunability that can be realized by varying the external magnetic field. The frequency tuning of the first wave in the backward-wave operation regime ranges from its cutoff frequency, i.e., $\omega_{1,\min} = \omega_{1,\text{cut}}$, up to the frequency that corresponds to the intersection of this beam with the second wave dispersion curve at the cutoff, i.e., $\omega_{1,\max} + |k_{z,1,\max}|v_{z0} = s\Omega_{0,\max} = \omega_{2,\text{cut}}$. Here $|k_{z,1,\max}| = (\omega_{1,\max}^2 - \omega_{1,\min}^2)^{1/2}/c$, since $\omega_{1,\min} = \omega_{1,\text{cut}}$. Using these simple formulas, one can easily determine the ratio of the frequency tuning,

$\Delta\omega = \omega_{1,\max} - \omega_{1,\min}$ to the separation of cutoff frequencies, $\Delta\omega_{\text{cut}} = \omega_{2,\text{cut}} - \omega_{1,\text{cut}}$, as

$$\delta\omega = \frac{\Delta\omega}{\Delta\omega_{\text{cut}}} = 1 - t\left(\sqrt{1 + \frac{2}{t}} - 1\right). \tag{13.1}$$

Here $t = \beta_{z0}^2/(\Delta\omega_{\text{cut}}/\omega_{1,\text{cut}})$. So, depending on the electron axial velocity and separation of cutoff frequencies, this parameter may be small or large. In the case of operation at very high-order modes, this parameter is large. Then, (13.1) reduces to $\delta\omega \approx 1/2t$, which shows that in this case the range of frequency tuning is much smaller than the separation between cutoff frequencies of the modes.

In a similar fashion we can also consider the effect of the mod-anode and beam voltage variations on the operating frequency of the gyro-BWO. The mod-anode variation affects the orbital-to-axial velocity ratio $\alpha = v_{\perp 0}/v_{z0}$, which determines the electron axial velocity present in the cyclotron resonance condition: $\beta_{z0}^2 = [1/(1+\alpha^2)][(\gamma_0^2 - 1)/\gamma_0^2]$. The variation of the beam voltage changes $\gamma_0 = 1 + eV_b/mc^2$, and hence affects not only the electron axial velocity, but also the electron cyclotron frequency. Typically, the voltage variation allows for smaller frequency tuning than the variation of the external magnetic field. Of course, the parasitic modes that we just discussed are dangerous only when electrons are coupled to these waves strongly enough. These modes can also be used for frequency step-tunable operation.

13.2 Gyrotwystron

The gyrotwystron schematic was already shown in Fig. 2.8 and discussed in Sec. 2.4. This device comprises the bunching stage of the gyroklystron with the output waveguide section in which a traveling wave is excited. So, its output stage is the same as the output stage of the gyro-TWT. Such a combination allows one to combine a high gain inherent in gyroklystrons with a large bandwidth typical for the gyro-TWT. (In gyrotwystrons with multistage prebunching, using the stagger-tuned cavities can enlarge the bandwidth.) Therefore these devices can be of interest for radar applications where a high gain should be accompanied with an appreciable bandwidth as well as for accelerator applications where high-power millimeter-wave sources are needed. In the latter case the use of the output waveguide instead of the cavity mitigates a critical issue of the RF breakdown. (This is why practically in all high-power klystrons used for accelerator applications the traveling-wave circuits are used in the output stage.)

Numerical studies of gyrotwystrons were started in the earlier 1970s. V. A. Zhurakhovsky (1972) reported some results of efficiency optimization of gyrotwystrons with one- and two-stage prebunching sections and constant external magnetic field. It was found that the maximum orbital efficiencies of the devices with one- and two-cavity prebunching sections are equal to 55% and 84%, respectively. Later, Ginzburg et al. (1981) and Nusinovich and Li (1992b) showed that in relativistic gyrotwystrons the efficiency enhancement is possible due to the recoil effect. This recoil effect leads to axial deceleration of electrons radiating EM waves with nonzero axial wavenumbers (see Chapter 1). This effect allows one to extract from electrons not only the energy of their gyration, but also the energy associated with their axial motion. It is interesting to note that the bunching and deceleration processes in gyrotwystrons with large and small bunching parameters are quite different. Examples of the phase bunching and corresponding efficiencies for the cases of small ($q = 0.2$) and large ($q = 1.84$) bunching parameters are shown in Fig. 13.6,

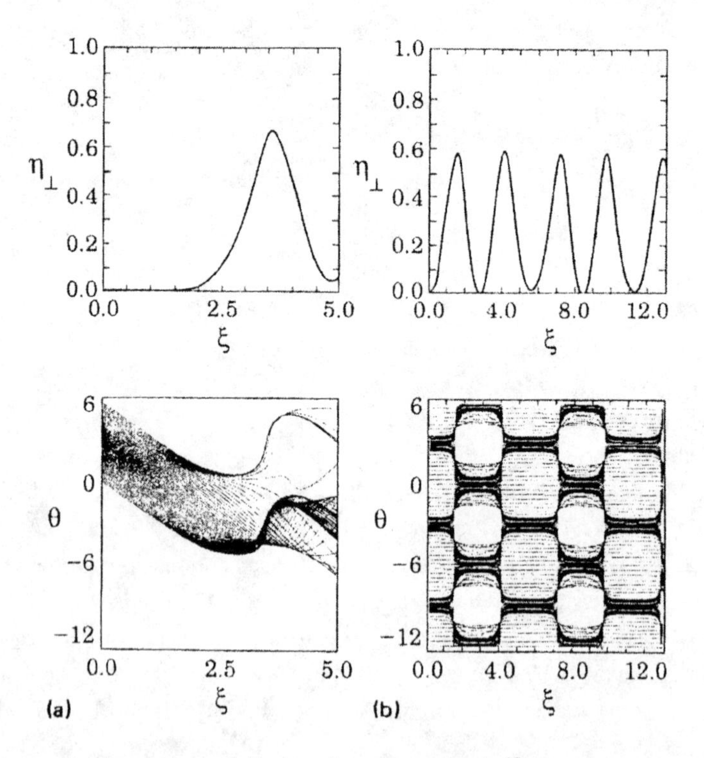

Fig. 13.6. Axial dependence of electron phases and orbital efficiency in gyrotwystrons. Cases (a) and (b) correspond to the small ($q = 0.2$) and large ($q = 1.84$) bunching parameters, respectively. (Reproduced from Nusinovich and Li 1992b)

reproduced from Nusinovich and Li (1992b). As one can see, in the case of a small prebunching the bunching process in a long output waveguide is very much the same as in gyro-TWTs considered in Chapter 7. On the contrary, in the case of a large bunching parameter the electron bunches can rapidly change their phase by a value on the order of π as in short cavities. The empty "eyes" in this case occur due to consideration of an idealized electron beam without a spread in electron velocities.

Attention was also paid to the bandwidth of gyrotwystrons. Moiseev (1977) studied analytically the bandwidth properties of a gyrotwystron consisting of input and output tapered waveguide sections separated by the cutoff drift region. (This device can also be treated as a two-stage gyro-TWT, discussed earlier in Sec. 12.2.) His estimates showed that the electron velocity spread typical for electron beams produced by magnetron-type electron guns limits the gyrotwystron bandwidth by a value of about 7%.

Later, an active development of gyrotwystrons was started at the Naval Research Laboratory (NRL) and at the University of Maryland. At NRL, first, modification of a 4.5 GHz three-cavity gyroklystron into a three-stage gyrotwystron allowed P. Malouff et al. (1995) to enhance the bandwidth from 0.4% to 1.5%. This device operated with a 42 dB small-signal gain (37 dB saturated gain), 80 kW output power, and 23% efficiency. Practically the same bandwidth (bandwidth of 1.6%, efficiency 27.5%) was also demonstrated in X-band gyrotwystrons (Blank, Zasypkin, and Levush 1998). Then, a one-percent bandwidth was demonstrated in a four-stage W-band gyrotwystron developed by M. Blank, B. G. Danly, and B. Levush (1999a). At the University of Maryland, relativistic gyrotwystrons have been studied both theoretically and experimentally for accelerator applications. In these devices the peak power reached 21 MW with 22% efficiency in the X-band experiments at the fundamental cyclotron resonance. Also, in the frequency-doubling regime with the output stage operating at the second cyclotron harmonic, 12 MW with 11% efficiency have been demonstrated; see P. Latham et al. (1994) and W. Lawson et al. (1995).

Recently, H. Guo proposed a concept of the inverted gyrotwystron (Guo et al. 1996). In this configuration, the input section is a waveguide followed by the drift region and the output section is a cavity. Theoretical study (Nusinovich and Walter 1997) revealed that electron modulation in a waveguide input section provides a better harmonic content in the electron current density than a standard ballistic bunching caused by a single-cavity modulation. Thus, it seems reasonable to use such a device as a frequency multiplier. Experiments with a frequency-doubling inverted gyrotwystron showed that such a device operating in rather high-order modes can demonstrate a 33 dB gain in about 1.3% bandwidth (Guo et al. 1997). This bandwidth, which significantly exceeds the width of a resonance curve of one output cavity mode, was later explained (Zhao et al. 2000) as a superposition of resonance curves of modes with different axial structures, as was discussed in Sec. 5.2.

13.3 Quasi-Optical Gyrotron

The quasi-optical gyrotron is a gyrotron in which a pair of mirrors forms an open resonator. Such a resonator is quite similar to highly selective open resonators used in lasers and masers. (We briefly discussed this selectivity in Sec. 2.3.) Since lasers and masers were invented in the late 1950s, it is not surprising that the concept of quasi-optical gyrotrons was considered right after the invention of the gyrotron (see, e.g., the schematic of a quasioptical gyrotron shown in Fig. 13.7). Since the transverse structure of the RF field in these resonators does not correspond to the axial symmetry of an annular electron beam formed by magnetron-type injection guns, the efficiency of quasi-optical gyrotrons is smaller than the efficiency of gyrotrons with axially symmetric resonators (Luchinin and Nusinovich 1984). This is why quasi-optical gyrotrons were not actively studied for a long time.

In addition to this disadvantage, the quasi-optical gyrotron has, however, an advantage that can be important for some applications. This advantage is the possibility of mechanically tuning the operating frequency by changing the distance between mirrors. Such a mechanical tuning was first realized by Antakov et al. (1975), who used such a gyrotron for high-resolution microwave

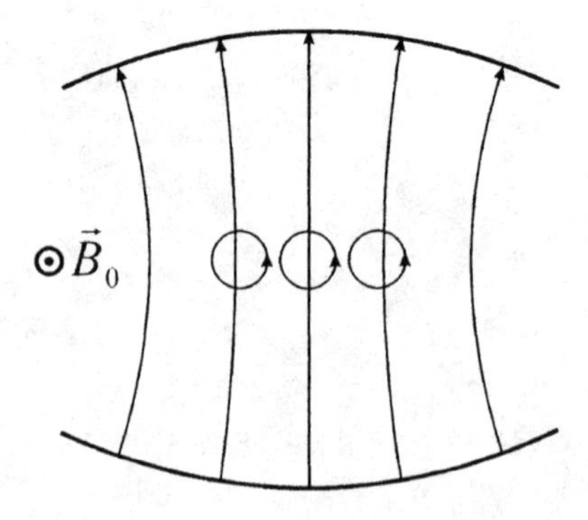

Fig. 13.7. Schematic of a quasioptical gyrotron. (After Rapoport, Nemak, and Zhurakhovsky 1967)

spectroscopy (Antakov et al. 1974). Later, similar programs based on the use of mechanically tunable gyrotrons for millimeter-wave spectroscopy and plasma scattering experiments were started first in Australia (G. F. Brand et al. 1982) and then in Japan (T. Idehara et al. 1991).

During the 1980s, an active development of quasi-optical gyrotrons took place at NRL (A. Fliflet et al. 1990). One of the motivations for this program was the ability of quasi-optical gyrotrons to operate at high power levels with a moderate density of ohmic losses in mirrors. Indeed, as in lasers, the ratio of the distance between mirrors to the skin depth can be very large, and hence, the ohmic Q-factor (see Sec. 3.3) can be much larger than the diffractive Q-factor, which is determined by the mirror width. Therefore, the power of ohmic losses, which relates to the output power as $P_{ohm} = (Q_{dif}/Q_{ohm})P_{out}$, can be small enough. However, later the progress in the mode selective operation in conventional high-power gyrotrons as well as concerns about low efficiency of quasi-optical gyrotrons led to the lack of interest in the development of quasi-optical gyrotrons for the electron-cyclotron plasma heating and current drive.

At the same time, recently, selective properties of such and similar microwave structures renewed an interest in using them in gyroamplifiers. A 140 GHz gyro-TWT with a confocal waveguide operating in a TE_{03}-like mode has been studied at MIT, where it was experimentally shown that the operation in, at least, 1% bandwidth (1.4 GHz) in gyro-TWTs with such microwave structures is possible (Sirigiri, Shapiro, and Temkin 2002).

Another novel microwave structure, whose use in gyrodevices was recently also studied at MIT, is a photonic-band-gap (PBG) structure (Sirigiri et al. 2001). In general, such a structure is a periodic array of varying dielectric or metallic elements. The structure that was used at MIT is shown in Fig. 13.8a. This structure consists of a set of metal rods placed parallel to one another and parallel to the gyrotron axis. A number of rods are omitted from the center just forming an area that is occupied by the field of an operating mode. A TE-like mode can exist in this central part of the structure if its resonant frequency lies in the band gap or stop band of the PBG structure. This band gap can be adjusted in such a way that resonant frequencies of all neighboring modes lie in the passband of the lattice and hence leak through the rods. Fig. 13.8b shows transverse distribution of the field of a TE_{041}-like high-Q mode in such a PBG structure. This structure, indeed, is quite similar to the structure of a TE_{04}-mode in a conventional cylindrical cavity. MIT experiments have demonstrated that in such a gyrotron it is possible to selectively excite only one mode in at least 30% frequency range.

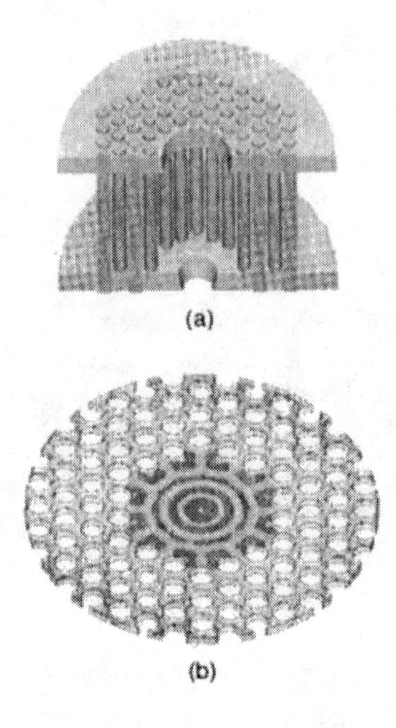

Fig. 13.8. (a) A section of the computer aided design of the PBG resonator used in the gyrotron experiment; (b) transverse distribution of the TE_{04}-like eigenmode of the PBG cavity. (Reproduced from Sirigiri et al. 2001)

13.4 Cyclotron Auto-Resonance Maser (CARM)

Development of high-current accelerators in the late 1960s provoked a strong interest in utilizing the intense relativistic electron beams generated by these accelerators for producing very high-power, coherent electromagnetic radiation. Various mechanisms of producing this radiation, including cyclotron radiation, were tested. The first attempts to experimentally demonstrate the generation of coherent cyclotron radiation from intense relativistic electron beams were very impressive; radiated microwave power reached a GW level. However, the efficiency demonstrated in these experiments was not very high (see, for example, Granatstein et al. 1974, 1975). Later on, Bratman, Ginzburg, and Nusinovich (1977) showed that the gyrotron theory could easily be generalized for the case of relativistic electron beams. Soon after that, first quite efficient experiments with relativistic gyrotrons were conducted at the Lebedev Institute in Moscow (Ginzburg et al. 1978). In parallel with this gyrotron activity, the study of cyclotron-auto-resonance masers began in the middle 1970s.

In 1974, M. I. Petelin proposed a concept of cyclotron auto-resonance masers (Petelin 1974). Already by that time two facts were clearly understood. (Both were discussed in Chapter 1.) First, in the case of operation in the gyrotron regime, i.e., at waves with small axial wavenumbers, the energy exchange between relativistic electrons and the waves disturbs the cyclotron resonance condition even when the electron kinetic energy changes slightly [see (1.8) in Chapter 1]. Second, it was known that in the case of wave propagation along the external magnetic field two bunching mechanisms that are associated with the electron orbital bunching and with the axial bunching exactly compensate each other (Gaponov 1960). So, in this limiting case the bunching responsible for the coherent radiation of electromagnetic waves disappears.

Petelin proposed to vary an angle of the wave propagation accordingly to the electron energy and orbital-to-axial velocity ratio. This means that this angle should provide such compensation of one bunching mechanism by another, which would allow one to maintain the cyclotron resonance during many electron orbits and, at the same time, will make it possible to shift an electron bunch into a decelerating phase with respect to the wave. Since in the ultra-relativistic limit of electron energies the optimal angle of the wave propagation is very small, which means that the device tends to the autoresonance, the device based on this concept of partial compensation of one bunching mechanism by another was called the cyclotron autoresonance maser (CARM).

With regard to the partial compensation of two bunching mechanisms, it is expedient to remind readers about the general dispersion equation (6.3). The last term in the LHS of this equation describes the O-type bunching effects that are the dominant effects in the interaction mechanisms we discussed so far. In the case of operation close to autoresonance, this term diminishes due to the partial compensation of two bunching mechanisms. Therefore, it may happen that in some cases of CARM operation the M-type bunching described by the preceding term in this equation can be important. (Typically, however, CARMs operate far enough from such a situation.)

Two comments should be made regarding the nature of CARMs and their relation to other sources of cyclotron radiation. First, there is no distinctive border that separates the CARM from the gyro-TWT. In the gyro-TWT, electrons also interact with the waves having nonzero axial wavenumbers. So, the distinction can be tentatively made in the following way: the gyro-TWTs that operate at waves having axial wavenumbers close to ω/c should be called CARMs, while others, in which the difference between k_z and ω/c is on the order of ω/c, belong to the main category of gyro-TWTs. In other words, the CARM can be considered as a limiting case of the gyro-TWT with a small angle of wave propagation.

The second comment is essentially a continuation of the first one. The cyclotron resonance condition (1.11) can be rewritten as

$$\omega \approx \frac{s\Omega_0}{1 - n\beta_{z0}}. \tag{13.2}$$

Thus, devices that utilize electrons with large axial velocities and waves propagating at small angles to the axis operate with a large Doppler frequency up-shift. Such devices are sometimes called *Dopplertrons*. It is quite obvious that CARMs driven by relativistic electron beams as well as FELs can be called Dopplertrons. It is known that the Doppler frequency up-shift in Dopplertrons increases with the electron energy. In FELs this frequency up-shift increases proportionally to γ_0^2 (see, e.g., Freund and Antonsen 1996). In CARMs, however, this up-shift of the operating frequency with respect to the nonrelativistic cyclotron frequency increases proportionally to γ_0 because the relativistic electron cyclotron frequency in (13.2) decreases with the electron energy. Comparison of frequency up-shifts that are possible in both classes of these devices has been done elsewhere (see, e.g., Nusinovich et al. 1994.)

The first stage of theoretical studies of CARMs was summarized in 1981 (Bratman et al, 1981). It was shown that CARMs operating with high-quality electron beams, in which the spread of electron velocities is negligibly small, can be very efficient. However, in a long series of experiments with CARMs it was found that electron beams produced by available electron guns and electron-optical systems have a spread that is too large for efficient interaction with waves of large axial wavenumbers, k_z. Thus, the Doppler broadening of the cyclotron resonance band, which is associated with the spread in k_z's, hindered realization of high efficiencies predicted by simple theories: typical efficiencies realized in these experiments did not exceed the 10% level (Bratman et al. 1983, Pendergast et al. 1992).

At the same time, microwave structures capable of selective operation in modes with large axial wavenumbers were successfully developed (Kovalev, Petelin, and Reznikov 1978; see also Bratman et al. 1983 and Chong et al. 1992). These structures were based on the concept of Bragg reflectors. In these resonators, a helical corrugation of an inner surface of a waveguide, which we already discussed in Sec. 10.2, couples two counterpropagating fast waves,

$$A_1 \propto \exp\{i(\omega t - m_1\varphi - k_{z,1}z)\}, \qquad A_2 \propto \exp\{i(\omega t - m_2\varphi + k_{z,2}z)\},$$

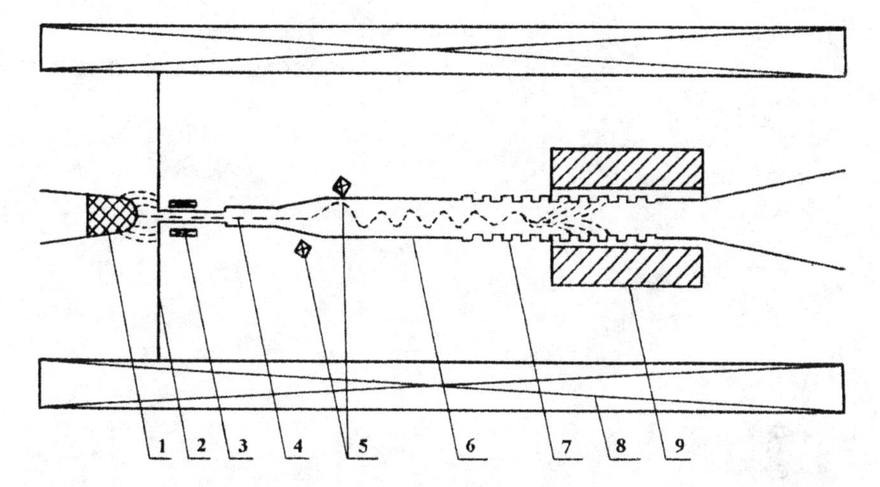

Fig. 13.9. Schematic of the CARM-experiment: 1, cathode; 2, anode; 3, correcting coil; 4, electron beam; 5, kicker; 6, operating section of cavity; 7, Bragg reflector; 8, solenoid; 9, magnetic screen. (Reproduced from Bratman et al. 1995)

for which Bragg's resonance conditions given by (10.2) and (10.3) are fulfilled. So, just these two waves form a mode with a high diffractive Q-factor, while all other waves have large diffractive losses. In addition to Bragg resonators, other possible configurations of cavities providing selective excitation of modes with large axial wavenumbers have been considered by Bratman et al. (1984).

It took a rather long time to develop an electron-optical system that formed electron beams with an acceptable spread. This work, done by the group led by Bratman, culminated in 1995 in the experiment with the CARM-oscillator, in which 26% efficiency was demonstrated (Bratman et al. 1995). A schematic of this experiment is shown in Fig. 13.9, reproduced from this paper.

13.5 Cyclotron Maser Based on the Anomalous Doppler Effect

As we discussed in Sec. 1.4, gyrating electrons can interact not only with fast but also with slow waves. In the latter case, when the phase velocity of a wave is smaller than the axial velocity of electrons, the anomalous Doppler effect takes place. A remarkable feature of the cyclotron radiation under the anomalous Doppler effect condition is the fact that a radiating electron in the process of radiating a wave loses its axial momentum but gains the orbital momentum. From this fact it follows that such radiation is possible even when an electron beam does not have initial orbital velocity at all. J. Pierce (1950) was, perhaps, the first to discover this possibility of radiation, which he treated as instability caused by interaction of an initially linear electron beam with purely transverse electromagnetic waves.

We already discussed some key features in the cyclotron radiation under the anomalous Doppler effect in Sec. 1.4. M. I. Petelin (1974) briefly considered such a CRM based on the anomalous Doppler effect and showed that its maximum efficiency approaches a 100% limit. This can be explained by the fact that, once an initially linear electron beam is injected into the interaction region with a transverse electromagnetic wave, all electrons have the same initial phase with respect to the wave. Therefore, there is no need for electron bunching, and all electrons entering the interaction region in a given instant of time interact with the wave in the same way. Later, N. Ginzburg (1979) carried out a much more detailed analysis of CRM-oscillators and amplifiers operating under the anomalous Doppler effect. In particular, he pointed out that in such devices a competition of the desired operating mode with conventional Cherenkov interaction can be a severe problem. To solve this problem, it is necessary to utilize those slow-wave structures in which the waves with large axial electric fields near the axis are eliminated by some means. Another interesting conclusion from his and Petelin's studies was the result that the high efficiency can be realized at small external magnetic fields and strong EM fields in the interaction region. Correspondingly, the interaction time should be rather short, i.e., electrons should make only about one orbit in the interaction space.

Since it is rather difficult to fulfill these requirements in the experiments, Nusinovich, Korol, and Jerby (1999) later studied the anomalous Doppler CRM with tapered parameters. It was shown that the efficiency of practical configurations of such CRM-amplifiers with tapered waveguide parameters can be about 30%. This efficiency was calculated for a device driven by a weakly relativistic electron beam. In the past, there were also several attempts to experimentally study anomalous-Doppler-CRMs driven by relativistic electron beams (see, e.g., Galuzo et al. 1982 and Didenko et al. 1983).

13.6 Large-Orbit Gyrotron

So far, we have considered the gyrotron utilizing a thin annular electron beam in which electrons gyrate around local guiding centers and assumed that the radius of these centers is much larger than the Larmor radius ($R_0 \gg a = v_\perp/\Omega$). Clearly, the cyclotron maser instability does not critically depend on the beam geometry and, hence, can also be present when all electrons gyrate

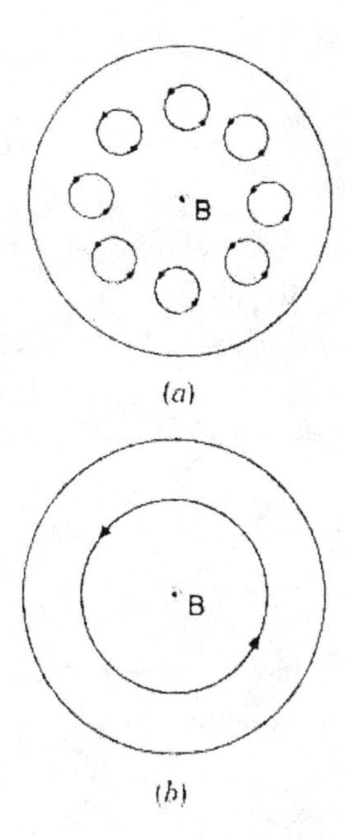

Fig. 13.10. Cross sections of the small-orbit (a) and large-orbit (b) gyrotrons.

about the common axis, which may coincide with the axis of a microwave circuit. Cross sections of both configurations are shown in Fig. 13.10. From their comparison it is obvious why, in contrast to the conventional gyrotron, which can be called a "small-orbit" gyrotron, the new device can be called a "large-orbit" gyrotron, or LOG. Often, it is also called a gyrotron with an axis-encircling electron beam.

There are several ways to generate such electron beams. One of them is to generate, first, a hollow nonrotating beam and let it pass through a narrow, nonadiabatic magnetic cusp where the axial magnetic field changes its sign. In accordance with the conservation law for the canonical angular momentum, when the absolute value of the field at both sides of the cusp is the same, electrons execute after the cusp the helical orbits with the same radius as the initial beam radius. Also, their guiding center coincides with the axis of the original beam. Another method is based on the initial injection of a thin, linear, low-energy electron beam into an accelerating cavity, in which a circularly polarized TE-mode is excited. The radial component of this rotating wave deflects electrons from the axis and accelerates them. This leads to formation of a high-energy spiral electron beam. The first method originated from the studies of microwave radiation from rotating electron layers in electron ring accelerators (Destler et al. 1988); the second method was proposed and realized by H. Jory 1968 (see also McDermott et al. 1983).

Since both the small- and large-orbit gyrotrons employ the same radiating mechanism based on the cyclotron maser instability, the theory described above can be readily applied to LOGs, as was pointed out by Nusinovich (1992b). From the formalism developed for small-orbit gyrotrons with cylindrical geometry of the interaction space, it immediately follows that electrons with the zero guiding center radius have a nonzero coupling impedance (see (3.59)) only to the modes whose azimuthal index is equal to the cyclotron harmonic number ($m = s$). This fact makes LOGs much more mode-selective devices than conventional gyrotrons. Although the LOG looks advantageous in regard to the mode selectivity, the problem is the transverse dimensions of its interaction space. Indeed, in conventional gyrotrons the beam radius is on the order of the wall radius of a microwave circuit and, as we just wrote above, $R_0 \gg a = v_\perp/\Omega$. In the LOG $R_0 = 0$, therefore, to avoid miniaturization of the interaction space while using the electrons with the same orbital velocity, one should operate at much lower magnetic fields. This means that in order to generate radiation at a given wavelength in the interaction space of the same volume, the LOG should operate at high cyclotron harmonics, $s \gg 1$.

The intensity of cyclotron radiation at harmonics, however, decreases as s increases. To compensate for this effect, a number of authors proposed to use slotted or vane, magnetron-type structures (Lau and Barnett 1982, Chu and Dialetis 1984). Indeed, the fields near the walls of such structures are strongly nonuniform. Therefore, in line with our general discussion in Sec. 1.1, the multipole components responsible for interaction at cyclotron harmonics are here larger than in a smooth-wall structure. Note that in such periodic structures the fields also contain slow space harmonics that make it possible to use M-type electron bunching for producing coherent radiation. This mechanism of radiation, which has no relation to the cyclotron maser instability, will be discussed in the next section. Also note that Gaponov and Yulpatov developed the linear theory of a device in which an electron layer rotates around a slotted microwave structure (Gaponov and Yulpatov 1962). They considered the two cases shown in Fig. 13.11: an electron ring rotates either inside (a) or outside (b) a vane structure. It was shown that in the first case the instability might occur even when only slow space harmonics are present in the wave. It was also shown that in the second case the system could be unstable even in the $\beta_\perp \to 0$ limit. This indicates that this instability is similar in nature to the cyclotron radiation under the anomalous Doppler effect, which was discussed in Sec. 13.5.

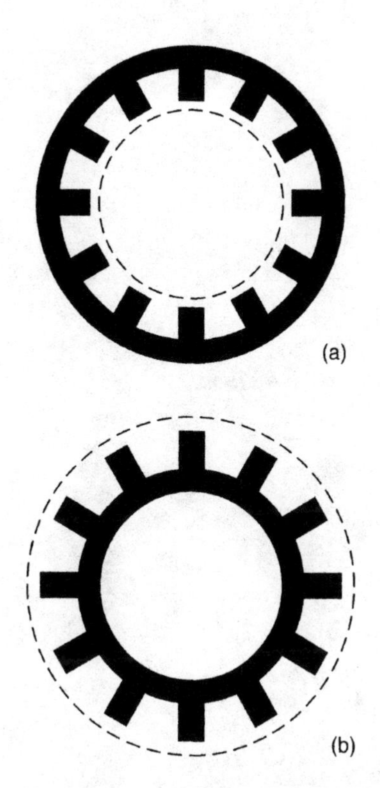

Fig. 13.11. Possible interaction circuits for axis-encircling electron beams: (a) electrons rotate inside a slow-wave vane structure; (b) electrons rotate outside a slow-wave vane structure (the latter structure can, for instance, be used in inverted magnetrons).

13.7 Peniotron, Gyropeniotron, and Autoresonant Peniotron

To explain the principles of operation of the peniotron and gyropeniotron, let us first come back to Eqs. (6.21), which describe perturbations in the motion of electrons gyrating in the external magnetic field under the action of an electromagnetic wave. Consider the perturbations in the radial and azimuthal electron motion given by the first two equations. Denominators of terms in the right-hand sides of these equations contain not only the cyclotron resonance mismatch $\Delta\omega_n$ for the cyclotron harmonic under study, but also resonance mismatches at neighboring harmonics, $\Delta\omega_{n-1}$ and $\Delta\omega_{n+1}$. In the dispersion equation (6.22) these terms result in what was called the M-type bunching. As follows from the analysis of electron motion, which is carried out in Appendix 1, these M-type effects caused by resonances at neighboring cyclotron harmonics can be associated with the transverse drift of electron guiding centers. In gyrotrons, as was shown in Sec. 3.2, this effect is negligibly small. However, when electrons are positioned in such a place where the gyrotron-type interaction is small while the transverse nonuniformity of the RF field, which is responsible for the transverse drift of guiding centers, is strong, the situation can be quite different.

In order to show when such a situation can happen, let us consider the right-hand side of (6.22) in more detail. For the sake of simplicity, assume that the device operates close to cutoff, since effects associated with electron axial bunching are typically not important for peniotron operation. The first term in figure brackets there, as was discussed in Sec. 6.2, describes the O-type electron bunching. As follows from (6.28), its numerator can be rewritten as

$$O_s = -\beta_{\perp 0}^2 [J_s'(\xi)]^2 \, |L_s|^2 \, . \tag{13.3}$$

Also, taking into account the definitions of the radial component of the resonance harmonic of the Lorentz force given in Appendix 1 by (A1.11) and (A1.13), the numerator of the second term in figure brackets of (6.22) determined by (6.23) can be rewritten (cf. Petelin and Yulpatov 1975) as

$$\begin{aligned} M_s = \frac{1}{2} \Big\{ & \left(1 - \beta_{\perp 0}^2\right) \left[J_{s-1}^2(\xi) - J_{s+1}^2(\xi)\right] |L_s|^2 \\ & + \beta_{\perp 0}^2 [J_s'(\xi)]^2 \left(\left|L_{s-1}\right|^2 - \left|L_{s+1}\right|^2\right) \Big\} . \end{aligned} \tag{13.4}$$

When the Larmor radius is much smaller than the transverse wavelength (it is enough to satisfy the condition $\beta_{\perp 0}^2 \ll 1$), Eqs. (13.3) and (13.4) reduce, respectively, to

$$O_s = -\left[\frac{s^{s-1}}{2^s (s-1)!}\right]^2 \beta_{\perp 0}^{2s} \, |L_s|^2 \tag{13.5}$$

and

$$M_s = \frac{1}{2}\left[\frac{s^{s-1}}{2^s (s-1)!}\right]^2 \beta_{\perp 0}^{2(s-1)} \left\{4\left(1-\beta_{\perp 0}^2\right) |L_s|^2 + \beta_{\perp 0}^2 \left(\left|L_{s-1}\right|^2 - \left|L_{s+1}\right|^2\right)\right\}. \tag{13.6}$$

By using these equations one can easily find when the M-type electron bunching can dominate the O-type bunching. Of course, to realize this situation it is not enough to have $|M_s| \gg |O_s|$, but the condition

$$|(\Delta\omega_s) M_s| \gg |2\Omega_0 O_s| \tag{13.7}$$

should be fulfilled. In a general case, all coupling impedances are of the same order ($|L_s|^2 \sim \left|L_{s\pm 1}\right|^2$), and hence the coefficients O_s and M_s, as follows from (13.5) and (13.6), are related as $O_s \sim (\beta_{\perp 0}^2/2) M_s$. Assume that, in accordance with the general principles formulated in Chapter 1, the cyclotron resonance detuning, $\Delta\omega_s$, is inversely proportional to the electron transit time, i.e., $\Delta\omega_s \sim \pi v_z/L$. Then, the condition given by (13.7) becomes equivalent to a small value of the parameter μ ($\mu \ll 1$). This μ is the same parameter that, as we discussed in Sec. 3.1 and later, is responsible for the electron phase bunching caused by the relativistic dependence of the electron cyclotron frequency on electron energy. So, we again came back to the conclusion made many times before (see Secs. 1.2 and 3.1) that the relationship between the M-type and O-type electron bunching mechanisms is determined by the parameter μ. In devices with large values of μ, the M-type bunching can dominate only in some special cases when the O-type effects are very small.

Let us illustrate this statement by considering a device with a cylindrical waveguide. In such a case the coupling impedance responsible for the O-type interaction is equal to $|L_s|^2 = J^2_{m\pm s}(k_\perp R_0)$, while the terms in the equations defining the coefficient M_s also contain $\left|L_{s-1}\right|^2 = J^2_{m\pm(s-1)}(k_\perp R_0)$ and $\left|L_{s+1}\right|^2 = J^2_{m\pm(s+1)}(k_\perp R_0)$. Assume now that we consider an interaction of a symmetric TE_{0p}-wave with electrons gyrating about the waveguide axis (i.e., as in large-orbit gyrotrons, the guiding center radius of electrons equals zero). Clearly, in such a case $|L_s|^2$ equals zero for interaction at any harmonic. However, for $s = 1$ the term $\left|L_{s-1}\right|^2$ equals one, thus M-type electron bunching will exist here.

Let us now evaluate for this example the restrictions on the finite radial thickness of an electron beam positioned on the waveguide axis. Assume that the beam (Larmor radii are not included into this consideration) can be characterized by the spread in the guiding center radii, ΔR_0. Thus, the coupling impedance responsible for the O-type interaction at the fundamental cyclotron resonance is now on the order of $|L_1|^2 = [k_\perp(\Delta R_0)/2]^2$. Substituting these values into (13.7) and using (13.6), one can readily find that the condition given by (13.7) can be fulfilled only when the spread in the electron guiding centers obeys the following restriction:

$$(k_\perp \Delta R_b)^2 \ll \left|\frac{\Delta\omega_s}{\Omega_0}\right|. \tag{13.8}$$

Here, again, we can assume $\Delta\omega_s \sim \pi v_z/L$, which reduces the RHS of (13.8) to $|\Delta\omega_s/\Omega_0| \sim 1/N$, where N is the number of electron orbits in the interaction space.

After these general remarks, we can start describing the peniotron and gyropeniotron operation in terms adopted in the literature devoted to this specific topic. The principles of the peniotron operation are well described by

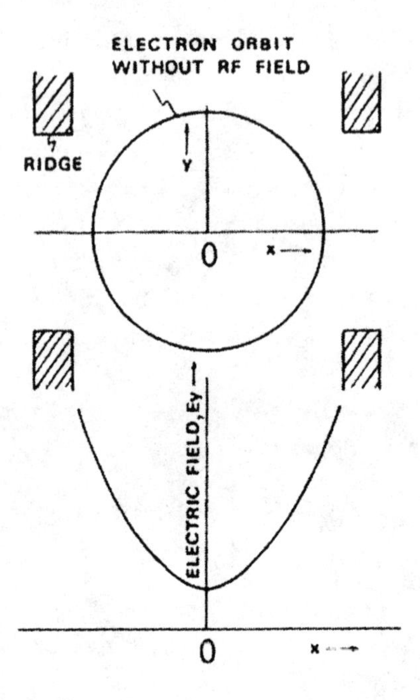

Fig. 13.12. Cross section of the interaction space in a traveling-wave peniotron and the transverse distribution of the RF electric field. (Reproduced from Ono et al. 1984)

S. Ono et al. (1984). Therefore, it makes sense to reproduce this description here. Consider a ring of electrons gyrating in a constant external magnetic field and interacting with a TE-wave of a waveguide containing four ridges as shown in Fig.13.12, reproduced from Ono et al. (1984). These ridges are positioned in such a manner that the nearer the electrons move to the ridges, the stronger they interact with the wave electric field. (The effect of the RF magnetic field on electrons, for simplicity, can be ignored.)

Assume now that the Doppler-shifted wave frequency is in resonance with the second cyclotron harmonic of electron gyration, and consider the motion of electrons with different entrance phases. This motion is shown in Fig.13.13. Here figures (a) and (b) show, respectively, the orbits of an electron that is initially accelerated by the RF electric field in the right-side space and another electron that entered the same point when the wave was shifted in phase by π, and hence the second electron is initially decelerated. The former electron increases its speed at first, and hence its orbit becomes larger. When an electron has moved half of an orbit, the electric field has made a complete cycle. Therefore, in the left side the first electron will be decelerated. However, since its orbit is now closer to the ridges than before, the electron deceleration will

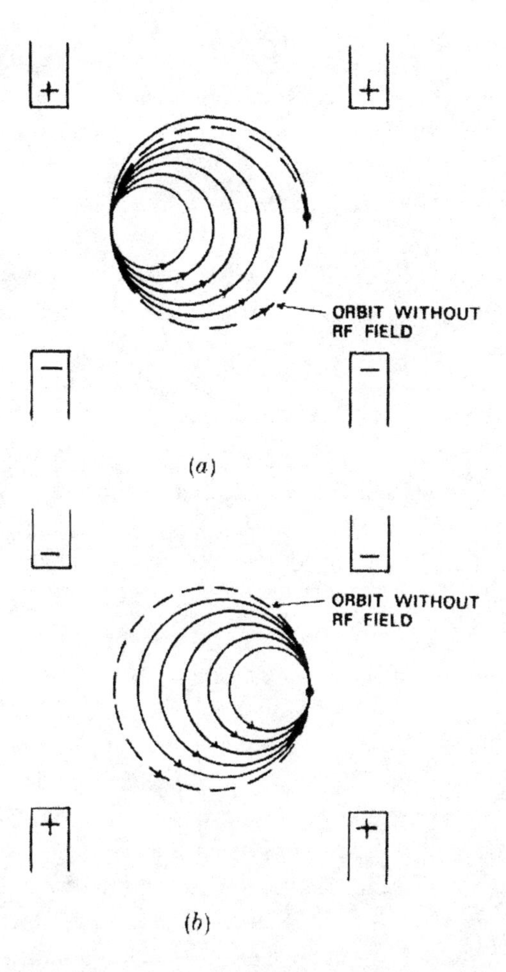

Fig. 13.13. Electron trajectories in a traveling-wave peniotron operating in the fundamental mode for (a) an initially accelerating electron, and (b) an initially decelerating electron. (Reproduced from Ono et al. 1984)

be greater than its previous acceleration. This results in an overall deceleration of this electron. So, it will continue to spiral into smaller and smaller orbits whose centers move to the left. In a similar fashion, the second electron, which was initially decelerated, will spiral into smaller orbits whose centers, however, move to the right. Hence, both sorts of electrons give up their rotational kinetic energy to the RF electric field of the wave. This fact makes the peniotron potentially a highly efficient device. Indeed, the analytical study done by S. Kuznetsov, Trubetskov, and Chetverikov (1980) confirmed that the peniotron efficiency limit is equal to 100%.

At the same time, Ono et al. (1984) correctly pointed out that there are two reasons for which the peniotron as a high-power microwave source must be inferior to the gyrotron. These reasons are: (1) an electron beam current, which can be used in the peniotron, is much smaller than that in the gyrotron, and (2) extremely high RF electric fields arise near the ridges during high-power operation.

Therefore, they proposed the gyropeniotron, whose schematic is shown in Fig. 13.14. In this device the same small-orbit electron beam is used as in conventional gyrotrons. So, the beam current, and hence, the beam power utilized in such a device can be the same as in conventional gyrotrons. Furthermore, there are no ridges, slots, or any other sharp elements in the microwave circuit. Thus, the second disadvantage of the peniotron can also be eliminated. (Of course, one can use slotted structures in the gyropeniotron as well.) The only limitation here is caused by the necessity to position the beam in such a place where gyrotron interaction with the electric field is weak while the peniotron interaction is strong. Such a case is shown in Fig. 13.14 for interaction with the TE_{02}-mode. The electric field of the mode at the beam radius shown is equal to zero, while its derivative, which is important for the peniotron interaction, is maximal.

This principle of operation offers some benefits in the case of utilizing low-order modes. However, discrimination of the gyrotron interaction with the operating mode may enhance the competition with other modes capable of gyrotron interaction with the beam of a given radius (Vitello and Ko 1985). Also, the necessity to precisely position the beam brings back the concerns about beam thickness and the ability to discriminate the gyrotron interaction in the case of beams with realistic thickness, which was discussed above.

Auto-Resonant Peniotron

In the beginning of this section, we compared contributions from O-type and M-type interactions for the case of operation near cutoff. When electrons interact with waves propagating at small Brillouin angles, the O-type interaction weakens due to the partial compensation of two bunching mechanisms (autoresonance effect). This effect, which we have discussed many times, makes the O-term in (13.5) $(1 - n^2)$ times smaller [compare (13.5) with (6.29) and (6.30)] and also allows one to extract electron energy not only from orbital, but also from axial motion. Reduction of the O-term enhances a role of the M-type interaction. Thus, a combination of two mechanisms (O-type auto-resonance effect and M-type interaction), both beneficial for efficient operation, is attractive for the development of high-efficiency cyclotron

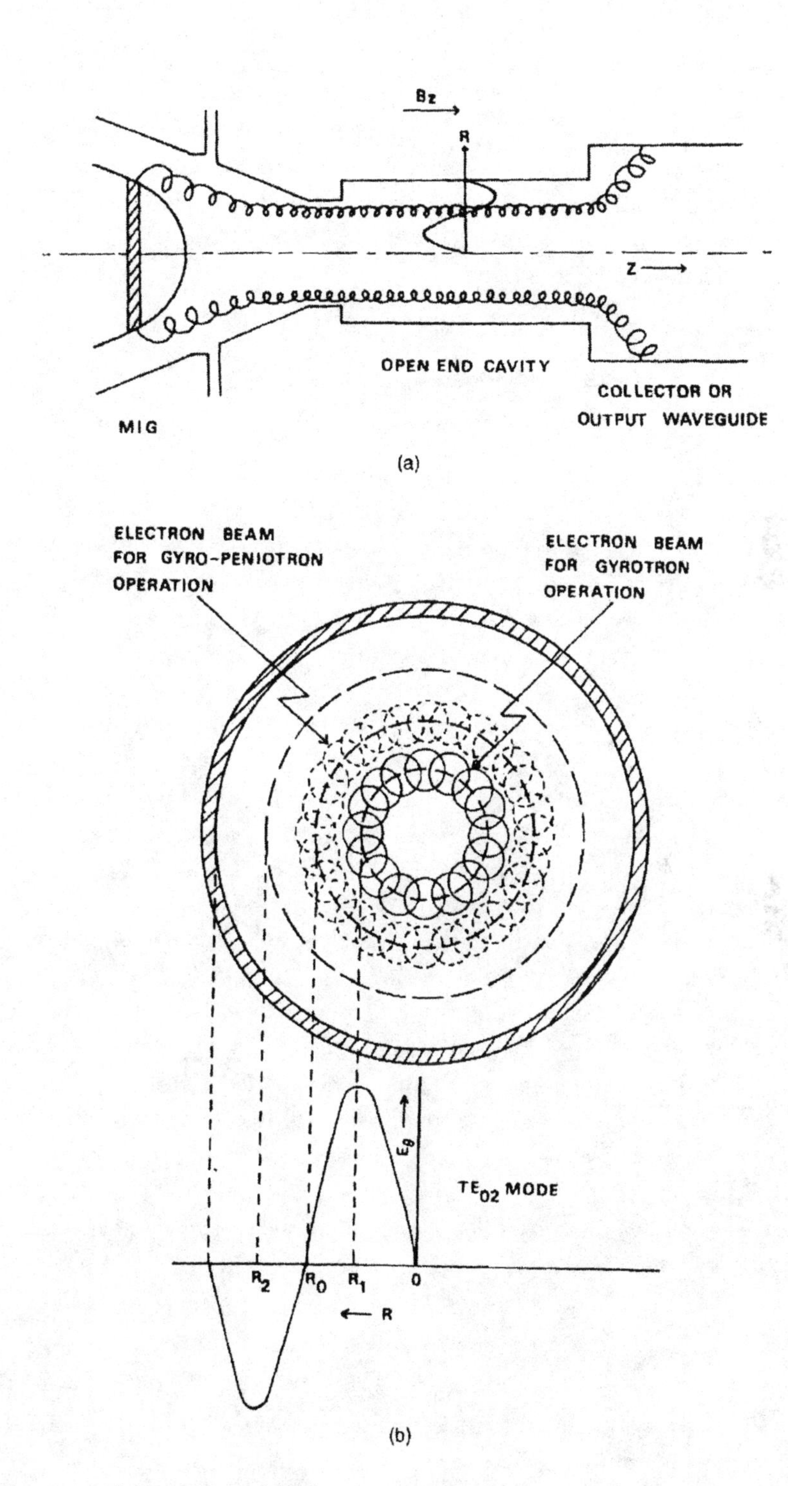

Fig. 13.14. (a) Schematic of a gyropeniotron; (b) electron beam positions in the cases of gyrotron and gyropeniotron operation. (Reproduced from Ono et al. 1984)

resonance masers. A device of this sort with combined autoresonant and peniotron properties is called an "autoresonant peniotron." This concept was originally proposed by J. Baird and coauthors in 1987 (Baird et al. 1987) and then actively studied by the group led by S. Ono (see N. Sato et al. 2002 for the latest progress and references).

13.8 Problems and Solutions

Problems

1. It was mentioned in Sec. 13.1 that the dispersion equation for the gyro-BWO can be written in a form similar to (6.27) for the gyro-TWT, namely, as $\gamma^2(\gamma - \delta) - 1 = 0$. From the small-signal theory of the linear-beam BWO (Johnson 1955) it is known that the starting conditions for the lossless BWO with negligibly small space charge effects yield the starting values for the normalized interaction length $\varsigma_{out} = C\,|k_z|\,L$ and normalized frequency detuning $\delta = (\omega + |k_z|\,v_{z0} - s\Omega_0)/\,|k_z|\,v_{z0}C$: $\varsigma_{out} = 1.9735$ and $\delta = 1.522$. Use these numbers and the expression for the Pierce gain parameter given by (6.30) (let (6.31) be also valid) for calculating the starting length and the oscillation frequency for the Ka-band gyro-BWO operating at the TE_{01}-wave at the fundamental cyclotron resonance. Consider an electron beam with the beam current of 5 A, beam voltage of 80 kV, and the orbital-to-axial velocity ratio of 1. Assume that a thin electron beam is positioned at the peak of the coupling impedance for the operating wave; the waveguide radius is 0.55 cm and the magnetic field is 12 kG.
2. In large-orbit gyrotrons, electrons gyrate around the system axis and interact with the rotating $TE_{m,p}$-wave at the $s = m$ cyclotron harmonic. Under properly chosen operating conditions, most of the electrons decrease their orbital momentum due to interaction with the wave, but some of them ("wrong-phase" electrons) can increase their momentum. Assume that electrons interact with the $TE_{m,1}$-wave and their initial clearance between the beam radius and the waveguide wall is $\lambda/4$. Estimate the changes in the orbital momentum that can lead to the electron interception by the wall, and analyze them.

Solutions

1. Let us start by determining the operating frequency in the zero-order approximation, which is the case of the exact cyclotron resonance: $\omega + |k_z|\,v_{z0} = s\Omega_0$. The values of ω and $k_z = (1/c)(\omega^2 - \omega_{cut}^2)^{1/2}$ that

correspond to this approximation we denote by the subscript (0). The cut-off frequency, f_{cut}, in accordance with given waveguide radius and known value of the eigennumber for the operating wave, is equal to 33.2637 GHz. The cyclotron frequency, f_{cycl}, for given values of the magnetic field and beam voltage is equal to 38.86 GHz. The normalized axial and orbital velocities for given values of the beam voltage and orbital-to-axial velocity ratio are equal to $\beta_{z0} = \beta_{\perp 0} = 0.35525$. Thus, the operating frequency in the zero-order approximation is equal to 35 GHz. Correspondingly, normalized transverse and axial wavenumbers are equal to $\kappa = f_{\mathrm{cut}}/f = 0.95$ and $n = \sqrt{1-\kappa^2} = 0.311$. The coupling impedance (3.59) for our choice of the beam position and operating mode is close to 0.145. Substituting all these values and the beam current into (6.30), one readily gets $C^3 = 5.22 \cdot 10^{-3}$ or $C = 0.17346$. Correspondingly, the starting length is equal to $L_{\mathrm{st}} = 5.82\lambda = 4.99\,\mathrm{cm}$. Now we should determine more accurately the operating frequency with the account for a finite cyclotron resonance mismatch. Denote this mismatch by $\Delta\omega = \omega + |k_z| v_{z0} - s\Omega_0$. Then, the detuning δ, which was given above, can be rewritten as $\delta = (\Delta\omega/\omega_{(0)})/n_{(0)}\beta_{z0}C = 1.522$, which, yields $\Delta\omega/\omega_{(0)} = 0.02917$. Representing the operating frequency as $\omega = \omega_{(0)} + \delta\omega$, one can easily find that the shift of the operating frequency caused by the frequency pulling effect of an electron beam is equal to $\delta f = 55.45$ MHz.

2. In a large-orbit gyrotron the beam radius is the Larmor radius $a = v_\perp/\Omega_0 = p_\perp c/eH_0$. Therefore, the changes in the electron position and orbital momentum are related as $\Delta a = (c/eH_0)\Delta p_\perp$. Since the critical increase in the beam radius is $(\Delta a)_{\mathrm{cr}} = \lambda/4$, the corresponding relative change in the orbital momentum is $(\Delta p_\perp/p_{\perp 0})_{\mathrm{cr}} = \lambda/4a = \pi/2s$. (Here we used the cyclotron resonance condition $\Omega_0 \approx \omega/s$ where for the LOG $s = m$.) So, this limitation becomes more important as the resonance harmonic number increases and, correspondingly, the guiding magnetic field decreases. Recall that the modulation of the orbital momentum of "wrong-phase" electrons depends on the initial cyclotron resonance mismatch. This is obvious from the phase portrait shown in Fig. 7.4: "wrong-phase" electrons are those particles that are initially located outside of the separatrix having the form of a cat's eye.

Summary

Operation of gyrotrons, as well as of any other source of coherent electromagnetic radiation (microwave tubes, lasers, etc.), is based on certain physical effects. These effects were the focus of the present book. It was shown how some specific physical effects are used for generating high-power, millimeter- and submillimeter-wave radiation in gyrotrons. In the last chapter, it was also shown that some new configurations of gyrodevices could be proposed that offer advantages over the well-developed ones.

Performance of gyrotrons, as well as performance of any other source of coherent electromagnetic radiation (microwave tubes, lasers, etc.), is mainly determined by various technological issues. Among them, to mention a few, are such issues as achievable current density from cathodes, intense cooling of microwave structures and collectors, and material properties of output windows. The book was not focused on these issues, but I hope that their importance was clearly shown.

Applications of gyrotrons, as well as of any other source of coherent electromagnetic radiation, focus interest on some specific performance characteristics that can be improved in the process of further development. In the book I discussed some important gyrotron applications, but not all of them. For instance, gyrotron applications to microwave sintering of ceramics and other kinds of material processing have not been mentioned. I also did not mention the use of gyrotrons for producing multicharge ions for heavy ion accelerators and just briefly mentioned the use of gyrotrons for active plasma diagnostics, various kinds of spectroscopy, and others. In this regard, it should be mentioned that several years ago the book *Applications of High-Power Microwaves* edited by A. V. Gaponov-Grekhov and V. L. Granatstein (1994), was published, which was focused on these issues. New applications of gyrotrons are also actively discussed at the annual International Conferences on Infrared and Millimeter Waves.

I hope that this book has helped the readers to better understand the physics of gyrotron operation and attain some ideas about gyrotron performance and problems in gyrotron development. I also hope that this book will be helpful to those who are just starting to work in the gyrotron field, and that it will be useful to those who already know how interesting and exciting are studies of the device called the *gyrotron*.

Appendix 1

Derivation of Gyro-Averaged Equations

Let us start from the equation for electron motion (3.21), in which the RF Lorentz force is determined by the electric and magnetic fields [see Eqs. (3.23)] expressed via the Hertz potential H_s. In our resonator formed by a slightly irregular waveguide, which is open in axial direction, this potential can be represented as $\Psi_s(x, y, z) f_s(z)$. Here the membrane function Ψ_s describes the transverse structure of the field. This function obeys the Helmholtz equation

$$\Delta_\perp \Psi_s + k_\perp^2 \Psi_s = 0 \tag{A1.1}$$

(here $k_\perp \approx \omega/c$) with the boundary condition $[[\nabla_\perp \Psi_s \times \vec{z}_0] \times \vec{n}] = 0$ at the side wall of the resonator; $\vec{n}$ is the unit vector perpendicular to its surface. The function $f(z)$ in this potential obeys the inhomogeneous string equation (2.13). (Strictly speaking, in (2.13) the axial wavenumber k_z should be determined not by the oscillation frequency ω, but by the mode eigenfrequency ω_s, which is assumed to be close to ω.)

Transverse coordinates of the electron obey the standard equation

$$\frac{d\vec{r}_\perp}{dz} = \frac{\vec{p}_\perp}{p_z}. \tag{A1.2}$$

Below we will assume that the EM field is much weaker than the external magnetic field. Thus, transverse momentum and coordinates of electrons can be represented as

$$p_x = -p_\perp \sin\theta, \qquad p_y = p_\perp \cos\theta, \tag{A1.3}$$
$$x = X + a\cos\theta, \qquad y = Y + a\sin\theta,$$

where $\theta = h_H z + \psi$ is the gyrophase, $a = p_\perp / h_H p_z$ is the Larmor radius, and X and Y are coordinates of the guiding center. In the absence of an EM field,

$p_\perp$, ψ, X, Y, and a are constant. In the presence of this field, in accordance with the Lagrange method, they can be considered as slowly varying during one period of gyration. (Van der Pol actively used this method of perturbations in his studies of radio-oscillators.)

Substituting this representation into (3.21) and (A1.2) yields

$$\frac{dp_\perp}{dz}\sin\theta + p_\perp\frac{d\psi}{dz}\cos\theta = \frac{e}{v_z}\text{Re}\left\{CG_x e^{i\omega t}\right\},$$
$$\frac{dp_\perp}{dz}\cos\theta - \frac{d\psi}{dz}p_\perp\sin\theta = -\frac{e}{v_z}\text{Re}\left\{CG_y e^{i\omega t}\right\},$$
$$\frac{dX}{dz} + \frac{da}{dz}\cos\theta - a\frac{d\psi}{dz}\sin\theta = 0,$$
$$\frac{dY}{dz} + \frac{da}{dz}\sin\theta + a\frac{d\psi}{dz}\cos\theta = 0.$$

After eliminating $\sin\theta$ and $\cos\theta$ from the left-hand sides of these equations, they can be rewritten as

$$\frac{dp_\perp}{dz} = -\frac{e}{v_z}\text{Re}\left\{CG_\theta e^{i\omega t}\right\},\tag{A1.4}$$
$$p_\perp\frac{d\psi}{dz} = \frac{e}{v_z}\text{Re}\left\{CG_r e^{i\omega t}\right\},$$
$$\frac{dX}{dz} = \frac{c}{v_z}\text{Re}\left\{\frac{C}{H_0}G_y e^{i\omega t}\right\},$$
$$\frac{dY}{dz} = -\frac{c}{v_z}\text{Re}\left\{\frac{C}{H_0}G_x e^{i\omega t}\right\}.$$

Here G_r and G_θ are components of the RF Lorentz force in the polar coordinate system with the origin located at the instantaneous position of the electron guiding center. These components are related to G_x and G_y as

$$G_x = G_r\cos\theta - G_\theta\sin\theta,\tag{A1.5}$$
$$G_y = G_r\sin\theta + G_\theta\cos\theta.$$

Since the Lorentz force is the periodic function of θ, it can be expanded in the Fourier series as

$$\vec{G} = \sum_{k=-\infty}^{\infty}\vec{G}_k e^{-ik\theta}, \qquad \text{where } \vec{G}_k = \frac{1}{2\pi}\int_0^{2\pi}\vec{G}e^{ik\theta}d\theta.$$

Substituting this expansion into (A1.4) and using the cyclotron resonance condition $\omega \approx s\Omega$ and the assumption that $p_\perp$, ψ, X, Y, and a vary slowly allow one to average these equations over fast gyrations. Reduced (or gyro-averaged) equations have a form quite similar to (A1.4):

$$\frac{dp_\perp}{dz} = -\frac{e}{v_z}\mathrm{Re}\left\{CG_{s\theta}e^{-i\vartheta}\right\},$$

$$p_\perp\left(\frac{d\vartheta}{dz} + \frac{\omega}{v_z} - sh_H\right) = \frac{e}{v_z}\mathrm{Re}\left\{sCG_{sr}e^{-i\vartheta}\right\},$$

$$\frac{dX}{dz} = \frac{c}{v_z}\mathrm{Re}\left\{\frac{C}{H_0}G_{sy}e^{-i\vartheta}\right\},$$

$$\frac{dY}{dz} = -\frac{c}{v_z}\mathrm{Re}\left\{\frac{C}{H_0}G_{sx}e^{-i\vartheta}\right\}. \quad \textbf{(A1.6)}$$

Here $\vartheta = s\theta - \omega t$ is the slowly variable gyrophase of the synchronous cyclotron harmonic with respect to the phase of the EM field.

In order to determine the resonant components of the Lorentz force in (A1.6), it is necessary to make use of the known integral representation (Whittaker and Watson 1963) for the membrane function

$$\Psi_s(x, y) = \int_0^{2\pi} \Psi_s^\dagger(\varphi)e^{ik_\perp(x\cos\varphi + y\sin\varphi)}d\varphi \quad \textbf{(A1.7)}$$

[One can easily verify the correctness of this representation by substituting it into (A1.1)]. By using representation of electron coordinates x and y given above by (A1.3), one can determine the angular harmonic of the membrane function as

$$\Psi_{s,k} = \frac{1}{2\pi}\int_0^{2\pi} \Psi_s e^{ik\theta}\,d\theta \quad \textbf{(A1.8)}$$

$$= \int_0^{2\pi} \Psi_s^\dagger(\varphi)e^{ik_\perp(X\cos\varphi + Y\sin\varphi) + ik\varphi}d\varphi \frac{1}{2\pi}\int_0^{2\pi} e^{ik_\perp a\cos\theta' + ik\theta'}d\theta',$$

where $\theta' = \theta - \varphi$. Here the last integral is the integral representation of the Bessel function (see, e.g., Gradshtein and Ryzhik 2000), and thus is equal to $2\pi i^k J_k(k_\perp a)$. The first integral for the case of membrane functions described by analytical functions can be represented with the use of the relation

$$\int_0^{2\pi} \Psi_s^\dagger e^{ik_\perp(X\cos\varphi + Y\sin\varphi)}\begin{Bmatrix}\cos\varphi \\ \sin\varphi\end{Bmatrix}d\varphi = \left(\frac{1}{ik_\perp}\right)\begin{Bmatrix}\frac{\partial\Psi_s}{\partial X} \\ \frac{\partial\Psi_s}{\partial Y}\end{Bmatrix}$$

as

$$\int_0^{2\pi} \Psi_s^{\dagger} e^{ik_\perp(X\cos\varphi+Y\sin\varphi)+ik\varphi}d\varphi = \left(\frac{1}{ik_\perp}\right)^k \left(\frac{\partial}{\partial X}+i\frac{\partial}{\partial Y}\right)^k \Psi_s(X,Y).$$

Finally, the angular harmonic of the membrane function can be written in a compact form as

$$\Psi_{s,k} = J_k(k_\perp a)L_k(X,Y,z), \tag{A1.9}$$

where

$$L_k = \left[\frac{1}{k_\perp}\left(\frac{\partial}{\partial X}+i\frac{\partial}{\partial Y}\right)\right]^k \Psi_s(X,Y). \tag{A1.10}$$

By using the definitions of the electric and magnetic fields given by Eqs. (3.23), one can readily express angular components of the RF Lorentz force in (A1.6) via the angular harmonic of the membrane function:

$$G_{k,r} = -i\left\{\frac{ck}{\omega a} - \frac{v_\perp}{c}\right\} f_s \Psi_{s,k}, \tag{A1.11}$$

$$G_{k,\theta} = -\frac{c}{\omega} f_s \frac{\partial \Psi_{s,k}}{\partial a}.$$

Here $v_\perp/c = \Omega a/c \approx \omega a/sc$, and therefore,

$$G_{k,r} = -i\left\{\frac{k}{\xi} - \frac{\xi}{n}\right\} f_s \Psi_{s,k}, \qquad G_{k,\theta} = -f_s \frac{\partial \Psi_{s,k}}{\partial \xi}, \tag{A1.12}$$

where $\xi = \omega a/c$. For the resonant harmonic ($k = s$), one can rewrite these expressions with the use of (A1.9) and Bessel equation

$$\frac{d}{d\xi}\left(\xi\frac{dJ_s}{d\xi}\right) + \left(\xi - \frac{s^2}{\xi}\right)J_s = 0,$$

as

$$G_{s,r} = -\frac{i}{s} f_s \frac{d}{d\xi}\left[\xi\frac{dJ_s(\xi)}{d\xi}\right] L_s, \tag{A1.13}$$

$$G_{s,\theta} = -f_s \frac{dJ_s(\xi)}{d\xi} L_s.$$

Correspondingly, resonant components of G_x and G_y, in accordance with (A1.5), can be given as

$$G_{s,x} = \frac{1}{2}[(G_{s+1,r} + iG_{s+1,\theta}) + (G_{s-1,r} - iG_{s-1,\theta})],$$

$$G_{s,y} = -\frac{1}{2}[(G_{s+1,r} + iG_{s+1,\theta}) - (G_{s-1,r} - iG_{s-1,\theta})].$$

Taking into account that $J_{s\pm1}(\xi) = \frac{s}{\xi}J_s(\xi) \mp \frac{dJ_s(\xi)}{d\xi}$, one can represent these combinations as

$$G_{s+1,r} + iG_{s+1,\theta} = -i\frac{\xi}{s} f_s \frac{dJ_s}{d\xi} L_{s+1},$$

$$G_{s-1,r} - iG_{s-1,\theta} = i\frac{\xi}{s} f_s \frac{dJ_s}{d\xi} L_{s-1}.$$

Since not only Ψ_s, but also L_s obeys Helmholtz equation, one can readily derive

$$L_k = \frac{1}{k_\perp}\left(\frac{\partial}{\partial X} - i\frac{\partial}{\partial Y}\right) L_{k+1}.$$

Indeed,

$$\frac{1}{k_\perp}\left(\frac{\partial}{\partial X} - i\frac{\partial}{\partial Y}\right) L_{k+1} = \frac{1}{k_\perp}\left(\frac{\partial}{\partial X} - i\frac{\partial}{\partial Y}\right)\left[\frac{1}{k_\perp}\left(\frac{\partial}{\partial X} + i\frac{\partial}{\partial Y}\right) L_k\right]$$
$$= \frac{1}{k_\perp^2}\left(\frac{\partial^2}{\partial X^2} + \frac{\partial^2}{\partial Y^2}\right) L_k = L_k$$

Therefore, in $G_{s,x}$ and $G_{s,y}$

$$L_{s+1} \pm L_{s-1} = \frac{1}{k_\perp}\left[\left(\frac{\partial}{\partial X} + i\frac{\partial}{\partial Y}\right) \mp \left(\frac{\partial}{\partial X} - i\frac{\partial}{\partial Y}\right)\right] L_s = \frac{2}{k_\perp}\begin{Bmatrix} i\frac{\partial}{\partial Y} \\ \frac{\partial}{\partial X} \end{Bmatrix} L_s,$$

and, correspondingly,

$$G_{s,x} = -i\frac{\xi}{s} f_s \frac{dJ_s}{d\xi}\frac{1}{k_\perp}\frac{\partial L_s}{\partial X} \simeq -i\frac{v_\perp}{c} f_s \frac{dJ_s}{d\xi}\frac{1}{k_\perp}\frac{\partial L_s}{\partial X},$$

$$G_{s,y} = -i\frac{\xi}{s} f_s \frac{dJ_s}{d\xi}\frac{1}{k_\perp}\frac{\partial L_s}{\partial Y} \simeq -i\frac{v_\perp}{c} f_s \frac{dJ_s}{d\xi}\frac{1}{k_\perp}\frac{\partial L_s}{\partial Y}.$$

With the use of these expressions, one can rewrite (A1.6) in a quasi-Hamiltonian form:

$$\frac{d\xi}{dz} = e\mathrm{Re}\left\{C\frac{\partial \mathcal{H}}{\partial \vartheta}\right\}, \tag{A1.14}$$

$$\frac{d\vartheta}{dz} + \frac{\omega}{v_z} - sh_H = -e\mathrm{Re}\left\{C\frac{\partial \mathcal{H}}{\partial \xi}\right\},$$

$$\frac{dX}{dz} = -\frac{1}{k_\perp}\mathrm{Re}\left\{\frac{C}{H_0}\frac{\partial \mathcal{H}}{\partial Y}\right\},$$

$$\frac{dY}{dz} = \frac{1}{k_\perp}\mathrm{Re}\left\{\frac{C}{H_0}\frac{\partial \mathcal{H}}{\partial X}\right\}.$$

Here we expressed $p_\perp$ via the energy ξ in accordance with (1.20) and used the fact that the axial momentum is constant. A role of Hamiltonian is played here by the function

$$\mathcal{H} = i\frac{p_\perp}{p_z} f_s \frac{dJ_s}{d\xi} L_s(X, Y, z)e^{-i\vartheta}. \qquad \textbf{(A1.15)}$$

In conclusion, let us also present the formula for L_s for the case of widely used cylindrical geometry. In such a case, it is expedient to replace Cartesian coordinates of the guiding center X and Y by the polar coordinates R, ψ, which are related as

$$X = R\cos\psi, \qquad Y = R\sin\psi.$$

Then, differential operator in L_k can also be expressed in polar coordinates with the use of relations

$$\frac{\partial}{\partial X} = \cos\psi\frac{\partial}{\partial R} - \frac{\sin\psi}{R}\frac{\partial}{\partial\psi},$$

$$\frac{\partial}{\partial Y} = \sin\psi\frac{\partial}{\partial R} + \frac{\cos\psi}{R}\frac{\partial}{\partial\psi},$$

as

$$L_k = \left\{e^{i\psi}\frac{1}{k_\perp}\left(\frac{\partial}{\partial R} + i\frac{\partial}{\partial\psi}\right)\right\}^k \Psi_s(R, \psi).$$

As is known, the membrane function for a $\mathrm{TE}_{\mathrm{m,p}}$-mode of the cylindrical resonator (waveguide) is equal to $J_m(k_\perp \mathrm{r})\mathrm{e}^{\mp im\psi}$. Here m is the azimuthal index of the mode and p is its radial index; the transverse wavenumber $k_\perp$ is equal to $\nu_{m,p}/R_w$, where $\nu_{m,p}$ is the pth root of the equation $dJ_m(\nu)/d\nu = 0$, which is the boundary condition for the TE-mode at the wall of the radius R_w. With the account for the known relation for Bessel functions $\frac{dJ_m(\xi)}{d\xi} = J_{m-1}(\xi) - \frac{m}{\xi}J_m(\xi) = \frac{m}{\xi}J_m(\xi) - J_{m+1}(\xi)$, differentiation in L_k results in

$$L_k = J_{m\mp k}(k_\perp R)e^{i(k\mp m)\psi}. \qquad \textbf{(A1.16)}$$

Here $m - k$ and $m + k$ correspond, respectively, to the co- and counterrotating waves with respect to electron gyration. This expression for L_k also follows from the Schwinger lemma discussed in Chapter 1 and coincides with the resonant term in the presentation of the Bessel function with the use of the Graff theorem (Whittaker and Watson 1963).

APPENDIX 2

Wave Excitation by Electron Beams in Waveguides

In this appendix, we will present the formalism that was developed a long time ago by L. A. Weinstein. Since his books (Weinstein 1988, Weinstein and Solntsev 1973) were published in Russian and not translated into English, we reproduce here this formalism with minor modifications for completeness of our treatment.

Below we shall consider a monochromatic wave whose electric and magnetic fields can be given as

$$\vec{E}_s = \mathrm{Re}\left\{\vec{E}_\omega e^{i\omega t}\right\}, \qquad \vec{H}_s = \mathrm{Re}\left\{\vec{H}_\omega e^{i\omega t}\right\}. \tag{A2.1}$$

In the absence of an electron beam, the complex amplitudes of the electric and magnetic fields can be represented as

$$\vec{E}_\omega = \sum_s (C_s \vec{E}_s + C_{-s}\vec{E}_{-s}), \qquad \vec{H}_\omega = \sum_s (C_s \vec{H}_s + C_{-s}\vec{H}_{-s}), \tag{A2.2}$$

where $C_{\pm s}$ are arbitrary constants. When a waveguide is uniform, the functions $\vec{E}_{\pm s}$ and $\vec{H}_{\pm s}$ depend on the axial coordinate as $\exp(-ik_{s,z}z)$. Here the index s designates a wave with a given transverse structure, the waves with negative indices $(-s)$ are the waves with the same transverse structure, but propagating in the $-z$-direction, so $k_{-s,z} = -k_{s,z}$. In the presence of losses in waveguide walls, for forward waves Im $k_{s,z} < 0$, while for backward waves Im $k_{-s,z} > 0$.

When there are no currents, the complex amplitudes of the electric and magnetic fields, $\vec{E}_\omega$ and $\vec{H}_\omega$, obey homogeneous Maxwell equations

$$\nabla \times \vec{H}_\omega = i\frac{\omega}{c}\vec{E}_\omega, \qquad \nabla \times \vec{E}_\omega = -i\frac{\omega}{c}\vec{H}_\omega. \tag{A2.3}$$

Here and below, for the sake of simplicity, we assume no dielectrics and magnetics: $\varepsilon = \mu = 1$. Below, we shall consider orthogonal nondegenerate waves, for which the orthogonality condition

$$\int_{S_\perp} \{[\vec{E}_s \times \vec{H}_{s'}] - [\vec{E}_{s'} \times \vec{H}_s]\}\vec{z}_0 ds_\perp = 0 \quad \text{for } s' \neq -s \tag{A2.4}$$

is fulfilled. Here $S_\perp$ is the waveguide cross-section area.

Now, let us assume that this wave can be excited by a monochromatic high-frequency component of an electron current with the density

$$\vec{j} = \text{Re}\left\{\vec{j}_\omega e^{i\omega t}\right\}. \tag{A2.5}$$

Then, the wave fields, $\vec{E}_\omega$ and $\vec{H}_\omega$, should obey inhomogeneous Maxwell equations

$$\nabla \times \vec{H}_\omega = i\frac{\omega}{c}\vec{E}_\omega + \frac{4\pi}{c}\vec{j}_\omega, \qquad \nabla \times \vec{E}_\omega = -i\frac{\omega}{c}\vec{H}_\omega. \tag{A2.6}$$

Let us represent the magnetic field of the wave as

$$\vec{H}_\omega = \sum_s (C_s\vec{H}_s + C_{-s}\vec{H}_{-s}), \tag{A2.7}$$

where C_s are now the wave amplitudes, which in the presence of an electron beam depend on the axial coordinate z. The next step is to determine the relation between representations of $\vec{H}$ and $\vec{E}$ in this case and to derive the equation for the axial dependence of C_s on the beam current. To do this, let us substitute the magnetic field representation given by (A2.7) into the first equation of (A2.6) and take into account that

$$\nabla \times (C_{\pm s}\vec{H}_{\pm s}) = C_{\pm s}(\nabla \times \vec{H}_{\pm s}) + [(\nabla \times C_{\pm s}) \times \vec{H}_{\pm s}].$$

This, in combination with the first equation in (A2.3), yields the following expression for the electric field

$$\vec{E}_\omega = \sum_s (C_s\vec{E}_s + C_{-s}\vec{E}_{-s}) + \frac{c}{i\omega}\sum_s \left(\frac{dC_s}{dz}[\vec{z}_0 \times \vec{H}_s] + \frac{dC_{-s}}{dz}[\vec{z}_0 \times \vec{H}_{-s}]\right) - \frac{4\pi}{i\omega}\vec{j}_\omega. \tag{A2.8}$$

Assume now that amplitudes C_s obey the additional condition

$$\sum_s \left(\frac{dC_s}{dz}[\vec{z}_0 \times \vec{H}_s] + \frac{dC_{-s}}{dz}[\vec{z}_0 \times \vec{H}_{-s}]\right) = \frac{4\pi}{c}\vec{j}_{\omega\perp}, \tag{A2.9}$$

where $\vec{j}_\perp$ is the transverse component of the complex current density. Then, the electric field becomes equal to

$$\vec{E}_\omega = \sum_s [C_s(z)\vec{E}_s + C_{-s}(z)\vec{E}_{-s}] + i\frac{4\pi}{\omega}j_{\omega,z}\vec{z}_0, \tag{A2.10}$$

where the last term makes the difference between the amplitudes of magnetic and electric fields.

Now we can substitute the expressions for $\vec{H}_\omega$ and $\vec{E}_\omega$ given, respectively, by Eqs. (A2.7) and (A2.10) into the second Maxwell equation. This results in

$$\sum_s \left(\frac{dC_s}{dz}[\vec{z}_0 \times \vec{E}_s] + \frac{dC_{-s}}{dz}[\vec{z}_0 \times \vec{E}_{-s}] \right) = -i\frac{4\pi}{\omega} \nabla \times (j_{\omega,z}\vec{z}_0). \quad \textbf{(A2.11)}$$

Let us now multiply Eq. (A2.9) by $\vec{E}_{-s}$ and Eq. (A2.11) by $\vec{H}_{-s}$, sum them, integrate over the waveguide cross section, and take into account the orthogonality condition given by (A2.4). This yields the following equation for the forward wave:

$$\frac{dC_s}{dz} = \frac{1}{N_s}\int_{S_\perp} \left\{ \vec{j}_{\omega,\perp}\vec{E}_{-s} + i\frac{c}{\omega} \nabla \times (j_{\omega,z}\vec{H}_{-s})\vec{z}_0 \right\} ds_\perp. \quad \textbf{(A2.12)}$$

Here we introduced the norm of the wave:

$$N_s = \frac{c}{4\pi}\int_{S_\perp} \{[\vec{E}_s \times \vec{H}_{-s}] - [\vec{E}_{-s} \times \vec{H}_s]\}\vec{z}_0 ds_\perp. \quad \textbf{(A2.13)}$$

Using the relation $\nabla \times (j_{\omega,z}\vec{H}_s) = j_{\omega,z}(\nabla \times \vec{H}_s) + [(\nabla \times j_{\omega,z}) \times \vec{H}_s]$ and Maxwell equation for $\nabla \times \vec{H}_s$ (the first equation in (A2.3)) allows one to rewrite (A2.12) as

$$\frac{dC_s}{dz} = \frac{1}{N_s}\int_{S_\perp} \left\{ \vec{j}_\omega\vec{E}_{-s} + i\frac{c}{\omega}\vec{z}_0 \nabla \times (j_{\omega,z}\vec{H}_{-s}) \right\} ds_\perp. \quad \textbf{(A2.14)}$$

As follows from the Stokes theorem, $\int_{S_\perp} \nabla \times (j_{\omega,z}\vec{H}_{-s})\vec{z}_0 ds_\perp = \oint_L j_{\omega,z}\vec{H}_{-s}d\vec{l}$, and, once the contour L is chosen inside the waveguide wall, the field $\vec{H}_{-s}$ at the depth exceeding the skin layer is equal zero, and therefore, (A2.14) reduces to

$$\frac{dC_s}{dz} = \frac{1}{N_s}\int_{S_\perp} \vec{j}_\omega \cdot \vec{E}_{-s} ds_\perp. \quad \textbf{(A2.15)}$$

A similar equation can readily be derived for waves propagating in the $-z$-direction, for which $k_{s,z}$ is negative.

The field components, as is known, can be expressed in terms of the vector Hertz potential $\vec{\Phi}$, which obeys the Helmholtz equation

$$\Delta\vec{\Phi} + k^2\vec{\Phi} = 0. \quad \textbf{(A2.16)}$$

In the case of TE-waves, the electric and magnetic fields can be given as

$$\vec{E}_\omega = \frac{k}{k_\perp^2} \nabla \times \vec{\Phi}_{TE}, \qquad \vec{H}_\omega = \frac{i}{k_\perp^2} \nabla \times (\nabla \vec{\Phi}_{TE}) + i\frac{k^2}{k_\perp^2}\vec{\Phi}_{TE}, \quad \textbf{(A2.17)}$$

where $k_\perp = \omega_{cut}/c$ is the transverse wavenumber. For a wave with a given axial wavenumber $k_z = (\omega^2 - \omega_{cut}^2)^{1/2}/c$, the vector Hertz potential can be presented as $\vec{\Phi} = H_s\vec{z}_0 = \Psi_s e^{-ik_z z}\vec{z}_0$, where the membrane function Ψ obeys the same membrane equation

$$\Delta_\perp \Psi + k_\perp^2 \Psi = 0. \tag{A2.18}$$

Correspondingly, the components of the electric and magnetic fields, $\vec{E}_{\pm s} = \vec{e}(\vec{r}_\perp)e^{-ik_{\pm s,z}z}$ and $\vec{H}_{\pm s} = \vec{h}(\vec{r}_\perp)e^{-ik_{\pm s,z}z}$, in the cylindrical coordinate system are equal to

$$e_r = \frac{k}{k_\perp^2 r}\frac{\partial \Psi}{\partial \varphi}, \qquad e_\varphi = -\frac{k}{k_\perp^2}\frac{\partial \Psi}{\partial r}, \qquad e_z = 0, \tag{A2.19}$$

$$h_r = \frac{k_z}{k_\perp^2}\frac{\partial \Psi}{\partial r}, \qquad h_\varphi = \frac{k_z}{k_\perp^2 r}\frac{\partial \Psi}{\partial \varphi}, \qquad h_z = i\Psi.$$

So, in the case of operation close to cutoff $k_z \to 0$, $k_\perp \to k$, and (A2.17) reduce to (3.23).

In the cylindrical waveguide, $\Psi = J_m(k_\perp r)e^{\pm im\varphi}$, where m is the azimuthal index of the $\text{TE}_{m,p}$-mode and the transverse wavenumber in a waveguide of a radius R is equal to $\nu_{m,p}/R$. Here $\nu_{m,p}$ is the pth root of the equation $J_m'(\nu) = 0$, which is the boundary condition at the waveguide wall. Correspondingly, the components of the electric and magnetic fields are equal to

$$e_r = \frac{k}{k_\perp^2 r}(\pm im)J_m(k_\perp r)e^{\pm im\varphi}, \quad e_\varphi = -\frac{k}{k_\perp}J_m'(k_\perp r)e^{\pm im\varphi}, \quad e_z = 0, \tag{A2.20}$$

$$h_r = \frac{k_z}{k_\perp}J_m'(k_\perp r)e^{\pm im\varphi}, \quad h_\varphi = \frac{k_z}{k_\perp^2 r}(\pm im)J_m(k_\perp r)e^{\pm im\varphi}, \quad h_z = iJ_m(k_\perp r)e^{\pm im\varphi}.$$

The fields $\vec{E}_{\pm s}$ and $\vec{H}_{\pm s}$ and their complex conjugates obey homogeneous Maxwell equations (A2.3). Therefore, the components of the complex conjugate forward wave and the components of the wave propagating in the opposite direction should be related either as $\vec{E}_{-s} = -\vec{E}_s^*$ and $\vec{H}_{-s} = \vec{H}_s^*$ or as $\vec{E}_{-s} = \vec{E}_s^*$ and $\vec{H}_{-s} = -\vec{H}_s^*$. In the former case, the norm of the wave given by (A2.13) can be rewritten as

$$N_s = \frac{c}{4\pi}\int_{S_\perp}\left\{\left[\vec{E}_s \times \vec{H}_s^*\right] + \left[\vec{E}_s^* \times \vec{H}_s\right]\right\}\vec{z}_0 ds_\perp. \tag{A2.21}$$

Substituting into (A2.21) the wave field components in the cylindrical waveguide given by (A2.20) yields the known expression

$$N_s = \frac{c}{2k_\perp^2}\frac{k_z k}{k_\perp^2}J_m^2(\nu_{m,p})\left(\nu_{m,p}^2 - m^2\right). \tag{A2.22}$$

In a similar fashion one can readily derive an expression for the wave norm in any waveguide by using the expressions for field components given by (A2.19).

Correspondingly, the wave excitation equation (A2.15) can be rewritten as

$$\frac{dC_s}{dz} = -\frac{1}{N_s} \int_{S_\perp} \vec{j}_\omega \vec{E}_s^* ds_\perp. \tag{A2.23}$$

Note that, if we choose the relation $\vec{E}_{-s} = \vec{E}_s^*$ and $\vec{H}_{-s} = -\vec{H}_s^*$, the result will be the same: there will be no minus in the RHS of (A2.23), but the norm will be negative.

Appendix 3

Derivation of the Self-Consistent Set of Equations for the Gyro-TWT

A self-consistent set of equations describing the gyro-TWT consists of the equations for electron motion in waveguide fields and the equation describing the wave excitation by an electron beam. For the case of electron motion in resonators, the equations for electron motion were derived in Appendix 1. It is shown there that, in general, equations for electron orbital momentum and phase (A1.4) can be written as

$$\frac{dp_\perp}{dz} = -\frac{e}{v_z}\mathrm{Re}\left\{CG_\theta e^{i\omega t}\right\}, \tag{A3.1}$$

$$p_\perp \frac{d\psi}{dz} = \frac{e}{v_z}\mathrm{Re}\left\{CG_r e^{i\omega t}\right\}. \tag{A3.2}$$

Also, the equation for the electron axial momentum can be written in a similar fashion as

$$\frac{dp_z}{dz} = -\frac{e}{v_z}\mathrm{Re}\left\{CG_z e^{i\omega t}\right\}. \tag{A3.3}$$

Above, in Appendix 1, we did not consider this equation because in the case of operation at TE-modes near cutoff the axial momentum remains practically constant. These three equations correspond to the representation of the wave fields as

$$\vec{E} = \mathrm{Re}\left\{C\vec{E}_s(\vec{r})e^{i\omega t}\right\}, \qquad \vec{H} = \mathrm{Re}\left\{C\vec{H}_s(\vec{r})e^{i\omega t}\right\}.$$

In the right-hand sides of (A3.1)–(A3.3) there are components of the function

$$\vec{G} = \vec{E}_s + \left[\frac{\vec{v}}{c} \times \vec{H}_s\right].$$

This function describes the spatial structure of the RF Lorentz force acting upon electrons. In waveguides, the electric and magnetic fields of the forward wave can be given by

$$\vec{E}_s(\vec{r}) = \vec{e}_s(\vec{r}_\perp)e^{-ik_z z}, \qquad \vec{H}_s(\vec{r}) = \vec{h}_s(\vec{r}_\perp)e^{-ik_z z},$$

so these equations can be rewritten as

$$\frac{dp_\perp}{dz} = -\frac{e}{v_z}\mathrm{Re}\left\{CG_\theta e^{i(\omega t - k_z z)}\right\}, \tag{A3.4}$$

$$p_\perp \frac{d\psi}{dz} = \frac{e}{v_z}\mathrm{Re}\left\{CG_r e^{i(\omega t - k_z z)}\right\}, \tag{A3.5}$$

$$\frac{dp_z}{dz} = -\frac{e}{v_z}\mathrm{Re}\left\{CG_z e^{i(\omega t - k_z z)}\right\}, \tag{A3.6}$$

where

$$\vec{G} = \vec{e}_s + \left[\frac{\vec{v}_\perp}{c} \times \vec{h}_s\right]. \tag{A3.7}$$

Below, the subscript s, which designates the type of the operating wave, will be omitted. Components of these wave fields $\vec{e}(\vec{r}_\perp)$ and $\vec{h}(\vec{r}_\perp)$ were determined in Appendix 2 via the membrane function $\Psi(\vec{r}_\perp)$ by (A2.19). This membrane function in the reference frame with the origin on the axis of electron gyration can be expanded in the Fourier series as

$$\Psi = \sum_l \Psi_l e^{il\theta}.$$

Correspondingly, the field components are equal to

$$e_r = \frac{k}{k_\perp^2 a}\sum_l (-il)\Psi_l e^{-il\theta}, \qquad e_\theta = -\frac{k}{k_\perp^2}\sum_l \left.\frac{d\Psi_l}{dr}\right|_{r=a} e^{-il\theta}, \qquad e_z = 0,$$

$$h_\theta = \frac{k_z}{k}e_r = ne_r, \qquad h_r = -\frac{k_z}{k}e_\theta = -ne_\theta, \qquad h_z = i\sum_l \Psi_l e^{-il\theta}.$$

Eq. (A1.9), as is shown in Appendix 1, determines angular harmonics of the membrane function. Correspondingly, the components of the Lorentz force (A3.7) can be determined as

$$G_r = i\sum_l \left\{\beta_\perp - \frac{\ell}{\kappa\xi}(1 - n\beta_z)\right\} J_l(\xi)L_l e^{-il\theta},$$

$$G_\theta = -\frac{1}{\kappa}(1 - n\beta_z)\sum_l J_l'(\xi)L_l e^{-il\theta},$$

$$G_z = -\frac{n}{\kappa}\beta_\perp \sum_l J_l'(\xi)L_l e^{-il\theta}.$$

So, now we can come back to (A3.4)–(A3.6), substitute into these equations the expressions for the Lorentz force components just derived, and make an averaging of these equations over fast gyrations. Introducing a slowly varying resonant harmonic of the gyrophase with respect to the phase of the traveling wave, $\vartheta = s\theta - \omega t + k_z z$, these gyro-averaged equations can be written as

$$\frac{dp_\perp}{dz} = \frac{e}{\kappa v_z}(1 - n\beta_z)\mathrm{Re}\left\{CJ_s'(\xi)L_s e^{-i\vartheta}\right\}, \tag{A3.8}$$

$$\frac{d\vartheta}{dz} + \frac{\omega - k_z v_z - s\Omega}{v_z} = s\frac{e}{v_z p_\perp}\left[\beta_\perp - \frac{s}{\kappa\xi}(1 - n\beta_z)\right]\mathrm{Re}\left\{iCJ_s(\xi)L_s e^{-i\vartheta}\right\} \tag{A3.9}$$

$$\frac{dp_z}{dz} = n\frac{ev_\perp}{\kappa c v_z}\mathrm{Re}\left\{CJ_s'(\xi)L_s e^{-i\vartheta}\right\}. \tag{A3.10}$$

As one can easily see, the equation for electron energy (1.2) can be rewritten as

$$\frac{d\xi}{dz} = -\frac{e}{v_z}(\vec{v}\vec{E}) = -\frac{e}{v_z}v_\perp E_\theta = e\frac{v_\perp}{v_z\kappa}\mathrm{Re}\left\{C\sum_l J_l'(\xi)L_l e^{i(\omega t - k_z z - i\theta)}\right\}.$$

After averaging over fast gyrations this yields

$$\frac{d\xi}{dz} = \frac{ev_\perp}{\kappa v_z}\mathrm{Re}\left\{CJ_s'(\xi)L_s e^{-i\vartheta}\right\}.$$

So, again, both nonaveraged and gyroaveraged equations for the electron energy and axial momentum yield the same autoresonance integral (1.17). Correspondingly, with the use of the general relation between electron energy and total momentum, we can express the axial and orbital components of the momentum via normalized electron energy as

$$p_z = p_{z0}(1 - bw), \qquad p_\perp = p_{\perp 0}(1 - w)^{1/2}, \tag{A3.11}$$

where $w = 2\frac{1-n\beta_{z0}}{\beta_{\perp 0}^2}\frac{\gamma_0 - \gamma}{\gamma_0}$. This derivation is given in Sec. 7.1. As a result, below we shall consider only equations for electron energy w and slowly variable gyrophase ϑ in which orbital and axial components of the momentum should be expressed in terms of w.

Let us start from a more complicated equation for the gyrophase (A3.9). In the LHS of this equation there is a variable detuning of the cyclotron resonance. In accordance with Sec. 1.3, this detuning can be represented as a sum of two terms, the first of which determines the initial cyclotron resonance detuning at the entrance to the interaction space, while the second one describes the effect of the changes in electron energy on the cyclotron resonance condition. So,

$$\begin{aligned}\frac{\omega - k_z v_z - s\Omega}{v_z} &= \frac{\gamma_0}{v_z\gamma}\left(\omega\frac{\gamma}{\gamma_0} - k_z\frac{p_z}{m\gamma_0} - s\Omega_0\right)\\ &= \frac{\gamma_0}{v_z\gamma}\left[\omega - k_z v_{z0} - s\Omega_0 - \omega(1 - n^2)\frac{\gamma_0 - \gamma}{\gamma_0}\right]\end{aligned}$$

(In transforming this detuning we used (A3.11).) Here we have in the RHS the ratio $\gamma_0/v_z\gamma$, which is equal to $1/\beta_{z0}(1-bw)$. So, finally this detuning can be represented as

$$\frac{\omega - k_z v_z - s\Omega}{v_z} = \frac{\omega}{c\beta_{z0}(1-bw)}\left(\frac{\omega - k_z v_{z0} - s\Omega_0}{\omega} - \mu w\right), \tag{A3.12}$$

where parameter $\mu = \beta_{\perp 0}^2(1-n^2)/2(1-n\beta_{z0})$ characterizes the effect of the changes in electron energy on the cyclotron resonance conditions. In the RHS of (A3.9) there is an expression in square brackets, which can be rewritten as

$$\beta_\perp - \frac{s}{\kappa\xi}(1-n\beta_z) = \frac{1}{\kappa}\left[\kappa\beta_\perp - \frac{s(1-n\beta_z)}{\xi}\right] = \frac{1-n\beta_z}{s\kappa}\left(\xi - \frac{s^2}{\xi}\right). \tag{A3.13}$$

Here we expressed the normalized orbital velocity $\beta_\perp$ via the normalized gyroradius ξ and used the cyclotron resonance condition $\Omega/\omega \approx (1-n\beta_z)/s$. Now, as in Appendix 1, we can use the Bessel equation, which in (A3.9) yields

$$\left(\xi - \frac{s^2}{\xi}\right)J_s(\xi) = -\frac{d}{d\xi}\left[\xi\frac{dJ_s(\xi)}{d\xi}\right]. \tag{A3.14}$$

The ratio $(1-n\beta_z)/\beta_z$, which appears in the RHS of (A3.9) after using (A3.13), can be rewritten as

$$\frac{1-n\beta_z}{\beta_z} = \frac{\gamma - np_z'}{p_z'} = \frac{1}{p_z'}\{\gamma - \gamma_0 + \gamma_0 - n[\gamma_0\beta_{z0} - n(\gamma_0 - \gamma)]\}$$

$$= \frac{\gamma_0(1-n\beta_{z0})}{p_z'}\left[1 - \frac{1-n^2}{1-n\beta_{z0}}\frac{\gamma_0-\gamma}{\gamma_0}\right].$$

Here the last term in square brackets, in accordance with (1.25), can be neglected. The primed axial momentum is here normalized to mc. All these steps allow us to rewrite (A3.9) as

$$\frac{d\vartheta}{dz'} = \frac{1}{\beta_{z0}(1-bw)}\left\{\mu w - \Delta + \frac{1-n\beta_{z0}}{\gamma_0\beta_{\perp 0}(1-w)^{1/2}}\mathrm{Im}\left[\frac{eC}{mc\omega\kappa}L_s e^{-i\vartheta}\right]\frac{d}{d\xi}(\xi J_s')\right\}. \tag{A3.15}$$

Here we denoted the initial cyclotron resonance detuning $(\omega - k_z v_{z0} - s\Omega_0)/\omega$ by Δ and introduced the normalized axial coordinate $z' = \omega z/c$.

Correspondingly, the equation for the normalized electron energy can be rewritten as

$$\frac{dw}{dz'} = -\frac{2(1-n\beta_{z0})}{\gamma_0\beta_{z0}\beta_{\perp 0}}\frac{(1-w)^{1/2}}{1-bw}J_s'(\xi)\mathrm{Re}\left\{\frac{eC}{mc\omega\kappa}L_s e^{-i\vartheta}\right\}. \tag{A3.16}$$

Introducing a set of normalized parameters $\mu' = \mu/\beta_{z0}$, $\Delta' = \Delta/\beta_{z0}$ and $C' = eC(1-n\beta_{z0})/mc\omega\kappa\gamma_0\beta_{z0}\beta_{\perp 0}$ allows one to reduce these equations to the following form:

$$\frac{d\vartheta}{dz} = \frac{1}{1-bw}\left\{\mu w - \Delta + \frac{1}{(1-w)^{1/2}}\text{Im}\left[CL_s e^{-i\vartheta}\right]\frac{d}{d\xi}(\xi J_s')\right\}, \quad \textbf{(A3.17)}$$

$$\frac{dw}{dz} = -2\frac{(1-w)^{1/2}}{1-bw}J_s'(\xi)\text{Re}\left\{CL_s e^{-i\vartheta}\right\}. \quad \textbf{(A3.18)}$$

Here primes are omitted (except for the derivative of the Bessel function).

Let us note that in the RHS of (A3.17) and (A3.18) we have the coupling coefficient $L_s(X, Y)$, which can be different for electrons with different guiding centers. So, we can integrate these equations for a beam with different operators L_s for different beamlets and then, in the equation for wave excitation (A2.23), average the source term over the beam distribution in guiding centers. This corresponds to the integration over the cross-section area of the interaction region in the RHS of (A2.23).

Let us now consider a simple case of a cylindrical waveguide with a thin annular electron beam, in which all beamlets, in accordance with (A1.16), have the same absolute value of L_s. In such a case we can introduce a new normalized amplitude of the wave,

$$F = CJ_{m\mp s}(k_\perp R)e^{i\Delta z}. \quad \textbf{(A3.19)}$$

This reduces (A3.17) and (A3.18) to, respectively,

$$\frac{d\vartheta}{dz} = \frac{1}{1-bw}\left\{\mu w + \frac{1}{(1-w)^{1/2}}\frac{d}{d\xi}(\xi J_s')\text{Im}[Fe^{-i\vartheta}]\right\}, \quad \textbf{(A3.20)}$$

$$\frac{dw}{dz} = -2\frac{(1-w)^{1/2}}{1-bw}J_s'(\xi)\text{Re}\{Fe^{-i\vartheta}\}. \quad \textbf{(A3.21)}$$

Note that, in accordance with the definition of the operator L_s given by (A1.16) and the amplitude F given by (A3.19), we use in (A3.20), (A3.21), and below a new slowly variable phase,

$$\vartheta' = \vartheta + \Delta z - (s \mp m)\psi \quad \textbf{(A3.22)}$$

$$= s(\theta - \Omega_0\tau) - \omega t_0 + k_z\int_0^\tau (v_z - v_{z0})d\tau' - (s \mp m)\psi.$$

So, at the entrance, this phase has an initial value $\vartheta'(0) = s\theta_0 - \omega t_0 - (s \mp m)\psi$, which for an unmodulated electron beam is uniformly distributed in all beamlets from 0 to 2π.

Now we can rewrite the equation for wave excitation (A2.23) in these new variables. To do this, let us use the representation of the beam current density by (A2.5) and representation of the function describing the spatial structure of the wave electric field given by (A2.19). Then, using in (A2.23) the charge conservation law discussed in Sec. 3.3, making the Fourier transform of the membrane function (see definition of e_θ after (A3.7) above), and averaging this equation over fast gyration result in the following equation:

$$\frac{dC}{dz} = \frac{1}{N\kappa} \int_{S_\perp} j_0 \left\{ \frac{1}{\pi} \int_0^{2\pi} \frac{p_\perp}{p_z} J_s'(\xi) L_s^* e^{i\vartheta} d\vartheta_0 \right\} ds_\perp. \tag{A3.23}$$

In this equation the phase ϑ is the same as that used in (A3.17)–(A3.18) and above. In the normalized variables used in (A3.17) and (A3.18) this equation can be rewritten as

$$\frac{dC}{dz} = -I_0 \int_{S_\perp} f(\vec{R}_{\perp 0}) \left\{ \frac{1}{2\pi} \int_0^{2\pi} \frac{(1-w)^{1/2}}{1-bw} J_s'(\xi) L_s^* e^{i\vartheta} d\vartheta_0 \right\} ds_\perp. \tag{A3.24}$$

Here the beam current density at the entrance was represented as $j_0 = -|I_b| f(\vec{R}_{\perp 0})$, where the function $f(\vec{R}_{\perp 0})$ describing the initial beam distribution over the guiding center radii is normalized to one: $\int_{S_\perp} f(\vec{R}_{\perp 0}) ds_\perp = 1$ and the normalized beam current parameter I_0 is equal to $2(e|I_b|/mc^3)[(1 - n\beta_{z0})/\gamma_0 \beta_{z0}^2 \kappa^2][c^3/\omega^2 N]$, where the wave norm is given by (A2.22). So, Eqs. (A3.17), (A3.18), and (A3.24) form a self-consistent set of equations describing the gyro-TWT with an arbitrary geometry of the waveguide and the beam. Of course, when it is necessary to take into account the velocity spread in the beam, we should add in the wave excitation equation the function describing the electron velocity spread. Correspondingly, there should be an additional averaging of the source term over the velocity distribution.

In the case of a thin annular electron beam in a cylindrical waveguide, for which the equations for electron motion were reduced above to (A3.20) and (A3.21), the wave excitation equation can also be reduced further. For the normalized amplitude given by (A3.19), this equation has the following form:

$$\frac{dF}{dz} - i\Delta F = -I_0 \frac{1}{2\pi} \int_0^{2\pi} \frac{(1-w)^{1/2}}{1-bw} J_s'(\xi) e^{i\vartheta} d\vartheta_0. \tag{A3.25}$$

So, the self-consistent set of equations for the axially symmetric gyro-TWT with a thin annular electron beam consists of Eqs. (A3.20), (A3.21), and (A3.25). In the last equation the normalized beam current parameter is equal to

$$I_0 = \frac{e|I_b|}{mc^3} \frac{1 - h\beta_{z0}}{\gamma_0 \beta_{z0}^2} \frac{4\kappa^2}{h} \frac{J_{m\mp s}^2(k_\perp R_0)}{(\nu^2 - m^2) J_m^2(\nu)}. \tag{A3.26}$$

In deriving this expression we used the expression for the norm of a wave in a cylindrical waveguide given by (A2.22). Note that the detuning Δ in the LHS of (A3.25) contains the axial wavenumber, which in the presence of the wave attenuation in the waveguide walls also has the imaginary part.

For typical operating conditions we can take one more step to further simplify this set of equations. Indeed, this set of equations contains Bessel functions and their derivatives, which have the argument $\xi = k_\perp a = \kappa(\omega/\Omega_0)\beta_{\perp 0}(1-w)^{1/2}$, which with the use of the cyclotron resonance condition can be rewritten as

$$\xi = \kappa \frac{s\beta_{\perp 0}}{1 - n\beta_{z0}}(1-w)^{1/2} = \xi_0(1-w)^{1/2}. \tag{A3.27}$$

This argument usually does not exceed the order of the Bessel function s. Therefore, Bessel function can be represented by the polynomial $J_s(\xi) \approx \frac{1}{s!}(\xi/2)^s$. Correspondingly,

$$dJ_s/d\xi \simeq [1/(s-1)!2^s]\xi^{s-1} \quad \text{and} \quad \frac{d}{d\xi}[\xi J_s'(\xi)] \simeq [s/(s-1)!2^s]\xi^{s-1}.$$

So, we can now introduce the final set of normalized variables:

$$\varsigma = \mu z = \frac{\beta_{\perp 0}^2(1-n^2)}{2\beta_{z0}(1-n\beta_{z0})}\frac{\omega z}{c}, \tag{A3.28}$$

$$\Delta' = \frac{\Delta}{\mu} = \frac{2(1-n\beta_{z0})}{\beta_{\perp 0}^2(1-n^2)}\frac{\omega - k_z v_{z0} - s\Omega_0}{\omega},$$

$$F' = \frac{\xi_0^{s-1}}{(s-1)!2^s}\frac{1}{\mu}F = \left(\frac{\kappa s\beta_{\perp 0}}{1-n\beta_{z0}}\right)^{s-1}\frac{1}{(s-1)!2^s}\frac{2\beta_{z0}(1-n\beta_{z0})}{\beta_{\perp 0}^2(1-n^2)}F,$$

$$I_0' = \left[\frac{\xi_0^{s-1}}{(s-1)!2^s}\right]^2\frac{I_0}{\mu^2},$$

and rewrite these equations as

$$\frac{d\vartheta}{d\varsigma} = \frac{1}{1-bw}\{w + s(1-w)^{(s/2)-1}\operatorname{Im}(Fe^{-i\vartheta})\}, \tag{A3.29}$$

$$\frac{dw}{d\varsigma} = -2\frac{(1-w)^{s/2}}{1-bw}\operatorname{Re}(Fe^{-i\vartheta}), \tag{A3.30}$$

$$\frac{dF}{d\varsigma} - i\Delta F = -I_0\frac{1}{2\pi}\int_0^{2\pi}\frac{(1-w)^{s/2}}{1-bw}e^{i\vartheta}\,d\vartheta_0. \tag{A3.31}$$

All primes are omitted here. The boundary conditions for the electron energy and phase were discussed above. The boundary condition for the forward wave amplitude at the entrance is $F(0) = F_0$. To establish the relation between this amplitude and the input power, recall that in our notations the wave power in the cylindrical waveguide is equal to

$$P = \frac{c}{8}|C|^2 \frac{1}{k^2} \frac{h}{\kappa^4} (\nu^2 - m^2) J_m^2(\nu). \tag{A3.32}$$

Here we used the field representation given after (A3.3) and the wave norm determined by (A2.22) in combination with the standard expression for the wave power in a waveguide. Now we should take into account that on the way from the amplitude C to the normalized amplitude F', which we wrote in (A3.31) without prime, we made three steps. First, after (A3.16) we introduced the normalized amplitude C' proportional to the original amplitude C. Then we introduced the normalized amplitude F given in terms of C' by (A3.19), and finally we introduced the new normalized amplitude F', which was expressed in terms of F by (A3.28). So, after making all the corresponding steps, one gets the following relation:

$$|F'| = 4 \left[2 \frac{e^2 P}{m^2 c^5} \frac{G}{n} \right]^{1/2} \frac{(s\kappa)^{s-1} \beta_{\perp 0}^{s-4} (1 - n\beta_{z0})^{3-s}}{\gamma_0 \kappa (s-1)! 2^s}. \tag{A3.33}$$

Here we used the coupling parameter G given by (3.59) and took into account that $1 - n^2 = \kappa^2$. The ratio m^2c^5/e^2 in square brackets is equal to $8.687 \cdot 10^6$ kW (this is the product of $mc^3/e = 17.04$ kA and $mc^2/e = 511$ kV). Therefore, for the power expressed in kW, (A3.33) can be rewritten as

$$|F'| = 1.92 \cdot 10^{-3} (PG/n)^{1/2} \frac{(s\kappa)^{s-1} \beta_{\perp 0}^{s-4} (1 - n\beta_{z0})^{3-s}}{\gamma_0 \kappa (s-1)! 2^s}. \tag{A3.34}$$

In the case of operation at the fundamental cyclotron resonance, (A3.34) reduces to

$$|F'| = 0.96 \cdot 10^{-3} (PG/n)^{1/2} \frac{(1 - n\beta_{z0})^2}{\kappa \gamma_0 \beta_{\perp 0}^3}. \tag{A3.35}$$

References

Abbreviations: RQE = Radiophysics and Quantum Electronics; REEP = Radio Eng. and Electron Phys.

Abramowitz, Milton, and I. M. Stegun. 1964. *Handbook of Mathematical Functions.* Ch. 9. New York: Dover.

Abrams, Richard H., B. Levush, A. A. Mondelli, and R. K. Parker. 2001. Vacuum electronics for the 21st century, *IEEE Microwave Magazine,* Sept. 2001, 61–72.

Adler, Robert. 1946. A study of locking phenomena in oscillators, *Proc. IRE,* 34: 351–57; this paper was also reprinted in *Proc. IEEE,* 61: 1380–5 (1973).

Airila, Markus I., O. Dumbrajs, A. Reinsfelds, and U. Strautins. 2001.Nonstationary oscillations in gyrotrons, *Phys. Plasmas,* 8: 4608–12.

Alikaev, Vladimir V., G. A. Bobrovskiy, M. M. Ofitserov, V. I. Poznyak, and K. A. Razumova. 1972. Electron cyclotron heating at the TM-3 tokamak, *Pis'ma ZhETF,* 15: 41–44 (*JETP Lett.*, 15: 27–29).

Alikaev, Vladimir V., V. A. Flyagin, V. I. Khizhnyak, A. G. Luchinin, G. S. Nusinovich, V. G. Usov, and S. N. Vlasov. 1978. Gyrotrons for electron-cyclotron plasma heating in large tokamaks. First Symposium on Heating in Toroidal Plasmas, Grenoble, France, July 3–7, 1978. *Proc. Symp.,* ed. T. Consoli, Association Euroaton-CEA, Pergamon Press, vol. 2, 339–49 (1979).

Andronov, Alexander A., and A. A. Vitt. 1934. To the mathematical theory of auto-oscillating systems with two degrees of freedom, *Zh. Tech. Fiz.*, 4: 122–43 (in Russian).

Andronov, Alexander A., V. A. Flyagin, A. V. Gaponov, A. L. Goldenberg, M. I. Petelin, V. G. Usov, and V. K. Yulpatov. 1978. The gyrotron: High-power source of millimetre and submillimetre waves, *Infrared Physics,* 18: 385–93.

Antakov, Igor I., V. M. Bokov, R. P. Vasilyev, and A. V. Gaponov. 1960. Interaction of the trochoidal electron beam with electromagnetic waves in a rectangular waveguide, *Izv. VUZov, Radiofiz.*, 3: 1033–44 (in Russian).

Antakov, Igor I., A. V. Gaponov, O. V. Malygin, and V. A. Flyagin. 1966. Application of induced cyclotron radiation of electrons for the generation and amplification of high-power electromagnetic waves, *Radiotekh. I Elektron.*, 11: 2254–57 (*REEP,* 11: 1995–97).

Antakov, Igor I., A. V. Gaponov, E. V. Zasypkin, et al. 1993. Gyroklystrons—millimeter wave amplifiers of the highest power, *Proc. 3rd Int. Workshop "Strong microwaves in plasmas,"* Moscow-N. Novgorod-Moscow, Aug. 15-22, 1993, ed. A. G. Litvak, Inst. Appl. Phys., Academy of Science of the USSR, Nizhny Novgorod, 1994, 2: 587–96.

Antakov, Igor I., E. V. Sokolov, and E. V. Zasypkin. 1993. Design and performance of 94-GHz high power multi-cavity gyroklystron amplifier, *Proc. 3rd Int. Workshop "Strong microwaves in plasmas,"* Moscow-N. Novgorod-Moscow, Aug. 15–22, 1993, ed. A. G. Litvak, Inst. Appl. Phys., Academy of Science of the USSR, Nizhny Novgorod, 1994, 2: 754–58.

Antakov, Igor I., S. P. Belov, L. I. Gershtein, V. A. Gintsburg, A. F. Krupnov, and G. S. Parshin. 1975. Use of high resonant-radiation powers to increase the sensitivity of microwave spectroscopes, *Pis'ma ZhETF,* 19: 634–37 (*JETP Lett.*, 19: 329–30).

Antakov, Igor I., S. N. Vlasov, V. A. Gintsburg, L. I. Zagryadskaya, and L. V. Nikolaev. 1975. CRM-oscillators with mechanical tuning of frequency, *Elektron. Tekhn.*, Ser. 1, *Elektron. SVCh*, no. 8: 20–25.

Antakov, Igor I., A. V. Gaponov, V. A. Gintsburg, A. L. Goldenberg, M. I. Petelin, and V. K. Yulpatov. 1975. An amplifier of electromagnetic oscillations in centimeter, millimeter and sub-millimeter wavelength regions. Copyright No. 302050 with priority of June 16, 1967. Official Bulletin KDIO of SM USSR, no. 41: 205 (1975).

Antakov, Igor I., L. A. Aksenova, E. V. Zasypkin, et al. 1990. Multi-cavity phase-locked gyrotrons for low-hybrid heating in toroidal plasmas, *Proc. Int. Workshop "Strong microwaves in plasmas,"* Suzdal, Russia, Sept. 18 - 23, 1990, ed. A. G. Litvak, Inst. Appl. Phys., Academy of Science of the USSR, Nizhny Novgorod, 1991, 2: 773–78.

Antonsen, Thomas M., Jr., and W. M. Manheimer. 1998. Shot noise in gyroklystrons, *IEEE-PS,* 26: 444–50.

Antonsen, Thomas M., Jr., B. Levush, and W. M. Manheimer. 1990. Stable single mode operation of a quasioptical gyrotron, *Phys. Fluids* B, 2: 419–26.

Antonsen, Thomas M., Jr., S. Y. Sai, and G. S. Nusinovich. 1992. Effect of window reflection on gyrotron operation, *Phys. Fluids* B, 4: 4131–39.

Appleton, E. V. 1924. On the anomalous behaviour of a vibration galvanometer, *Phil. Mag.* S.6, 47: 609–19.

Arjona, Melany R., and W. G. Lawson. 2000. Design of a 34-GHz second-harmonic coaxial gyroklystron experiment for accelerator applications, *IEEE-PS,* 28: 700–705.

Baird, J. Mark, and W. Lawson. 1986. Magnetron injection gun (MIG) design for gyrotron applications, *Int. J. Electron.*, 61: 953–67.

Baird, J. Mark, L. R. Barnett, R. W. Grow, and R. C. Freudenberger. 1987. Harmonic auto-resonant peniotron (HARP) interactions, *IEDM*–87: 913–16.

Barker, Robert J., and E. Schamiloglu, eds. 2001. *High-Power Microwave Sources and Technologies*. New York: IEEE Press.

Barnett, Larry R., Y. Y. Lau, K. R. Chu, and V. L. Granatstein. 1981. An experimental wide-band gyrotron traveling-wave amplifier, *IEEE-ED,* 28: 872–75.

Basu, B. N. 1995. *Electromagnetic Theory and Applications in Beam-Wave Electronics,* Sec. 8.4. Singapore: World Scientific.

Bates, D. J., and E. L. Ginzton. 1957. A traveling-wave frequency multiplier, *Proc. IRE,* 45: 938–44.

Belousov, Vladimir I., V. S. Ergakov, and M. A. Moiseev. 1978. Two-cavity CRM at harmonics of the electron cyclotron frequency, *Elektron. Tekhnika, Ser. I, Elektronika* SVCh, no. 9: 41–51 (in Russian).

Bezruchko, Boris P., S. P. Kuznetsov, and D. I. Trubetskov. 1979. Experimental observation of stochastic self-oscillations in the electron beam—backscattered electromagnetic wave dynamic system, *Pis'ma Zh. Eksp. Teor. Fiz.*, 29: 180–84 (*JETP Lett.*, 29: 162–65).

Blank, Monica, B. G. Danly, and B. Levush, 1999a. Experimental demonstration of a W-band (94 GHz) gyrotwystron amplifier, *IEEE-PS,* 27: 405–11.

———. et al. 1999b. Demonstration of a 10 kW average power 94 GHz gyroklystron amplifier, *Phys. Plasmas,* 6: 4405–9.

Blank, Monica, E. V. Zasypkin, and B. Levush. 1998. An investigation of X-band gyrotwystron amplifiers, *IEEE-PS,* 26: 577-81.

Blank, Monica, B. G. Danly, B. Levush, P. E. Latham, and D. E. Pershing, 1997. Experimental demonstration of a W-band gyroklystron amplifier, *Phys. Rev. Lett.*, 79: 4485–88.

Bollen, W. M., A. H. McCurdy, B. Arfin, R. K. Parker, and A. K. Ganguly. 1985. Design and performance of a three-cavity gyroklystron amplifier, *IEEE-PS,* 13: 417–23.

Borie, Edit, and B. Jodicke. 1987. Startup and mode competition in a 150 GHz gyrotron, *Int. J. IR and MM Waves,* 8: 207–26.

Bott, Ian B. 1964. Tunable source of millimeter and submillimeter electromagnetic radiation, *Proc. IEEE,* 52: 330–32.

Botton, Moti, T. M. Antonsen, Jr., B. Levush, K. T. Nguyen, and A. N. Vlasov. 1998. MAGY: A Time-Dependent Code for Simulation of Slow and Fast Microwave Sources, *IEEE-PS,* 26: 882–92.

Bovsheverov, V. M. 1936. On some oscillation problems leading to functional equations, *Zh. Tech. Fiz.,* 6: 1480–89 (in Russian).

Brand, G. Ferguson, N. G. Douglas, M. Gross, J. Y. L. Ma, L. C. Robinson, and Chen Zhiyi. 1982. Tuneable millimeter-wave gyrotrons, *Int. J. Infrared and Millimeter Waves,* 3: 725–34.

Bratman, Vladimir L. 1975. Instability of orbital motion in a layer of electrons rotating in a uniform magnetic field. I, *Zh. Tech. Fiz.,* 45: 1591–96 (*Sov. Phys. Tech. Phys.,* 20: 1017–19).

Bratman, Vladimir L. 1976. Instability of orbital motion in a layer of electrons rotating in a uniform magnetic field. II, *Zh. Tech. Fiz.,* 46: 2030–36 (*Sov. Phys. Tech. Phys.,* 21: 1188–92).

Bratman, Vladimir L., and M. A. Moiseev. 1975. Conditions for self-excitation of a cyclotron-resonant maser with a non-resonant electrodynamic system, *Izv. VUZov Radiofiz.,* 18: 1045–55 (*RQE,* 18: 772–79).

Bratman, Vladimir L., and M. I. Petelin. 1975. Optimizing the parameters of high-power gyromonotrons with RF field of nonfixed structure, *Izv. VUZov, Radiofiz.,* 18: 1538–43 (*RQE,* 18: 1136–40).

Bratman, Vladimir L., and A. E. Tokarev. 1974. On the theory of the relativistic cyclotron-resonance maser, *Izv. VUZov, Radiofiz.,* 17: 1224–28 (*RQE,* 17: 932–35).

Bratman, Vladimir L., N. S. Ginzburg, and G. S. Nusinovich. 1977. Theory of the relativistic gyrotron, *Pis'ma Zh. Tekh. Fiz.,* 3: 961–65 (*Sov. Tech. Phys. Lett.,* 3: 395–96).

Bratman, Vladimir L., M. A. Moiseev, M. I. Petelin, and R. E. Erm. 1973. Theory of gyrotrons with a nonfixed structure of the high-frequency field, *Izv. VUZov, Radiofiz.,* 16: 622–30 (*RQE,* 16: 474–80).

Bratman, Vladimir L., N. S. Ginzburg, G. S. Nusinovich, M. I. Petelin, and V. K. Yulpatov. 1979. Cyclotron and synchrotron masers, in *Relativistic High-Frequency Electronics,* 157–216. Gorkiy, USSR: IAP (in Russian).

Bratman, Vladimir L., N. S. Ginzburg, G. S. Nusinovich, M. I. Petelin, and P. S. Strelkov. 1981. Relativistic gyrotrons and cyclotron autoresonance masers, *Int. J. Electron.,* 51: 541–68.

Bratman, Vladimir L., G. G. Denisov, N. S. Ginzburg, and M. I. Petelin. 1983. FEL's with Bragg reflection resonators: Cyclotron autoresonance masers versus ubitrons, *IEEE-QE,* 19: 282–96.

Bratman, Vladimir L., G. G. Denisov, S. D. Korovin, M. M. Ofitserov, S. D. Polevin, and V. V. Rostov. 1984. Relativistic millimeter-wave generators, in *Relativistic High-Frequency Electronics,* ed. A. V. Gaponov-Grekhov, 4: 119–76. Nizhny Novgorod, USSR: IAP (in Russian).

Bratman, Vladimir L., G. G. Denisov, B. D. Kol'chugin, S. V. Samsonov, and A. B. Volkov. 1995. Experimental demonstration of high-efficiency cyclotron-autoresonance-maser operation, *Phys. Rev. Lett.,* 75: 3102–5.

Bratman, Vladimir L., A. W. Cross, G. G. Denisov, et al. 2000. High-gain wide-band gyrotron traveling wave amplifier with a helically corrugated waveguide, *Phys. Rev. Lett.,* 84: 2746–49.

Briggs, Richard J. 1964. Electron-stream interaction with plasmas, *Research Monograph,* no. 29, Cambridge, MA: MIT Press.

Bykov, Yuri V., A. L. Goldenberg, L. V. Nikolaev, M. M. Ofitserov, and M. I. Petelin. 1975. Experimental investigation of a gyrotron with whispering-gallery modes, *Izv. VUZov, Radiofiz.*, 18: 1544-47 (*RQE*, 18: 1141–43).

Calame, Jeff P., B. G. Danly, and M. Garven. 1999. Measurements of intrinsic shot noise in a 35 GHz gyroklystron, *Phys. Plasmas,* 6: 2914–25.

Calame, Jeff P., M. Garven, J. J. Choi, et al. 1999. Experimental studies of bandwidth and power production in a three-cavity, 35 GHz gyroklystron amplifier, *Phys. Plasmas,* 6: 285–97.

Charbit, P., A. Herscovici, and G. Mourier. 1981. A partly self-consistent theory of the gyrotron, *Int. J. Electron,* 51: 303–30.

Chen, S. H., K. R. Chu, and T. H. Chang. 2000. Saturated behavior of the gyrotron backward-wave oscillator, *Phys. Rev. Lett.*, 85: 2633–36.

Chodorow, Marvin, and C. Susskind. 1964. *Fundamentals of Microwave Electronics.* 17. New York-San Francisco-London: McGraw-Hill.

Chong, C. K., D. B. McDermott, M. M. Razeghi, N. C. Luhmann, Jr., J. Pretterebner, D. Wagner, M. Thumm, M. Caplan, and B. Kulke. 1992. Bragg Reflectors, *IEEE-PS,* 20: 393–402.

Chu, Kwo R., and D. Dialetis. 1984. Theory of harmonic gyrotron oscillator with slotted resonant structure, *Int. J. IR and MM Waves,* 5: 37–56.

Chu, Kwo R., and J. Hirshfield. 1978. Comparative study of the axial and azimuthal bunching mechanisms in electromagnetic instabilities, *Phys. Fluids,* 21: 461–66.

Chu, Kwo R., and A. T. Lin. 1988. Gain and bandwidth of the gyro-TWT and CARM amplifier, *IEEE-PS,* 16: 90–104.

Chu, Kwo R., H. Guo, and V. L. Granatstein. 1997. Theory of the harmonic multiplying gyrotron traveling wave amplifier, *Phys. Rev. Lett.*, 78: 4661–64.

Chu, Kwo R., M. E. Read, and A. K. Ganguly. 1980. Methods of efficiency enhancement and scaling for the gyrotron oscillator, *IEEE-MTT,* 28: 318-25.

Chu, Kwo R., Y. Y. Lau, L. R. Barnett, and V. L. Granatstein. 1981a. Theory of a wide-band distributed gyrotron traveling-wave amplifier, *IEEE-ED,* 28: 866–71.

Chu, Kwo R., A. K. Ganguly, V. L. Granatstein, J. L. Hirshfield, S. Y. Park, and J. M. Baird. 1981b. Theory of a slow wave cyclotron amplifier, *Int. J. Electron.*, 51: 493–502.

Chu, Kwo R., V. L. Granatstein, P. E. Latham, W. Lawson, and C. D. Striffler. 1985. A 30-MW gyroklystron-amplifier design for high-energy linear accelerator, *IEEE-PS,* 13: 424–34.

Chu, Kwo R., P. E. Latham, and V. L. Granatstein. 1988. Penultimate cavity tuning of the gyroklystron amplifier, *Int. J. Electron.*, 65: 419–28.

Chu, Kwo R., L. R. Barnett, W. K. Lau, L. H. Chang, and H. Y. Chen. 1990. A wide-band millimeter-wave gyrotron traveling-wave amplifier experiment, *IEEE-ED,* 37: 1557–60.

Chu, Kwo R., L. R. Barnett, H. Y. Chen, et al. 1995. Stabilizing of absolute instabilities in gyrotron traveling-wave amplifier, *Phys. Rev. Lett.*, 74: 1103–6.

Chu, Kwo R., et al. 1999. Theory and experiment of ultrahigh-gain gyrotron traveling wave amplifier, *IEEE-PS,* 27: 391–404.

Collin, Robert E. 1966. *Foundations for Microwave Engineering.* New York: McGraw-Hill.

Cooke, Simon J., and G. G. Denisov. 1998. Linear theory of a wide-band gyro-TWT amplifier using spiral waveguide, *IEEE-PS,* 26: 519–30.

Dammertz, Gunter, O. Braz, A. K. Chopra, et al. 1999. Recent results of the 1-MW, 140-GHz, $TE_{22,6}$-mode gyrotron, *IEEE-PS,* 27: 330–39.

Danly, Bruce G., and R. J. Temkin. 1986. Generalized nonlinear harmonic gyrotron theory, *Phys. Fluids*, 29: 561–67.

Davydovskii, Vladimir Ya. 1962. On the possibility of accelerating charged particles by electromagnetic waves in a constant magnetic field, *Zh. Eksp. Teor. Fiz.*, 13: 886–88 (*Soviet Physics JETP*, 16 (1963): 629–30).

Denisov, Gregory G., A. N. Kuftin, V. I. Malygin, N. P. Venedictov, D. V. Vinogradov, and V. E. Zapevalov. 1992. 110 GHz gyrotron with a built-in high-efficiency converter, *Int. J. Electronics*, 72: 1079–92.

Denisov, Gregory G., V. L. Bratman, A. D. R. Phelps, and S. V. Samsonov. 1998. Gyro-TWT with a helical operating waveguide: New possibilities to enhance efficiency and frequency bandwidth, *IEEE-PS*, 26: 508–18.

Destler, William W., E. Chojnacki, R. F. Hoeberling, W. Lawson, A. Singh, and C. D. Striffler. 1988. High-power microwave generation from large-orbit devices, *IEEE-PS*, 16: 71–89.

Didenko, Andrei N., A. R. Borisov, G. P. Fomenko, A. S. Shlapakovskii, and Yu. G. Shtein. 1983. Cyclotron maser using the anomalous Doppler effect, *Pis'ma ZhETF*, 9: 1331–32 (*Sov. Lett. Phys. Lett.*, 9: 572–73).

Drobot, Adam T., and K. Kim. 1981. Space charge effects on the equilibrium of guided electron flow with gyromotion, *Int. J. Electron.*, 51: 351–68.

Dumbrajs, Olgierd, and G. S. Nusinovich. 1992. Theory of a frequency-step-tunable gyrotron for optimum plasma ECRH, *IEEE-PS*, 20: 452–57.

Dumbrajs, Olgierd, and G. S. Nusinovich. 1997. Effect of technical noise on radiation linewidth in free-running gyrotron oscillators, *Phys. Plasmas*, 4: 1413–23.

Dumbrajs, Olgierd, V. I. Khizhnyak, A. B. Pavelyev, B. Pyosczyk, and M. K. Thumm. 2000. Design of rapid-frequency step-tunable powerful coaxial-cavity harmonic gyrotrons, *IEEE-PS*, 28: 681–87.

Eckstein, J. N., D. W. Latshaw, and D. S. Stone. 1983. 95GHz Gyro-Traveling Wave Tube, Varian Associates Final Report, Contract DASG60-79-C-005 MOD P003 (BMDACT), November 1983.

Erckmann, Volker, G. Dammertz, D. Dorst, et al. 1999. ECRH and ECCD with high power gyrotrons at the stellarators W7-AS and W7-X, *IEEE-PS*, 27: 538–46.

Ergakov, Vsevolod S., and M. A. Moiseev. 1975. Theory of synchronization of oscillations in a cyclotron-resonance maser monotron by an external signal, *Izv. VUZov Radiofiz.*, 18: 120–31 (*RQE*, 18: 89–97).

Ergakov, Vsevolod S., M. A. Moiseev, and R. E. Erm. 1980. Effect of the electron velocity spread on gyrotron characteristics, *Elektron. Tekhn., Ser. I, Elektron. SVCh*, no.3, 20–27 (in Russian).

Ergakov, Vsevolod S., M. A. Moiseev, and A. A. Shaposhnikov. 1977. Signal fluctuations in a CRM-monotron caused by electron beam shot noise, *Radiotekhnika I Elektron.*, 22: 2154–61 (*REEP*, 22:107).

———. 1983. Low-frequency fluctuations in oscillations in a gyrotron due to thermal noise, *Radiotekhnika I Elektron.*, 28: 1347–53 (*REEP*, 28: 85–90).

Felch, Kevin, and R. Temkin. 2003. CPI 140 GHz gyrotron for Wendelstein 7X achieves 154 second pulsed operation at 500 kW power level, *VLT News*, 6, no.1, April 2003.

Felch, Kevin, R. Bier, L. Fox, et al. 1984. A 60 GHz, 200 kW CW gyrotron with a pure output mode, *Int. J. Electron.*, 57: 815–20.

Felch, Kevin, T. S. Chu, J. Feinstein, et al. 1992. Long-pulse operation of a gyrotron with beam/RF separation, Paper T4.1 at the 17th Int. Conf. on IR and MM Waves, Dec. 14–17, 1992, Pasadena, CA.

Felch, Kevin, M. Blank, P. Borchard, et al. 1996. Long-pulse and CW tests of a 110-GHz gyrotron with an internal quasi-optical converter, *IEEE-PS*, 24: 558–69.

Felch, Kevin, M. Blank, P. Borchard, et al. 2002. Progress update on CPI 500 kW and 1 MW, multi-second-pulsed gyrotrons, 3rd Int. Conf. IVEC-2002, 332–333, April 23–25, 2002, Monterey, CA.

Ferguson, Patrick E., G. Valier, and R. S. Symons. 1981. Gyrotron-TWT Operating Characteristics, *IEEE-MTT,* 29: 794–99.

Fliflet, Arne W. 1986. Linear and non-linear theory of the Doppler-shifted cyclotron resonance maser based on TE and TM waveguide modes, *Int. J. Electron.,* 61: 1049–80.

Fliflet, Arne W., and W. M. Manheimer. 1989. Nonlinear theory of phase-locking gyrotron oscillators driven by an external signal, *Phys. Rev.* A, 39: 3432–43.

Fliflet, Arne W., A. J. Dudas, M. E. Read, and J. M. Baird, 1982a. Use of electrode synthesis technique to design MIG-type guns for high power gyrotrons, *Int. J. Electron.,* 53: 743–54.

Fliflet, Arne W., M. E. Read, K. R. Chu, and R. Seeley. 1982b. A self-consistent field theory for gyrotron oscillators: Application to a low Q gyromonotron, *Int. J. Electron.,* 53: 505–22.

Fliflet, Arne W., T. A. Hargreaves, R. P. Fischer, W. M. Manheimer, and P. Sprangle. 1990. Review of quasi-optical gyrotron development, *J. Fusion Energy,* 9: 31–58.

Flyagin, Valeriy A., and G. S. Nusinovich. 1988. Gyrotron oscillators, *Proc. IEEE,* 76: 644–56.

Flyagin, Valeriy A., A. L. Goldenberg, and G. S. Nusinovich. 1984. Powerful gyrotrons, in *Infrared and Millimeter Waves,* ed. K. J. Button, vol.11, 179–226. New York: Academic Press.

Flyagin, Valeriy A., A. Goldenberg, V. Zapevalov. 1992. Investigation and development of gyrotrons at the Institute of Applied Physics, Paper T4.4 at the 17th Int. Conf. on IR and MM Waves, Dec. 14–17, 1992, Pasadena, CA.

Flyagin, Valeriy A., A. G. Luchinin, and G. S. Nusinovich. 1983. Submillimeter-wave gyrotrons: Theory and experiment, *Int. J. of Infrared and Millimeter Waves,* 4: 629–38.

Flyagin, Valeriy A., A. V. Gaponov, M. I. Petelin, and V. K. Yulpatov. 1977. The gyrotron, *IEEE-MTT,* 25: 514–21.

Freund, Henry P., and T. M. Antonsen, Jr. 1996. *Principles of Free-electron Lasers,* Ch. 6. 2d ed. New York: Chapman and Hall.

Friedman, Moshe, D. A. Hammer, W. M. Manheimer, and P. Sprangle. 1973. Enhanced microwave emission due to the transverse energy of a relativistic electron beam, *Phys. Rev. Lett.,* 31: 752–55.

Galuzo, Sergei Yu., V. I. Kanavets, A. I. Slepkov, and V. A. Pletyushkin. 1982. Relativistic cyclotron accelerator exploiting the anomalous Doppler effect, *Zh. Tekh. Fiz.,* 52: 1681–83 (Sov. Phys. Tech. Phys., 27: 1030–32).

Ganguly, Achinta K., and S. Ahn. 1982. Self-consistent large signal theory of the gyrotron traveling wave amplifier, *Int. J. Electron.,* 53: 641–58.

———. 1989. Non-linear analysis of the gyro-BWO in three dimensions, *Int. J. Electron.,* 67: 261–76.

Ganguly, Achinta K., and K. R. Chu. 1981. Analysis of two-cavity gyroklystron, *Int. J. Electron.,* 51: 503–20.

———. 1984. Limiting currents in gyrotrons, *Int. J. Infrared and Millimeter Waves,* 5: 103–21.

Ganguly, Achinta K., A. W. Fliflet, and A. H. McCurdy. 1985. Theory of multi-cavity gyroklystron amplifier based on a Green's function approach, *IEEE-PS,* 13: 409–16.

Gantenbein, Gunter, H. Zohm, G. Giruzzi, et al. 2000. Complete suppression of neoclassical tearing modes with current drive at the electron-cyclotron-resonance frequency in ASDEX upgrade tokamak, *Phys. Rev. Lett.,* 85: 1242–45.

Gaponov, Andrei V. 1959a. Interaction between electron fluxes and electromagnetic waves in waveguides, *Izv. VUZov, Radiofiz.*, 2: 450–62 (in Russian).

———. 1959b. Interaction of irrectilinear electron beams with electromagnetic waves in transmission lines, *Izv. VUZov, Radiofiz.*, 2: 836–37 (in Russian).

———. 1959c. Report at the Popov Society Meeting, Moscow.

———. 1960. Instability of a system of excited oscillators with respect to electromagnetic perturbations, *ZhETF*, 39: 326–31 (*Soviet Physics JETP*, 12 (Feb.1961): 232–36).

———. 1961. Relativistic dispersion equations for waveguides with helical and trochoidal electron beams, *Izv. VUZov, Radiofiz.*, 4: 547–60 (in Russian).

Gaponov, Andrei V., and Yulpatov V. K. 1962. Interaction of electron beams with electromagnetic field in resonators, *Radiotekhn. I Elektron.*, 7: 631–42.

———. Interaction of helical electron beams with the electromagnetic field in a waveguide, *Radiotekh. I Elektron.*, 12: 627–32 (*REEP*, 12: 582–87).

Gaponov, Andrei V., M. I. Petelin, and V. K. Yulpatov. 1967. The induced radiation of excited classical oscillators and its use in high-frequency electronics, *Izv. VUZov, Radiofiz.*, 10: 1414–53 (*RQE*, 10: 794–813).

Gaponov, Andrei V., A. L. Goldenberg, D. P. Grigoryev, I. M. Orlova, T. B. Pankratova, and M. I. Petelin. 1965. Induced synchrotron radiation of electrons in cavity resonators, *Pis'ma JETP*, 2: 430–35 (*JETP Letters*, 2: 267–69).

Gaponov, Andrei V., A. L. Goldenberg, D. P. Grigoryev, T. B. Pankratova, M. I. Petelin, and V. A. Flyagin. 1975. Experimental investigation of centimeter-band gyrotrons, *Izv. VUZov, Radiofiz.*, 18: 280–89 (*RQE*, 18: 204–10).

Gaponov, Andrei V., A. L. Goldenberg, G. N. Rapoport, and V. K. Yulpatov. 1975. Multicavity cyclotron resonance maser, Copyright no. 273001 with priority of June 16, 1967, Official Bulletin KDIO of SM USSR, no. 40, p. 185.

Gaponov, Andrei V., A. L. Goldenberg, M. I. Petelin, and V. K. Yulpatov. 1976. A device for cm, mm and submm wave generation, Copyright no. 223931 with priority of March 24, 1967, Official Bulletin KDIO of SM USSR, no. 11, p. 200.

Gaponov, Andrei V., V. A. Flyagin, A. L. Goldenberg, G. S. Nusinovich, Sh. E. Tsimring, V. G. Usov, and S. N. Vlasov. 1981. Powerful millimetre-wave gyrotrons, *Int. J. Electron.*, 51: 277–302.

Gaponov-Grekhov, Andrei V., and V. L. Granatstein, eds. 1994. *Applications of High-Power Microwaves*. Boston-London: Artech House.

Garven, Morag, J. P. Calame, B. G. Danly, K. T. Nguyen, B. Levush, F. N. Wood, and D. E. Pershing. 2002. A gyrotron-traveling-wave tube amplifier experiment with a ceramic loaded interaction region, *IEEE-PS*, 30: 885–93.

Gause, G. F. 1934. *The Struggle for Existence,* Baltimore: Williams and Wilkins.

Giguet, Eric, Ph. Thouvenin, C. Tran, et al. 1997. Status of the 118 GHz–0.5 MW-quasi CW gyrotron for Tore Supra and TCV tokamaks, Conf. Digest, 22nd Int. Conf. on IR & MM Waves, Wintergreen, VA, July 20–25, 1997, pp. 104–5.

Gilmour, A. S., Jr. 1994. *Principles of Traveling Wave Tubes*, Chs. 5, 15. Boston, London: Artech House.

Ginzburg, Naum S. 1979. Nonlinear theory of amplification and generation of electromagnetic waves based on the anomalous Doppler effect, *Izv. VUZov, Radiofiz.*, 22: 470–79 (*RQE*, 22: 323–29).

Ginzburg, Naum S. 1987. On the theory of relativistic CRMs operating under synchronous adiabatic deceleration of the electron beam by the electromagnetic wave, *Izv. VUZov, Radiofiz.*, 7: 1181–87 (*RQE*, 30: 865–70).

Ginzburg, Naum S., S. P. Kuznetsov, and T. N. Fedoseeva. 1978. Theory of transients in relativistic backward-wave tubes, *Izv, VUZov, Radiofiz.*, 21: 1037–52 (*RQE*, 21: 728–39).

Ginzburg, Naum S., V. I. Krementsov, M. I. Petelin, P. S. Strelkov, and A. G. Shkvarunets. 1978. Cyclotron resonance maser with a relativistic high-current electron beam, *Pis'ma ZhTF*, 4: 149–53 (*Sov. Tech. Phys. Lett.*, 4: 61–62).

Ginzburg, Naum S., G. S. Nusinovich, and N. A. Zavolsky. 1986. Theory of non-stationary processes in gyrotrons with low Q resonators, *Int. J. Electronics*, 61: 881–94.

Ginzburg, Naum S., I. G. Zarnitsyna, and G. S. Nusinovich. 1979. Theory of relativistic maser at cyclotron self-resonance with the counter wave, *Radiotekhn. I Elektron.*, 24: 1146–52 (*REEP*, 24: 113–18).

———. 1981. Theory of relativistic CRM amplifiers, *Izv. VUZov, Radiofiz.*, 24: 481–90 (*RQE*, 24: 331–38).

Ginzburg, Vitaliy L. 1959. Certain theoretical aspects of radiation due to superluminal motion in a medium, *Usp. Fiz. Nauk*, 69: 537–64 (*Sov. Phys. Uspekhi*, 2 (1960): 874–93).

Glushenko, V. N., S. V. Koshevaya, and V. A. Prus. 1970. Improved efficiency of a gyrotron at the fundamental gyroresonance by means of a corrected distribution of magnetostatic fields, *Izv. VUZov, Radioelectron.*, 13: 12–17 (*Radioelectron. Commun. Systems*, 13: 10–15).

Glyavin, Mikhail Yu, V. E. Zapevalov, and M. L. Kulygin. 1999. Influence of the microwave-signal reflection on the generation efficiency of tunable gyrotrons, *Izv. VUZov, Radiofiz.*, 42: 1092–96 (*RQE*, 42: 962–66).

Golant, Victor E., M. G. Kaganskiy, L. P. Pakhomov, K. A. Podushnikova, and K. G. Shakovets. 1972. Experiments on microwave heating of plasmas at Tuman-2 installation, *J. Tech. Phys.*, 42: 488–96.

Goldenberg, Arkadiy L., and M. I. Petelin. 1973. The formation of helical electron beams in an adiabatic gun, *Izv. VUZov, Radiofiz.*, 16: 141–49 (*RQE*, 16: 106–11).

Goldenberg, Arkadiy L., G. S. Nusinovich, and A. B. Pavelyev. 1980. Diffractive Q of a resonator with a helical gopher, in *Gyrotrons*, 91–97. Gorky, USSR: IAP (in Russian).

Golubyatnikova, Elena R., and M. I. Petelin. 1994. Wave-beam passage through dielectric plate, *Izv. VUZov, Radiofiz.*, 37: 535–38 (*RQE*, 37:335–37).

Gradshtein, Israel S., and I. M. Ryzhik. 2000. *Tables of Integrals, Series, and Products*. San Diego: Academic Press, 2000.

Granatstein, Victor L., M. Herndon, R. K. Parker, and P. Sprangle. 1974. Coherent synchrotron radiation from an intense relativistic electron beam, *IEEE-QE*, 10: 651–54.

Granatstein, Victor L., M. Herndon, P. Sprangle, Y. Carmel, and J. Nation. 1975. Gigawatt microwave emission from an intense relativistic electron beam, *Plasma Phys.*, 17: 23–28.

Gryaznova, Tatyana A., S. V. Koshevaya, and G. N. Rapoport. 1969. Analysis of possibilities to enhance the efficiency of CRM-devices by the phase method, *Izv. VUZov, Radioelektron.*, 9: 998–1005 (*Radioelectron. Commun. Systems*, 12: 27–32).

Guo, Hezhong, J. Rodgers, S. Chen, M. Walter, and V. L. Granatstein. 1996. *Experimental Investigation of a Phase-Locked Harmonic Multiplying Inverted Gyrotwystron*, 1996 Int. Conf. on Plasma Science, June 3–5, 1996, Boston, IEEE Conference Records-Abstracts (Institute of Electrical and Electronics Engineers, Piscataway, NJ, 1996), 304.

Guo, Hezhong, S. H. Chen, V. L. Granatstein, J. Rodgers, G. Nusinovich, M. Walter, B. Levush, and W. J. Chen. 1997. Operation of a highly overmoded, harmonic-multiplying, wideband gyrotron amplifier, *Phys. Rev. Lett.*, 79: 515–18.

Heidinger, R., R. Schwab, R. Sporl, and M. Thumm. 1997. Dielectric loss measurements in CVD diamond windows for gyrotrons, 22nd Int. Conf. on Infrared and Millimeter Waves, July 20–25, 1997, Wintergreen, VA, Conf. Digest, ed. H. P. Freund, 142–43.

Herrmannsfeldt, William. 1979. Electron trajectory program, SLAC Report No. 226, Stanford, CA; also *Nucl. Instrum. Methods Phys. Res.* 187 (1981): 245.

Hirshfield, Jay L., and V. L. Granatstein. 1977. The electron cyclotron maser—An historical survey, *IEEE-MTT,* 25: 522–27.

Hirshfield, Jay L., M. A. LaPointe, A. K. Ganguly, R. B. Yoder, and C. Wang. 1996. Multimegawatt cyclotron autoresonance accelerator, *Phys. Plasmas,* 3: 2163–68.

Idehara, Toshitaka, and Y. Shimizu. 1994. Mode cooperation in a submillimeter wave gyrotron, *Phys. Plasmas,* 1: 3145–47.

Idehara,Toshitaka, T. Tatsukawa, H. Tanabe, S. Matsumoto, K. Kunieda, K. Hemmi, and T. Kanemaki. 1991. High-frequency, step tunable, cyclotron harmonic gyrotron, *Phys. Fluids* B, 3: 1766–72.

Ilyinskiy, A. S., and A. G. Sveshnikov. 1968. Methods for studying irregular waveguides, *J. Comp. Math. and Math. Phys.*, 8: 363–73.

Jackson, John D. 1962. *Classical Electrodynamics.* New York and London: John Wiley and Sons.

James, Bill G., and M. Kreutzer. 1995. *High-power millitron^{TM} TWTs in W-band,* 20th Int. Conf. on IR and MM Waves, Dec. 11–14, 1995, Orlando, FL, Conf. Digest, 13–14.

Johnson, H. R. 1955. Backward-wave oscillators, *Proc. IRE,* 43: 684–97.

Jory, Howard. 1968. Investigation of electronic interaction with optical resonators for microwave generation and amplification, Final Report, Contract ECOM-01873-F, Varian Associates, July 1968 (unpublished).

———. 1978. Development of gyrotron power sources in the millimeter wavelength range, *Proc. of the Int. Symp. "Heating in Toroidal Plasmas,"* Grenoble, France, July 3–7, 1978, ed. T. Consoli, vol. 2, 351–62 New York and Oxford: Pergamon Press.

Jory, Howard R., F. Friedlander, S. J. Hegji, J. F. Shively, and R. S. Symons. 1977. Gyrotrons for high-power millimeter-wave generation, *IEDM Digest,* 234–37.

Katsenelenbaum, Boris Z., L. Mercader del Rio, M. Pereyaslavets, M. Sorolla Ayza, and M. Thumm. 1998. *Theory of Nonuniform Waveguides (the Cross-Section Method).* London; Institution of Electrical Engineers.

Kolomenskiy, Andrei A., and A. N. Lebedev. 1959. Stability of a charged beam in storage systems, *Atomic Energy,* 7: 549–50 (in Russian).

———. 1963. Resonance effects associated with particle motion in a plane electromagnetic wave, *Zh. Eksp. Teor. Fiz.*, 44: 261–69 (*Soviet Physics JETP,* 17: 179–84).

Koshevaya, Svetlana V. 1970. Private communication.

Kou, C. S., S. H. Chen, L. R. Barnett, H. Y. Chen, and K. R. Chu. 1993. Experimental study of an injection-locked gyrotron backward-wave oscillators, *Phys. Rev. Lett.*, 20: 924–27.

Kovalev, Nikolay F., I. M. Orlova, and M. I. Petelin. 1968. Wave transformation in a multimode waveguide with corrugated walls, *Izv. VUZov, Radiofiz.*, 11: 783–86 (*RQE,* 11: 449–50).

Kovalev, Nikolay F., M. I. Petelin, and M. G. Reznikov. 1978. *Resonator,* Author's Certificate 720591 with priority of Nov. 14, 1978, *Bull. KDIO,* no. 9, 1980.

Kreischer, Kenneth E., R. J. Temkin, H. R. Fetterman, and W. J. Mulligan. 1984. Multimode oscillation and mode competition in high frequency gyrotrons, *IEEE-MTT,* 32: 481–90.

Kreischer, Kenneth E., B. G. Danly, J. B. Schutkeker, and R. J. Temkin. 1985. The design of megawatt gyrotron, *IEEE-PS,* 13: 364–73.

Kroll, Norman M. 1978. The free-electron laser as a traveling wave amplifier, *in Novel Sources of Coherent Radiation: Physics of Quantum Electronics,* eds. S. F. Jacobs, M. Sargent III, and M. O. Scully, vol. 5, 115–57. Reading, MA: Addison-Wesley.

Kroll, Norman M., P. L. Morton, and M. N. Rosenbluth. 1980. Variable parameter free-electron laser, in *Physics of Quantum Electronics,* ed. S. F. Jackobs et al., vol.7, 81–112. Reading, MA: Addison-Wesley.

Kuraev, Alexander A. 1979. *Theory and Optimization of Electron Microwave Devices.* 50, Minsk, USSR: Nauka I Tekhnika (in Russian).

Kuraev, Alexander A., I. S. Kovalev, and S. V. Kolosov. 1975. *Numerical Optimization Methods in Microwave Electronics.* Minsk, USSR: Nauka I Tekhnika (in Russian).

Kuraev, Alexander A., V. A. Stepukhovich, and V. A. Zhurakhovskiy. 1970. Stimulated synchrotron radiation of electrons in a piecewise homogeneous magnetic field, *Pis'ma ZhETF,* 11: 429–32 (*JETP Letters,* 11: 289–90).

Kuznetsov, Sergei P., D. I. Trubetskov, and A. P. Chetverikov. 1980. Nonlinear analytic theory of the peniotron, *Pis'ma Zh. Tekh. Fiz.,* 6: 1164–68 (*Sov. Tech. Phys. Lett.,* 6: 498–99).

Lamb, William E., Jr. 1965. Lectures in "Quantum Optics and Electronics," Summer School of Theoretical Physics, Les Houches, University of Grenoble, 1964, ed. C. DeWitt, A. Blandin, and C. Cohen-Tannoudji. New York, London, and Paris: Gordon and Breach. 1965.

Landau, Lev D., and E. M. Lifshitz. 1958. *Quantum Mechanics: Non-Relativistic Theory.* London-Paris: Pergamon Press; Reading, MA: Addison-Wesley.

———. 1960. *Electrodynamics of Continuous Media,* sec. 77. Oxford-London: Pergamon Press; Reading, MA: Addison-Wesley.

———. 1975. *The Classical Theory of Fields.* 4th rev. English ed. Oxford: Pergamon Press.

Latham, Peter E., W. Lawson, V. Irwin, et al. 1994. High power operation of an X-band gyrotwistron, *Phys. Rev. Lett.,* 72: 3730–33.

Lau, Y. Y. 1987. Effects of cathode surface roughness on the quality of electron beam, *J. Appl. Phys.,* 61: 36–44.

Lau, Y. Y., and L. R. Barnett. 1982. Theory of a low magnetic field gyrotron (gyromagnetron), *Int. J. IR and MM Waves,* 3: 619–44.

Lau, Y. Y., and K. R. Chu. 1981. Gyrotron traveling-wave amplifier: III. A proposed wide band fast wave amplifier, *Int. J. Infrared MM Waves,* 2: 415–21.

Lau, Y. Y., K. L. Jensen, and B. Levush. 1998. A comparison of flicker noise and shot noise on a hot cathode, *IEEE-PS,* 28: 794–97.

Lawson, Wesley G. 1990. Theoretical evaluation of nonlinear tapers for a high-power gyrotron, *IEEE-MTT,* 38: 1617–22.

Lawson, Wesley G., J. Calame, B. Hogan, et al. 1991. Efficient operation of a high power X-band gyroklystron. *Phys. Rev. Lett.,* 67: 520–23.

Lawson, Wesley G., and V. Specht, 1993. Design comparison of single-anode and double-anode 300-MW magnetron injection gun, *IEEE-ED,* 40: 1322–28.

Lawson, Wesley G., H. W. Matthews, M. K. E. Lee, et al. 1993. High-Power Operation of a K-Band Second-Harmonic Gyroklystron, *Phys. Rev. Lett.,* 71: 456–59.

Lawson, Wesley, P. E. Latham, J. P. Calame, et al. 1995. High power operation of first and second harmonic gyrotwystrons, *J. Appl. Phys.,* 78: 550-59.

Lawson, Wesley G., J. Cheng, J. P. Calame, et al. 1998. High-power operation of a three-cavity X-band coaxial gyroklystron, *Phys. Rev. Lett.,* 81: 3030–33.

Lawson, Wesley G., B. Hogan, S. Gouveia, B. Huebschman, and V. L. Granatstein. 2002. Development of Ku-band frequency-doubling coaxial gyroklystrons for accelerator applications, 3rd Int. Vacuum Electronics Conf., April 23–25, 2002, Monterey, CA, *Conf. Abstracts,* 81–82.

Leupold, Herbert A., E. Potenziani II, and A. S. Tilak. 1992. Tapered fields in cylindrical and spherical spaces, *IEEE Trans. Magn.,* MAG-28: 3045–48.

Levush, Baruch, and T. M. Antonsen, Jr. 1990. Mode competition and control in high-power gyrotron oscillators, *IEEE-PS,* 18: 260–72.

Levush, Baruch, T. M. Antonsen, Jr., A. Bromborsky, W.-R. Lou, and Y. Carmel. 1992. Theory of relativistic backward-wave oscillators with end reflections, *IEEE-PS,* 20: 263–80.

Li, H., and T. M. Antonsen, Jr. 1994. Space charge instability in gyrotron beams, *Phys. Plasmas,* 1: 714–29.

Lichtenberg, Allan J. *Phase-Space Dynamics of Particles.* New York-London-Sydney-Toronto: John Wiley and Sons.

Lin, Anthony T., K. R. Chu, C. C. Lin, C. S. Kou, D. B. McDermott, and N. C. Luhmann, Jr. 1992. Marginal stability design criterion for gyro-TWTs and comparison of fundamental with second harmonic operation, *Int. J. Electron.,* 72: 873–86.

Liu, Chunbo, T. M. Antonsen, Jr., and B. Levush. 1996. Simulation of the velocity spread in magnetron injection guns, *IEEE-PS,* 24: 982–91.

Lohr, John, R. W. Callis, W. P. Cary, et al. 2000. Commissioning of the 110 GHz ECH system on DIII-D for physics applications, 42nd Annual Meeting of the APS-DPP and the 10th Int. Congress on Plasma Physics, Oct. 23–27, 2000, Quebec City, Quebec, Canada; *Bull. APS,* 45 no.7 (Oct. 2000); 220.

Louisell, William H. 1960. *Coupled Mode and Parametric Electronics,* 25. New York, London: John Wiley and Sons.

Lubyako, Lev V., E. V. Suvorov, A. B. Burov, et al. 1998. System for measuring collective scattering spectra for thermonuclear plasma diagnostics, *Zh. Tekh. Fiz.,* 68: 54–62 (*Tech. Physics,* 43: 926–33).

Luchinin, Alexei G., and G. S. Nusinovich. 1984. An analytical theory for comparing the efficiency of gyrotrons with various electrodynamic systems, *Int. J. Electron.,* 57: 827–34.

Lygin, Vladimir K., V. N. Manuilov, A. N. Kuftin, A. B. Pavelyev, and B. Piosczyk. 1995. Inverse magnetron injection gun for a coaxial 1.5 MW, 140 GHz gyrotron, *Int. J. Electronics,* 79: 227–35.

Ma, J. Y. L., M. M. Blanco, and L. C. Robinson. 1983. Night moth eye broadband waveguide window, 7th Int. Conf. on IR and MM Waves, Marseille, France, Feb. 14–18, 1983, *Conf. Digest,* paper J5-1, 247.

Makowski, Mike. 1996. ECRF systems for ITER, *IEEE-PS,* 24: 1023–32.

Malakhov, Askold N. 1968. *Fluctuations in Auto-Oscillating Systems,* Moscow: "Nauka" (in Russian).

Malouf, Perry, V. L. Granatstein, S. Y. Park, Gun-Sik Park, and C. M. Armstrong. 1995. Performance of a wideband, three-stage, mixed geometry gyrotwystron amplifier, *IEEE-ED,* 42: 1681–85.

Malygin, Sergei A. 1986. A high-power gyrotron operating at the third harmonic of the cyclotron frequency, *Radiotekh. I Elektron.,* 31: 334–36 (*Sov. J. Commun. Techn. Electron,* 31: 106–8).

Manheimer, Wallace M. 1987. Theory of the multi-cavity phase locked gyrotron oscillator, *Int. J. Electron.,* 63: 29–47.

Manheimer, Wallace M., H. P. Freund, B. Levush, and T. M. Antonsen, Jr. 2001. Theory and simulation of ion noise in microwave tubes, *Physics of Plasmas,* 8: 297–320.

Manuilov, Vladimir N., and Sh. E. Tsimring. 1978. Synthesis of axially-symmetric systems of the formation of intense helical electron beams, *Radiotekh. Elektron.,* 23: 1486–96 (*REEP,* 23 (1978): 111–19).

McCurdy, Alan H., A. K. Ganguly, and C. M. Armstrong. 1989. Operation and theory of a driven single-mode electron cyclotron maser, *Phys. Rev.* A, 40: 1402–21.

McDermott, David B., N. C. Luhmann, Jr., D. S. Furuno, A. Kupiszewski, and H. Jory. 1983. Operation of a millimeter-wave harmonic gyrotron, Int. J. Infrared and Millimeter Waves, 4: 639–64.

Meeker, John, and J. E. Rowe. 1962. Phase focusing in linear-beam devices, *IRE-ED,* 9: 257–66.

Messiah, A. *Quantum Mechanics*, vol.1. New York, Chichester, Brisbane, Toronto: John Wiley and Sons.

Mobius, Arnold, and M. Thumm. 1993. Gyrotron output launchers and output tapers, in *Gyrotron Oscillators: Their Principles and Practice*, ed. C. J. Edgcombe, Ch. 7. London: Taylor and Francis.

Moiseev, Mark A. 1977. Maximum amplification band of a CRM-twystron, *Izv. VUZov, Radiofiz.*, 20: 1218–23 (*RQE*, 20: 846–49).

Moiseev, Mark A., and G. S. Nusinovich. 1974. Concerning the theory of multimode oscillations in a gyromonotron, *Izv. VUZov, Radiofiz.*, 17:1709–17 (*RQE*, 17: 1305–11).

Moiseev, Mark A., G. G. Rogacheva, and V. K. Yulpatov. 1968. Theoretical study of the effect of the axial inhomogeneity of the EM field in a resonator on the efficiency of the CRM-monotron, Abstracts of reports presented at the Popov's NTORES Session, Gorkiy, 68 (in Russian).

Myasnikov, Vadim E., A. G. Litvak, S. V. Usachev, et al. 2002. Development of 170 GHz gyrotrons for ITER, 334–35. 3rd Int. Conf. IVEC-2002, April 23–25, 2002, Monterey, CA.

Neilson, Jeff M., R. L. Ives, M. Read, et al. 2002. Update on the development of a 10 MW, 91 GHz gyroklystron, 2002 IEEE Int. Conf. on Plasma Science, May 26–30, 2002, Banff, Alberta, Canada, *IEEE Conf. Records-Abstracts*, 183.

Nezhevenko, Oleg A., V. P. Yakovlev, A. K. Ganguly, and J. L. Hirshfield. 1998. High power pulsed magnicon at 34 GHz, in *High Energy Density Microwaves*, ed. R. M. Phillips, 195–206. AIP Conf. Proc. 474, Woodbury, New York.

Nezlin, Mikhail V. 1976. Negative-energy waves and the anomalous Doppler effect, *Usp. Fiz. Nauk*, 120: 481–95 (*Sov. Phys. Uspekhi*, 19: 946–54).

Ngo, M. T., B. G. Danly, R. Myers, D. E. Pershing, V. Gregers-Hansen, and G. Linde. 2002. High-power millimeter-wave transmitter for the NRL WARLOC radar, 3rd Int. Vacuum Electronics Conf., April 23–25, 2002, Monterey, CA, *Conf. Abstracts*, 363–64.

Nguyen, Khan T., J. P. Calame, D. E. Pershing, B. G. Danly, M. Garven, B. Levush, and T. M. Antonsen, Jr. 2001. Design of a Ka-band gyro-TWT for radar applications, *IEEE-ED*, 48: 108–15.

Nielsen, C., and A. Sessler. 1959. Longitudinal space charge effects in particle accelerators, *Rev. Sci. Instrum.*, 30: 80–89.

Nordsieck, A. 1953. Theory of the large signal behavior of traveling-wave amplifier, *Proc. IRE*, 41: 630–37.

Nortrop, T. 1963. *The Adiabatic Motion of Charged Particles*. New York: Wiley Interscience.

Nusinovich, Gregory S. 1974. Methods of voltage feeds for a pulsed gyromonotron which ensure high efficiency in a single-mode operation, *Elektron. Tekh., Ser. I, Elektron. SVCh*, no. 3, 44–49 (in Russian).

———. 1975. Theory of synchronization of multimode electron microwave oscillators, *Izv. VUZov, Radiofiz.*, 18: 1689–98 (*RQE*, 18: 1246–52).

———. 1981. Mode interaction in gyrotrons, *Int. J. Electron.*, 51: 457–74.

———. 1984. Some perspectives on operating frequency increase in gyrotrons, Ch.6 in *Infrared and Millimeter Waves*, ed. K. J. Button, vol. 11, Ch. 6, 227–38. New York: Academic Press.

———. 1986. Theory of mode interaction in the gyrotron, Preprint KfK 4111. Lectures given at the Kernforschungszentrum Karlsruhe, October 1984.

———. 1988. Linear theory of a gyrotron with weakly tapered external magnetic field, *Int. J. Electronics*, 64: 127–36.

———. 1992a. Cyclotron resonance masers with inhomogeneous external magnetic fields, *Phys. Fluids* B, 4: 1989–97.

———. 1992b. Non-linear theory of a large-orbit gyrotron, *Int. J. Electron.*, 72: 959–67.
———. 1992c. Parametric instabilities in gyro-devices at cyclotron harmonics, *Int. J. Electron.*, 72: 795–805.
———. 1999. Review of the theory of mode interaction in gyrodevices, *IEEE-PS*, 27: 313–26.
Nusinovich, Gregory S., and O. Dumbrajs. 1996. Theory of gyro-backward wave oscillators with tapered magnetic field and waveguide cross section, *IEEE-PS*, 24: 620–29.
Nusinovich, Gregory S., and R. E. Erm. 1972. Efficiency of the CRM-monotron with a Gaussian axial distribution of the high-frequency field, *Elektronnaya Tekhnika, Ser. I, Elektronika SVCh*, no. 8, 55–60 (in Russian).
Nusinovich, Gregory S., and H. Li. 1992a. Theory of gyro-travelling-wave tubes at cyclotron harmonics, *Int. J. Electron.*, 72: 895–907.
———. 1992b. Theory of the relativistic gyrotwystron, *Phys. Fluids* B, 4: 1058–65.
Nusinovich, Gregory S., and T. B. Pankratova. 1981. Theory of submillimeter-wave gyrotrons, in a book of collected papers, *Gyrotrons*, 169–84 Gorkiy, USSR: IAP (in Russian).
Nusinovich, Gregory S., and M. Walter. 1997. Theory of the inverted gyrotwystron, *Phys. Plasmas*, 4: 3394–3402.
———. 1999. Linear theory of multistage forward-wave amplifiers, *Phys. Rev.* E, 60: 4811–22.
Nusinovich, Gregory S., B. G. Danly, and B. Levush. 1997. Gain and bandwidth in stagger-tuned gyroklystrons, *Phys. Plasmas*, 4: 469–78.
Nusinovich, Gregory S., M. Korol, and E. Jerby. 1999. Theory of the anomalous Doppler cyclotron-resonance-maser amplifier with tapered parameters, *Phys. Rev.* E, 59: 2311–21.
Nusinovich, Gregory S., P. E. Latham, and O. Dumbrajs. 1995. Theory of relativistic cyclotron masers, *Phys. Rev.* E, 52: 998–1012.
Nusinovich, Gregory S., B. Levush, and B. G. Danly. 1999. Gain and bandwidth in stagger-tuned gyroklystrons and gyrotwystrons, *IEEE-PS*, 27: 422–28.
Nusinovich, Gregory S., A. N. Vlasov, and T. M. Antonsen, Jr. 2001. Nonstationary phenomena in tapered gyro-backward-wave oscillators, *Phys. Rev. Lett.*, 87: 218301.
Nusinovich, Gregory S., A. B. Pavelyev, and V. I. Khizhnyak. 1989. Restrictions on the choice of optimum parameters of gyrotrons for mode competition conditions, *Radiotekh. Elektron.*, 33: 649–52 (*Sov. J. Comm. Tech. Electron.*, 33: 114–17).
Nusinovich, Gregory S., T. M. Antonsen, Jr., V. L. Bratman, and N. S. Ginzburg. 1994. Principles and capabilities of high-power microwave generators, in *Applications of High-Power Microwaves*, ed. A. V. Gaponov-Grekhov and V. L. Granatstein, Ch. 2. Boston-London: Artech House.
Nusinovich, Gregory S., G. P. Saraph, and V. L. Granatstein. 1997. Scaling law for ballistic bunching in multicavity harmonic gyroklystrons, *Phys. Rev. Lett.*, 78: 1815–18.
Nusinovich, Gregory S., J. Rodgers, W. Chen, and V. L. Granatstein, 2001. Phase stability in gyro-traveling-wave-tubes, *IEEE-ED*, 48: 1460–68.
Ogawa, I., M. Iwata, T. Idehara, et al. 1997. Plasma scattering measurements using a submillimeter wave gyrotron (Gyrotron FU II) as a power source, *Fusion Eng. and Design*, 34-35: 455–58.
Ono, Shoichi, K. Tsutaki, and T. Kageyama. 1984. Proposal of a high efficiency tube for high power millimetre or submillimetre wave generation. The gyro-peniotron, *Int. J. Electron.*, 56: 507–20.

Orlova, Inessa M., and M. I. Petelin. 1968. Private communication.

Osepchuk, John M. 1978. Life begins at forty: Microwave tubes, *Microwave Journal,* 51–60 (Nov. 1978).

Ott, Edward, and W. M. Manheimer. 1975. Theory of microwave emission by velocity-space instabilities of an intense relativistic electron beam, *IEEE-PS,* 3: 1–5.

Pangonis, L. I., and M. V. Persikov. 1971. Analysis of wave modes based on radiation from waveguides with an inclined aperture, *Radiotekh. I Electron.*, 16: 2300–2 (*REEP,* 16: 2108–10).

Pantell, Richard H. 1959. Electron beam interaction with fast waves, in *Proc. of the Symposium on Millimeter Waves,* 301–11 New York, March 31, April 1 and 2, 1959, Brooklyn, NY: Polytechnic Press of the Polytechnic Institute, 1960.

Park, Gun-Sik, V. L. Granatstein, P. E. Latham, C. M. Armstrong, A. K. Ganguly, and S. Y. Park. 1991. Phase stability of gyroklystron amplifier, *IEEE-PS,* 19: 629–40.

Park, Gun-Sik, S. Y. Park, R. H. Kyser, et al. 1994. Broadband operation of a Ka-band tapered gyro-traveling wave amplifier, *IEEE-PS,* 22: 536–43.

Park, Gun-Sik, et al. 1995. Gain broadening of two-stage tapered gyrotron traveling wave tube amplifier, *Phys. Rev. Lett.*, 74: 2399–2402.

Park, S. Y., R. H. Kyser, C. M. Armstrong, R. K. Parker, and V. L. Granatstein. 1990. Experimental study of a Ka-band gyrotron backward-wave oscillator, *IEEE-PS,* 18: 321–25.

Parshin, Vladimir V., R. Heidinger, B. A. Andreev, A. V. Gusev, and V. B. Shmagin. 1995. Silicon with extra low losses for megawatt output gyrotron windows, 20th Int. Conf. IR and MM Waves, Lake Buena Vista, Orlando, FL, *Conf. Digest,* 22–23, paper M2.4.

Pendergast, K. D., B. G. Danly, W. L. Menninger, and R. J. Temkin. 1992. A long-pulse, CARM oscillator experiment, *Int. J. Electron.*, 72: 983–1004.

Petelin, Mikhail I. 1974. On the theory of ultrarelativistic autoresonance masers, *Izv. VUZov, Radiofiz.*, 17: 902–8 (*RQE,* 17: 686–90).

———. 2001. Private communication.

——— 2002. Private communication.

Petelin, Mikhail I., and W. Kasparek, 1991. Surface corrugation for broadband matching of windows in powerful microwave generators, *Int. J. Electronics,* 71: 871–73.

Petelin, Mikhail I., and V. K. Yulpatov. 1975. Linear theory of a CRM-monotron, *Izv. VUZov, Radiofiz.*, 18: 290–99 (*RQE,* 18: 212–19).

Phillips, Robert M., and D. W. Sprehn. 1999. High-power klystrons for the next linear collider, *Proc. IEEE,* 87: 738–51.

Pierce, John R. 1950. *Traveling-Wave Tubes*. Toronto: D. Van Nostrand.

———. 1954. *Theory and Design of Electron Beams,* 35. Toronto: D. Van Nostrand.

Piosczyk, Bernhard. 1993. Electron guns for gyrotron applications, in *Gyrotron Oscillators: Their Principles and Practice,* ed. C. J. Edgcombe, Ch. 5, 123–46. London: Taylor and Francis.

Piosczyk, Bernhard, A. Arnold, G. Dammertz, et al. 2000. Step-frequency operation of a coaxial cavity gyrotron from 134 to 169.5 GHz, *IEEE-PS,* 28: 918–23.

Rapoport, Grigoriy N., A. K. Nemak, and V. A. Zhurakhovskiy. 1967. Interaction between helical electron beams and strong electromagnetic cavity fields at cyclotron-frequency harmonics, *Radiotekh. I Elektron.*, 12: 633–41 (*REEP,* 12: 587–95).

Read, Michael E., K. R. Chu, and A. J. Dudas. 1982. Experimental examination of the enhancement of gyrotron efficiencies by use of profiled magnetic fields, *IEEE-MTT,* 30: 42–46.

Read, Michael E., R. L. Ives, G. Miram, et al. 2002. Inverted magnetron-injection electron gun for the relativistic coaxial gyroklystron at the University of Maryland, in *Advanced Accelerator Concepts,* ed. C. E. Clayton and P. Muggli, 10th

Workshop, June 2002, Mandalay Beach, CA, AIP Conf. Proc., vol. 647, Melville, NY, 2002.

Roberts, Charles S., and S. J. Buchsbaum. 1964. Motion of a charged particle in a constant magnetic field and a transverse electromagnetic wave propagating along the field, *Phys. Rev.*, 135: A381–89.

Rodgers, John, H. Guo, G. S. Nusinovich, and V. L. Granatstein. 2001. Experimental study of phase deviation and pushing in a frequency doubling, second harmonic gyro-amplifier, *IEEE-ED*, 48: 2434–41.

Rowe, Joseph E. 1959. Theory of the crestatron: A forward-wave amplifier, *Proc. IRE*, 47: 536–45.

Sakamoto, Keishi, M. Tsuneoka, A. Kasugai, et al. 1994. Major improvement in gyrotron efficiency with beam energy recovery, *Phys. Rev. Lett.*, 73: 3532–35.

Sakamoto, Keishi, A. Kasugai, Y. Ikeda, et al. 2002. Development of 170 GHz and 110 GHz gyrotrons for fusion application, 336–37, 3rd Int. Conf. IVEC-2002, April 23–25, 2002, Monterey, CA.

Sangster, A. J. 1980. Small-signal analysis of the traveling-wave gyrotron using Pierce parameters, *IEEE-MTT*, 28: 313–17.

Saraph, Girish P., T. M. Antonsen, Jr., G. S. Nusinovich, and B. Levush. 1993. Nonlinear theory of stable, efficient operation of a gyrotron at cyclotron harmonics, *Phys. Fluids* B, 5: 4473–85.

Saraph, Girish P., T. M. Antonsen, Jr., G. S. Nusinovich, and B. Levush. 1995. A study of parametric instability in a harmonic gyrotron: Designs of third harmonic gyrotrons at 94 GHz and 210 GHz, *Phys. Plasmas*, 2: 2839–46.

Sarcione, M., et al. 1996. The design, development and testing of the THAAD solid state phase array, IEEE Int. Symp. on Phased Array Systems and Technology, 1996, 260–65.

Sato, H., T. M. Tran, K. E. Kreischer, and R. J. Temkin. 1986. Analytical treatment of linearized self-consistent theory of a gyromonotron with a non-fixed structure, *Int. J. Electron.*, 61: 895–904.

Sato, Nobuyuki, O. Kamohara, K. Sagae, and K. Yokoo. 2002. Experiments of autoresonant peniotron using a parallel transmission line with double pairs of ridges, *IEEE-PS*, 30: 859–64.

Savelyev, Vladimir Ya. 1940. On the theory of a klystron, *J. Tech. Phys.*, 10: 1365–71 (in Russian).

Schneider, Jurgen. 1959. Stimulated emission of radiation by relativistic electrons in a magnetic field, *Phys. Rev. Lett.*, 2: 504–5.

Schottky, Walter. 1926. Small-shot effect and flicker effect, *Phys. Rev.*, 28: 74–103.

Schriever, R. L., and C. C. Johnson. 1966. A rotating beam waveguide oscillator, *Proc. IEEE*, 54: 2029–30.

Schwinger, Julian, and D. S. Saxon. 1968. *Discontinuities in Waveguides*, 39. New York: Gordon and Breach.

Shafranov, V. D. 1958. Refractive index of magnetized plasma in the vicinity of the ion cyclotron resonance, in *Plasma Physics and Controlled Thermonuclear Fusion*, vol. 4, 426–29. Moscow: AN SSSR.

Shapiro, Vitaliy D., and V. I. Shevchenko. 1976. Wave-particle interaction in non-equilibrium media, *Izv. VUZov, Radiofiz.*, 19:767–91 (*RQE*, 19: 543–60).

Singh, Amarjit, S. Rajapatirana, Y. Men, et al. 1999. Design of a multistage depressed collector system for 1-MW CW gyrotrons—Part I: Trajectory control of primary and secondary electrons in a two-stage depressed collector, *IEEE-PS*, 27: 490–502.

Sinitsyn, Olexander V., G. S. Nusinovich, K. T. Nguyen, and V. L. Granatstein. 2002. Nonlinear theory of the gyro-TWT: Comparison of analytical method with numerical code data for the NRL gyro-TWT, *IEEE-PS*, 30: 915–21.

Sirigiri, Jagadishvar R., M. A. Shapiro, and R. J. Temkin. 2002. Initial experimental results from the MIT 140 GHz quasioptical gyro-TWT, 3rd IEEE Int. Vacuum Electronics Conf., April 23–25, 2002, Monterey, CA, *Conf. Digest,* 83–84.

Sirigiri, Jagadishvar R., K. E. Kreischer, J. Machusak, I. Mastovsky, M. A. Shapiro, and R. J. Temkin. 2001. Photonic-band-gap resonator gyrotron, *Phys. Rev. Lett.,* 86: 5628–31.

Slater, John C. 1950. *Microwave Electronics*. New York: D. Van Nostrand.

Smith, J. Maynard. 1974. *Models in Ecology*, Ch. 5. Cambridge, UK: Cambridge University Press.

Smullin, Louis D., and H. A. Haus, eds. 1959. *Noise in Electron Devices*. New York: Wiley.

Solymar, L. 1959. Spurious mode generation in nonuniform waveguides, *IRE-MTT,* 7: 379–83.

Sprangle, Phillip, and A. T. Drobot. 1977. The linear and self-consistent nonlinear theory of the electron cyclotron maser instability, *IEEE-MTT,* 25: 528–44.

Sprangle, Phillip, and R. A. Smith. 1980. The nonlinear theory of efficiency enhancement in the electron cyclotron maser (gyrotron), *J. Appl. Phys.,* 51: 3001–7.

Sprangle, Phillip, C. M. Tang, and W. M. Manheimer. 1980. Nonlinear theory of free-electron lasers and efficiency enhancement, *Phys. Rev.* A, 21: 302–18.

Stratonovich, R. L. 1963. *Topics in the Theory of Random Noise,* vol. 1, *General Theory of Random Processes,* 21. New York: Gordon and Breach.

Sukhorukov, Anatoliy P., and A. V. Sheludchenkov. 2002. Gyrotrons and electromagnetic compatibility, in book of review papers *Vacuum Microwave Electronics,* ed. M. I. Petelin, 118–20. Nizhniy Novgorod: Inst. Appl. Phys. Russian Acad. Sci. (in Russian).

Sushilin, Pavel B., A. Sh. Fix, and V. V. Parshin. 1989. Perspectives of increasing the gyrotron output window capacity, In the book of collected papers *Gyrotrons,* ed. V. A. Flyagin, 181–94. Inst. Appl. Phys. Gorkiy USSR: (in Russian).

Symons, Robert S., and H. R. Jory. 1981. Cyclotron resonance devices, in *Advances in Electronics and Electron Physics,* vol. 55, 1–75. New York: Academic Press.

Tantawi, Sami G., W. T. Main, P. E. Latham, et al. 1992. Amplification from an overmoded three-cavity gyroklystron with a tunable penultimate cavity, *IEEE-PS,* 20: 205–15.

Temkin, Richard J., K. Kreischer, S. M. Wolfe, D. R. Cohn, and B. Lax. 1979. High frequency gyrotrons and their application to tokamak plasma heating, *J. Magnetism and Magnetic Materials,* 11: 368–71.

Thumm, Manfred. 1997. Modes and mode conversion in microwave devices, in *Generation and Application of High Power Microwaves,* ed. R. A. Cairns and A. D. R. Phelps, 121–72. The Scottish Universities Summer School in Physics and Institute of Physics Publishing, Bristol and Philadelphia.

———. 1998. Development of output windows for high-power long-pulse gyrotrons and EC wave applications, *Int. J. Infrared and Millimeter Waves,* 19: 3–14.

Tolkachev, Alexei A. 2002. Some tendencies in the development of radar and communication systems, in a book of collected papers *Vacuum Microwave Electronics*, ed. M. I. Petelin, 7–12. Nizhny Novgorod, Inst. Appl. Phys. Russian Academy of Sci. (in Russian).

Tolkachev, Alexei A., B. A. Levitan, G. K. Solovyev, V. V. Veytsel, and V. E. Farber. 2000. A megawatt power millimeter-wave phase-array radar, *IEEE AES Systems Magazine,* July 2000, 25–31.

Tran, T. M., B. G. Danly, K. E. Kreischer, J. B. Schutkeker, and R. J. Temkin. 1986. Optimization of gyroklystron efficiency, *Phys. Fluids,* 29: 1274–81.

Tsimring, Shulim E. 1974. Formation of helical electron beams, Lectures on Microwave Electronics, Saratov School of Engineers, 4: 3–94, Saratov University Publishers (in Russian).

Twiss, R. Q. 1958. Radiation transfer and the possibility of negative absorption in radio astronomy, *Austr. J. Phys.* 11: 564–79.

Van der Pol, Balthasar. 1922. On oscillation hysteresis in a triode generator with two degrees of freedom, *Phil. Mag.*, S.6, 43: 700–719.

———. 1927. Forced oscillations in a circuit with non-linear resistance, *Phil. Mag.* S.7, 3: 65–80; see also review paper, The nonlinear theory of electric oscillations, *Proc. IRE*, 22 (1934): 1051–86.

Van Sciver, Steven, and K. R. Marken. 2002. Superconducting magnets above 20 tesla, *Physics Today,* 55 (August 2002): 37–43.

Vaughan, J. Rodney M. 1988. Multipactor, *IEEE-ED*, 35: 1172–80.

Vitello, Peter, and K. Ko. 1985. Mode competition in the gyro-peniotron oscillator, *IEEE-PS*, PS-13: 454–63.

Vlasov, Alexander N., T. M. Antonsen, Jr., J. C. Rodgers, G. S. Nusinovich, M. Walter, M. Botton, and B. Levush. 2001. Simulations of multifrequency processes in fast wave devices by MAGY, in *Pulsed Power Plasma Science-2001,* IEEE Conference Record- Abstracts, paper O3G8, 455. June 17–22, 2001, Las Vegas, Nevada.

Vlasov, Sergei N., and K. M. Likin. 1980. Geometrical optics theory of wave transformers in oversized waveguides, in a book of collected papers *Gyrotrons,* ed. V. A. Flyagin, 125–38. Gorky, USSR: Inst. Appl. Phys. Acad. Sci. USSR (in Russian).

Vlasov, Sergei N., and I. M. Orlova. 1974. Quasi-optical transformer which transforms the waves in a waveguide having a circular cross-section into a highly-directional wave beam, *Izv. VUZov, Radiofiz.*, 17: 148–54 (*RQE*, 17: 115–19).

Vlasov, Sergei N., I. M. Orlova, and M. I. Petelin. 1972. Quasioptical transformation of natural waves of waveguides having a circular cross section by means of axially symmetrical reflectors, *Izv. VUZov, Radiofiz.*, 15: 1913–18 (*RQE*, 15: 1466–70).

———. 1981. Gyrotron cavities and electrodynamic mode selection, in a book of collected papers *Gyrotron,* ed. A. V. Gaponov-Grekhov, 62–76. Gorky, USSR: Inst. Appl. Phys., Acad. Sci. USSR (in Russian).

Vlasov, Sergei N., L. I. Zagryadskaya, and M. I. Petelin. 1975. Transformation of a whispering gallery mode, propagating in a circular waveguide into a beam of waves, *Radiotekhn. I Elektron.*, 20: 2026–30 (*REEP,* 20: 14–17).

Vlasov, Sergei N., L. I. Zagryadskaya, and I. M. Orlova. 1976. Open coaxial resonators for gyrotrons, *Radiotekhn. I Elektron.*, 21: 1485–92 (*REEP,* 21: 96–102).

Vlasov, Sergei N., G. M. Zhislin, I. M. Orlova, M. I. Petelin, and G. G. Rogacheva. 1969. Irregular waveguides as open resonators, *Izv. VUZov, Radiofiz.*, 12: 1236–44 (*RQE*, 12: 972–78).

Vomvoridis, John L. 1997. Cyclotron resonance effects on a rectilinear electron beam for the generation of high-power microwaves, in *Generation and Application of High Power Microwaves,* ed. R. A. Cairns and A. D. R. Phelps, 183–200. The Scottish Universities Summer School in Physics and Institute of Physics Publishing, Bristol and Philadelphia.

Wachtel, Jonathan M., and J. L. Hirshfield. 1966.Interference beats in pulse-simulated cyclotron radiation, *Phys. Rev. Lett.*, 17: 348–51.

Walsh, John E., G. L. Johnston, R. C. Davidson, and D. J. Sullivan. 1989. Theory of phase-locked regenerative oscillators with nonlinear frequency shift effects, SPIE vol. 1061, *Microwave and Particle Beam Sources and Directed Energy Concepts,* ed. H. E. Brandt, 161–69.

Wang, Q. S., D. B. McDermott, and N. C. Luhmann, Jr. 1995. Demonstration of marginal stability theory by a 200-kW second-harmonic gyro-TWT amplifier, *Phys. Rev. Lett.*, 75: 4322–25.

Webster, David L. 1939. Cathode-ray bunching, *J. Appl. Phys.*, 10: 501–8.

Weibel, E. S. 1959. Spontanously growing transverse waves in a plasma due to an anisotropic velocity distribution, *Phys. Rev. Lett.*, 2: 83–84.

Weinsten, Lev A. 1957. Nonlinear theory of the TWT: Part II, Numerical results, *Radiotekhnika I Elektronika*, 2: 1027–47 (in Russian).

———. 1966. *Open Resonators and Open Waveguides*. Boulder, CO: Golem Press, 1969 (translation).

———. 1969. General theory of resonant electron auto-oscillators, in *High-Power Electronics*, ed. P. L. Kapitsa and L. A. Weinstein, vol. 6, 84–129. Moscow: "Nauka,"(in Russian).

———. 1988. *Electromagnetic Waves*, 2nd ed. Moscow: "Radio I Svyaz"(in Russian).

Weinstein, Lev A., and V. A. Solntsev. 1973. *Lectures on Microwave Electronics*, 40. Moscow: "Sov. Radio"(in Russian).

Whaley, David R., M. Q. Tran, T. M. Tran, and T. M. Antonsen, Jr. 1994. Mode competition and startup in cylindrical cavity gyrotrons using high-order operating modes, *IEEE-PS*, 22: 850–60.

Whittaker, E. T., and G. N. Watson. 1979. *A Course of Modern Analysis: An Introduction to the General Theory of Infinite Processes and of Analytical Functions, with an Account of the Principal Transcendental Functions*. New York: AMS Press.

Wilson, Perry B. 1994. Application of high-power microwave sources to TeV linear colliders, in *Applications of High-Power Microwaves*, eds. A. V. Gaponov-Grekhov and V. L. Granatstein, Ch. 7, 229–317. Boston, London: Artech House.

Woskoboinikov, Pavel, D. R. Cohn, and R. J. Temkin. 1983. Application of advanced millimeter/far-infrared sources to collective thomson scattering plasma diagnostics, *Int. J. IR and MM Waves*, 4: 205–29.

Yariv, Amnon. 1975. *Quantum Electronics*, 2nd ed., Ch. 11. New York: John Wiley and Sons.

Yokoo, K., N. Sato, J. Sakuraba, et al. 1996. Fabrication of liquid helium-free superconducting magnet for millimeter wave gyrotron, 21st Int. Conf. IR and MM Waves, July 14–19, 1996, Berlin, Germany, *Conf. Digest*, paper AW7.

Yulpatov, Valeriy K. 1960. Report at the 4th All-Union Conference on Radioelectronics, Kharkov.

———. 1965. The nonlinear theory of interaction of a curvilinear periodic electron beam with an electromagnetic field, *Voprosy radioelektroniki, Ser. I, Elektronika*, 12: 15–32 (in Russian).

———. 1967. The nonlinear theory of interaction of periodic electron beam with an electromagnetic wave, *Izv. VUZov, Radiofiz.*, 10: 846–56 (*RQE*, 10: 471–76).

———. 1968. Private communication.

———. 1970. Private communication.

———. 1972. Private communication.

———. 1974. Averaged equations for oscillations in the CRM-monotron, 3rd Winter School of Engineers, Saratov, IV, 144–78 (in Russian).

Zapevalov, Vladimir E., and G. S. Nusinovich. 1985. Self-modulation instability of gyrotron radiation, *Radiotekh. Elektron.*, 30: 563–70 (*REEP*, 30:101–8).

———. 1989. The theory of amplitude-phase mode interaction in electron masers, *Izv. VUZov, Radiofiz.*, 32: 347–55 (*RQE*, 32: 269–76).

Zapevalov, Vladimir E., S. A. Malygin, and S. E. Tsimring. 1993. High-power gyrotron at second harmonic of cyclotron frequency, *Izv. VUZov, Radiofiz.*, 36: 543–52 (*RQE*, 36: 346–53).

Zarnitsyna, Irina G., and G. S. Nusinovich. 1974. Stability of single-mode self-excited oscillations in a gyromonotron, *Izv. VUZov, Radiofiz.*, 17: 1858–67 (*RQE*, 17: 1418–24).

———. 1975a. Competition of modes having arbitrary frequency separation in a gyromonotron, *Izv. VUZov, Radiofiz.*, 18: 303–6 (*RQE*, 18: 223–25).

———. 1975b. Concerning the stability of locked one-mode oscillations in a multimode gyromonotron, *Izv. VUZov, Radiofiz.*, 18: 459–62 (*RQE*, 18: 339–42).

———. 1977. Competition of modes resonant with different harmonics of cyclotron frequency in gyromonotrons, *Izv. VUZov, Radiofiz.*, 20: 461–67 (*RQE*, 20: 313–17).

Zasypkin, Evgeniy V., M. A. Moiseev, E. V. Sokolov, and V. K. Yulpatov. 1995. Effect of penultimate cavity position and tuning on three-cavity gyroklystron amplifier performance, *Int. J. Electron.*, 78: 423–33.

Zaytsev, Nikolay I., T. B. Pankratova, M. I. Petelin, and V. A. Flyagin. 1974. Millimeter and submillimeter wave gyrotrons, *Radiotekh. I Elektron.*, 19: 1056–60 (*REEP*, 19: 103–7).

Zhao, Jingjun, G. S. Nusinovich, H. Guo, J. C. Rodgers, and V. L. Granatstein. 2000. Axial mode locking in a harmonic-multiplying, inverted gyrotwystron, *IEEE-PS*, 28: 597–605.

Zheleznyakov, Vladimir V. 1960–61. On the instability of magnetoactive plasma with respect to electromagnetic perturbations. *Izv. VUZov, Radiofiz.*, 3: 57–66, part 1, 3: 180–91, part 2, 4: 619–29, part 3, 4: 849–60, part 4 (in Russian).

Zhurakhovskiy, Valeriy A. 1972. *Nonlinear Oscillations of Electrons in Magnetically Directed Flows*. Kiev: "Naukova Dumka" (in Russian).

Index